Surfactants
in
Tribology

Surfactants
in
Tribology

Edited by
Girma Biresaw
K.L. Mittal

CRC Press
Taylor & Francis Group
Boca Raton London New York

CRC Press is an imprint of the
Taylor & Francis Group, an **informa** business

CRC Press
Taylor & Francis Group
6000 Broken Sound Parkway NW, Suite 300
Boca Raton, FL 33487-2742

First issued in paperback 2019

ISBN-13: 978-1-4200-6007-2 (hbk)
ISBN-13: 978-0-367-38724-2 (pbk)

Library of Congress Cataloging-in-Publication Data

Surfactants in tribology / Girma Biresaw and K.L. Mittal, editors.
    p. cm.
  Includes bibliographical references and index.
  ISBN 978-1-4200-6007-2 (alk. paper)
  1. Surface active agents. 2. Tribology. I. Biresaw, Girma, 1948- II. Mittal, K. L., 1945- III. Title.

TP994.S8835 2008
621.8'9--dc22
                                                2008011646

Visit the Taylor & Francis Web site at
http://www.taylorandfrancis.com

and the CRC Press Web site at
http://www.crcpress.com

Contents

PART IV   Polymeric and Bio-Based Surfactants

Preface

This book presents recent developments and research activities that highlight the importance of surfactants in tribological phenomena. Even a cursory look at the literature will evince the high tempo of research in both fields: tribology and surfactants.

In light of these research trends, we organized a symposium on the topic of "The Role of Surfactants in Tribology" as a part of the 16th International Symposium on Surfactants in Solution (SIS) in Seoul, Korea, June 4–9, 2006. The SIS series is a biennial event started in 1976. Since then, these meetings have been held in many corners of the globe, attended by a "who's who" list of members of the surfactant community. These meetings are widely recognized as the premier forum for discussing the latest research findings on surfactants in solution.

In keeping with the SIS tradition, the leading researchers in the fields of surfactants and tribology were invited to present their latest findings at the SIS event in 2006. In essence, this symposium represented a nexus between the research arenas of surfactants and tribology. The participants were invited to submit chapters to this book based on their presentations. In addition, we solicited and obtained chapters from other leading researchers who were not able to participate in the 2006 symposium. Thus this book represents the cumulative wisdom of many active and renowned scientists and technologists engaged in the study of surfactants in variegated tribological phenomena.

Surfactants play a variety of critical roles in tribology, which subsumes the phenomena of three processes: friction, wear, and lubrication. The most widely recognized role of surfactants deals with their ability to control friction and wear. Surfactants also allow for control of a wide range of properties of lubricants, such as emulsification/demulsification, bioresistance, oxidation resistance, rust/corrosion prevention, etc. This book is a compendium of recent advances dealing with the role of surfactants in all three subjects within the purview of tribology. The book comprises chapters dealing with theoretical, experimental, and technological advances. Topics covered in the book include the role of surfactants in the tribological aspects of self-assembled monolayers (SAMs), microelectromechanical systems (MEMS), and nanoelectromechanical systems (NEMS). The book also addresses recent advances in fundamental tribological issues such as friction, wear, adsorption, tribochemical reactions, and surface/interfacial interactions.

The chapters in this book are grouped into five parts, each comprising chapters with a common theme. Part I consists of a single chapter dealing with the fundamentals of surfactants. Part II contains three chapters dealing with tribological aspects of micro- and nanodevices. Topics covered in Part II include tribological aspects of micropatterns of two-dimensional asperity arrays, MEMS, NEMS, and magnetic recording devices. Part III has six chapters dealing with self-assembled monolayers and ultrathin films relevant to tribological phenomena. Topics covered in Part III include tribological aspects of organosilane monolayers, ultrathin self-assembled films, superhydrophobic films, MoDTC/ZDDP tribofilms, and surfactant-coated copper nanoparticles. Part IV has five chapters dealing with polymeric and biobased surfactants. Topics dealt with in Part IV include various tribological aspects related to polymeric gels, elastomers sliding against hydrophilic and hydrophobic surfaces, agriculture-based amphiphiles, vegetable oils, and biobased greases. Part V contains six chapters dealing with surfactant adsorption and aggregation relevant to tribological phenomena. Topics covered in Part V include the design of surfactants for lubrication, aqueous nonionic surfactant-based lubricants, adsorption and aggregation kinetics, surfactant and polymer nanostructures, and engine oils.

This book is the first to comprehensively treat the relevance of surfactants in tribology and brings together researchers from both the tribology and surfactants arenas. Consequently, this book is a valuable repository of information for a wide range of individuals engaged in research, development, and manufacturing. Tribological phenomena play a crucial role in the performance of a legion of consumer, industrial, and high-tech products, including MEMS, NEMS, magnetic recording media, metalworking, cars, aircraft, etc. Surfactants can be beneficially used to reduce wear and friction, with significant economic implications.

We sincerely hope that the bountiful information in this book will be a valuable resource for chemists, chemical engineers, petroleum engineers, automotive engineers, lubricant formulators, materials scientists, and tribologists. Indeed, the book should be of special interest to those engaged in the study of MEMS, NEMS, and biodevices or, more broadly, to anyone who seeks to understand and solve tribological issues.

Girma Biresaw, Ph.D.
K. L. Mittal, Ph.D.

Editors

Girma Biresaw received a Ph.D. in physical-organic chemistry from the University of California–Davis and spent 4 years as a postdoctoral research fellow at the University of California–Santa Barbara, investigating reaction kinetics and products in surfactant-based organized assemblies. Biresaw then joined the Aluminum Company of America as a scientist and conducted research in tribology, surface/colloid science, and adhesion for 12 years. Biresaw joined the Agricultural Research Service of the U.S. Department of Agriculture in Peoria, IL, in 1998 as a research chemist, and became a lead scientist in 2002. To date, Biresaw has authored/coauthored more than 150 scientific publications, including more than 40 peer-reviewed manuscripts, 6 patents, more than 25 proceedings and book chapters, and more than 80 scientific abstracts.

Kashmiri Lal Mittal received his Ph.D. degree from the University of Southern California in 1970 and was associated with the IBM Corp. from 1972 to 1993. He is currently teaching and consulting worldwide in the areas of adhesion and surface cleaning. He is the editor of 89 published books, as well as others that are in the process of publication, within the realms of surface and colloid science and of adhesion. He has received many awards and honors and is listed in many biographical reference works. Mittal is a founding editor of the international *Journal of Adhesion Science and Technology* and has served on the editorial boards of a number of scientific and technical journals. Mittal was recognized for his contributions and accomplishments by the worldwide adhesion community by organizing on his 50th birthday, the First International Congress on Adhesion Science and Technology in Amsterdam in 1995. In 2002 he was honored by the global surfactant community, which inaugurated the Kash Mittal Award in the surfactant field in his honor. In 2003 he was honored by the Maria Curie-Sklodowska University, Lublin, Poland, which awarded him the title of doctor *honoris causa*.

Contributors

Yasuhisa Ando
National Institute of Advanced
  Industrial Science and Technology
Ibaraki, Japan

Girma Biresaw
Cereal Products and Food Science
  Research Unit
National Center for Agricultural
  Utilization Research
Agricultural Research Service, USDA
Peoria, Illinois

Sophie Bistac
UHA–CNRS
Institut de Chimie des Surfaces et
  Interfaces
Mulhouse, France

Carlo Carraro
Department of Chemical Engineering
University of California
Berkeley, California

Wei Dai
State Key Laboratory of Nonlinear
  Mechanics
Institute of Mechanics
Chinese Academy of Sciences
Beijing, China

Kenneth M. Doll
USDA/NCAUR/ARS
Food and Industrial Oil Research
Peoria, Illinois

Peter J. Dowding
Infineum UK Ltd.
Abingdon, Oxfordshire, U.K.

Miao Du
Laboratory of Soft and Wet Matter
Graduate School of Science
Hokkaido University
Sapporo, Japan
and
Department of Polymer Science and
  Engineering
Zhejiang University
Hangzhou, China

Julian Eastoe
School of Chemistry
University of Bristol
Bristol, U.K.

Sevim Z. Erhan
USDA/NCAUR/ARS
Food and Industrial Oil Research
Peoria, Illinois

Daniel A. Fischer
Materials Science and Engineering
  Laboratory
National Institute of Standards and
  Technology
Gaithersburg, Maryland

Joelle Frechette
Chemical and Biomolecular
  Engineering Department
Johns Hopkins University
Baltimore, Maryland

Michael L. Free
Department of Metallurgical
  Engineering
University of Utah
Salt Lake City, Utah

Jian Ping Gong
Laboratory of Soft and Wet Matter
Graduate School of Science
Hokkaido University
Sapporo, Japan

Ömer Gül
The Pennsylvania State University
The Energy Institute
University Park, Pennsylvania

Li-Ya Guo
State Key Laboratory of Nonlinear
  Mechanics
Institute of Mechanics
Chinese Academy of Sciences
Beijing, China

Shuchen Hsieh
Department of Chemistry and Center
  for Nanoscience and Nanotechnology
National Sun Yat-Sen University
Kaohsiung, Taiwan, Republic of China

Stephen M. Hsu
Materials Science and Engineering
  Laboratory
National Institute of Standards and
  Technology
Gaithersburg, Maryland

Hideomi Ishida
Graduate School of Engineering
Kyushu University
Fukuoka, Japan

Masataka Kaido
Toyota Motor Corporation
Aichi, Japan

T. E. Karis
Hitachi Global Storage Technologies
San Jose Research Center
San Jose, California

Motoyasu Kobayashi
The Institute for Materials Chemistry
  and Engineering
Kyushu University
Fukuoka, Japan

Bengt Kronberg
Institute for Surface Chemistry
Stockholm, Sweden

Roya Maboudian
Department of Chemical Engineering
University of California
Berkeley, California

Joseph M. Perez
Tribology Group, Chemical
  Engineering Department
The Pennsylvania State University
University Park, Pennsylvania

D. Pocker
Hitachi Global Storage Technologies
San Jose Materials Analysis Laboratory
San Jose, California

Parag Purohit
NSF Industry/University Cooperative
  Research Center for Advanced Studies
  in Novel Surfactants
Columbia University
New York, New York

Leslie R. Rudnick
Ultrachem Inc.
New Castle, Delaware

Sharadha Sambasivan
Suffolk Community College
Selden, New York

N. Satyanarayana
Department of Mechanical Engineering
National University of Singapore
Singapore

Brajendra K. Sharma
USDA/NCAUR/ARS
Food and Industrial Oil Research
Peoria, Illinois
and
Department of Chemical Engineering
Pennsylvania State University
University Park, Pennsylvania

S. K. Sinha
Department of Mechanical Engineering
National University of Singapore
Singapore

P. Somasundaran
NSF Industry/University Cooperative
 Research Center for Advanced Studies
 in Novel Surfactants
Columbia University
New York, New York

M. P. Srinivasan
Department of Chemical and
 Biomolecular Engineering
National University of Singapore
Singapore

Marian Wlodzimierz Sulek
Technical University of Radom
Department of Physical Chemistry
Radom, Poland

Lei Sun
Laboratory for Special Functional
 Materials
Henan University
Kaifeng, China

Atsushi Suzuki
Toyota Motor Corporation
Aichi, Japan

Rico F. Tabor
School of Chemistry
University of Bristol
Bristol, U.K.

Atsushi Takahara
The Institute for Materials Chemistry
 and Engineering
Kyushu University
Fukuoka, Japan

Tomasz Wasilewski
Technical University of Radom
Faculty of Materials Science,
 Department of Physical Chemistry
Radom, Poland

Zhishen Wu
Laboratory for Special Functional
 Materials
Henan University
Kaifeng, China

Jiping Ye
Research Department
NISSAN ARC, Ltd.
Yokosuka, Japan

Jun Yin
State Key Laboratory of Nonlinear
 Mechanics (LNM)
Institute of Mechanics
Chinese Academy of Sciences
Beijing, China

Zhijun Zhang
Laboratory for Special Functional
 Materials
Henan University
Kaifeng, China

Ya-Pu Zhao
State Key Laboratory of Nonlinear
 Mechanics
Institute of Mechanics
Chinese Academy of Sciences
Beijing, China

Part I

General

1 Introduction to Surfactants

Michael L. Free

ABSTRACT

Surfactants play an important role in tribological applications. They have properties and characteristics that make them ideal in many lubrication applications. This chapter discusses the general properties and characteristics of surfactants that form the foundation for their role in tribological applications.

1.1 INTRODUCTION

Surfactants are important to many technologically important areas that vary from detergency and cleaning to manufacturing of metal parts and, of course, tribology. Details regarding their general properties are found in a variety of fundamental literature [1–7]. More specific information regarding their application can be found in the fundamental literature as well as in literature that is more specifically related to tribology [8–14].

Surfactants are molecules that have both hydrophilic and hydrophobic sections that impart partial affinity toward both polar and nonpolar surfaces. The hydrophobic section of surfactant molecules consists of nonpolar moieties such as repeating carbon-hydrogen (CH_2) units or carbon-fluorine (CF_2) units. The hydrophilic sections of surfactant molecules contain ionic functional groups such as ammonium (NH_4^+) or carboxylate (CO_2^-), or they consist of nonionic polar groups such as the hydroxyl portion of alcohol molecules. Thus, surfactants are at least partially soluble in polar liquids such as water as well as nonpolar media such as hexane.

The surfactant affinity toward both polar and nonpolar media makes surfactants more likely to be found at boundaries as well as to associate with other surfactant molecules. Consequently, surfactant molecules are particularly active at interfacial boundaries between solids and liquids, liquids and liquids, and liquids and gasses. The word "surfactant" is a contraction of the words "surface active agent." An example of a common surfactant, sodium dodecyl sulfate, is shown in fig. 1.1.

1.2 TYPES OF SURFACTANTS

The three main types of surfactants are ionic, zwitterionic, and nonionic. Zwitterionic surfactants have anionic and cationic constituents. Examples of zwitterionic surfactants include betaines and dimethyl amine oxides. Ionic surfactants are generally classified as cationic if they are positively charged or anionic if they are negatively

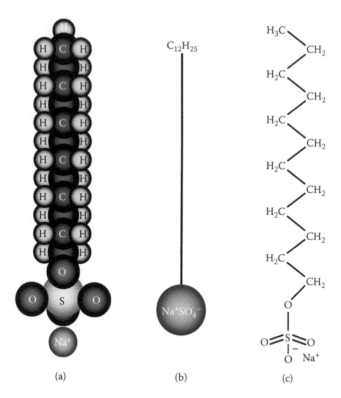

FIGURE 1.1 (See color insert following page 80.) Comparison of (a) space-filling atomic, (b) simplified, and (c) simplified atomic views of sodium dodecyl sulfate. The letters represent the atomic symbols. These views represent the lowest energy state. However, due to molecular vibrations, the actual shape varies with time, and the average shape is not a straight chain as shown. Note that the angle for these views does not reveal the fact that the repeating carbon-hydrogen section of adsorbed sodium dodecyl sulfate is not perpendicular to the surface.

charged when dissolved in aqueous media from a neutral salt form. Most common cationic surfactants are made with amine salts such as cetyl trimethyl ammonium bromide and cetyl pyridinium chloride. Anionic surfactant examples include carboxylate, sulfonate, and sulfate functional groups in salt forms such as sodium oleate, sodium dodecyl benzene sulfonate, and sodium dodecyl sulfate. Nonionic surfactants are not ionic and are, therefore, not found in salt forms. Nonionic surfactants generally consist of alcohols, polyglucosides, and poly(ethylene oxide)s that have attached hydrocarbon chains.

Nearly all surfactants contain between 4 and 18 carbon atoms that are connected in a continuous sequence as shown for sodium dodecyl sulfate in fig. 1.1. In some cases, hydrocarbon sections form branches from a central hydrocarbon chain. In other cases, the hydrophobic portion of the surfactant may contain an aromatic ring structure. Most common surfactants are near the middle of this range (10–14) for the number of carbon atoms in the central chain, which generally determines the molecular length.

The length of the hydrocarbon chain is the primary factor that determines the surfactant solubility. Long-chain surfactant molecules have very low solubility in aqueous media. In contrast, long-chain surfactant molecules have high solubility in nonpolar media such as oils. Structural features such as double bonds or aromatic structures also alter their solubilities.

1.3 ASSOCIATION AND AGGREGATION

The dual nature or amphiphilicity of surfactant molecules provides a thermodynamic driving force for adsorption and aggregation of surfactant molecules. In aqueous media the hydrophobic sections of surfactant molecules are attracted to the hydrophobic section of adjacent surfactant molecules. The association of adjacent hydrophobic sections of surfactant molecules reduces the less favorable interactions between water molecules and individual hydrophobic sections of surfactant, thereby reducing system free energy. The effects of association between adjacent hydrophobic sections of surfactant molecules are enhanced in aggregate structures such as adsorbed layers of surfactant and solution micelles.

Micelles are spherical aggregates of surfactant molecules that can be represented by fig. 1.2a. The concentration at which micelles form in solution is known as the critical micelle concentration (CMC). The concentration at which surfactants aggregate at surfaces to form monolayer-level surface coverage (see fig. 1.2b) is referred to as the surface aggregation concentration (SAC). The SAC is usually very similar to the CMC, although the SAC is usually lower due to interactions with immobile lattice atoms. Other aggregate structures such as bilayers and cylindrical micelles can also form above the CMC or SAC.

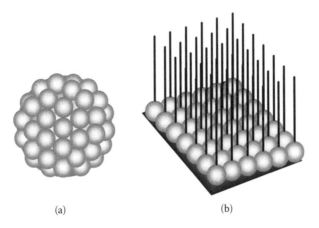

(a)   (b)

FIGURE 1.2   (a) Illustration of the outer portion of a spherical micelle in which the spheres represent the hydrophilic functional group of surfactant molecules; (b) illustration of surfactant molecules adsorbed at a surface at the monolayer coverage level in which the spheres represent the hydrophilic functional group of surfactant molecules and the lines represent the hydrocarbon chain.

1.3.1 FACTORS AFFECTING AGGREGATION

The aggregation process is a strong function of the length of the hydrophobic section of the surfactant molecule. Longer hydrocarbon chains lead to a greater tendency for aggregation. Other factors that affect aggregation strongly include ionic strength, which is indicative of the dissolved charge concentration, and temperature. The effect of these factors is mathematically considered in the following equation [15]:

$$\text{CMC} \cong \text{SAC} \cong \exp(\frac{1}{RT}[(L-x)\Delta G_{c.l.} + k(L-x)RT \ln(\alpha)]) \quad (1.1)$$

in which R is the gas constant (J/mole·K), T is the absolute temperature (K), L is the total number of consecutive CH_2 units in the surfactant molecule, $\Delta G_{c.l.}$ is the free energy increment for each CH_2 unit of chain length as denoted by the subscript "c.l." (J/mole), x is the number of CH_2 units needed to initiate aggregation, k is a solvent polarity factor, and α is the traditional ion activity coefficient that characterizes the interaction between hydrocarbon chains and the polar aqueous media based on a single-charged ion. Although the equation accounts for temperature effects, the Krafft point, which is the temperature above which surfactant solubility rises sharply, must also be considered. Surfactants are generally used at or above the Krafft point.

Aggregate structures such as micelles can be used to deliver organic or aqueous media. Micelles can easily accommodate oils in their interior. As the micelles swell with oil in their interior volume, they form emulsions. Micelles can also be inverted in organic solvents (hydrophobic portions forming the outer perimeter) to form reverse micelles that can accommodate small droplets of water to form oil-based emulsions. Many products containing water and oil are produced in emulsified form.

1.4 ADSORPTION

Interfaces between one solid phase and another phase, such as between solids and liquids, often provide an ideal location for adsorption of surfactant. The solid substrate can provide specific sites at which the hydrophilic head group of a surfactant molecule may bond chemically or physically. Chemical bonding or chemisorption involves a chemical reaction at the surface between the surfactant and the substrate that produces a chemical bond. Physical bonding may involve electrostatic and van der Waals forces between the surfactant and the substrate as well as between surfactant molecules. The immobile nature of a solid substrate, combined with bonding between surfactant and substrate, results in aggregation and solidification of the adsorbed surfactant at concentrations and temperatures that are, respectively, lower and higher than would be needed for aggregation or solidification in bulk media. The adsorption of a monolayer of surfactant causes hydrophilic substrates to change to hydrophobic surfaces due to the orientation of hydrophobic functional head groups toward the substrate, as shown in fig. 1.3a. Conversely, monolayer adsorption on hydrophobic substrates converts them to hydrophilic substrates, as illustrated in fig. 1.3b, which makes it easier to retain polar lubricants at the surface.

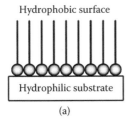
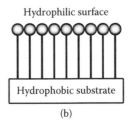

Hydrophobic surface   Hydrophilic surface

Hydrophilic substrate   Hydrophobic substrate

(a)   (b)

FIGURE 1.3 (a) Surfactant molecules, represented by spheres and lines, adsorbed at a hydrophilic substrate, and (b) surfactant molecules, represented by spheres and lines, adsorbed at a hydrophobic substrate

1.5 SURFACE TENSION

Surface tension is the force that holds an interface together. Molecules in the bulk of a medium can form bonds with molecules in all directions. In contrast, those at an interface with another phase are more restricted in bonding options. The lack of balanced bonding opportunities at the interface gives rise to surface tension. The addition of surfactant to aqueous media results in a higher concentration of surfactant molecules at the air–water interface than in the bulk. The presence of the surfactant at the interface provides an opportunity for water molecules to have more bonding opportunities at the interface and reduces the surface tension. The higher the concentration of surfactant at the interface, the lower the surface tension becomes until a close-packed monolayer of surfactant forms. Concentrations that exceed the level needed for close-packed monolayers result in the formation of micelles, as discussed previously. Consequently, surface tension provides useful information regarding the adsorption of a surfactant at an interface.

The concentration at an interface above the bulk concentration level is known as the surface excess, and it is often calculated using the Gibbs equation [1], which is rearranged to

$$\Gamma = -\frac{1}{RT}\left(\frac{d\gamma}{d\ln c}\right) \tag{1.2}$$

in which Γ is the surface excess (mole/m^2), R is the gas constant (J/mole·K), T is the absolute temperature (K), γ is the surface tension (N/m), and c is the concentration (mole/m^3). The Gibbs equation indicates that the surface excess concentration is linearly proportional to the change in surface tension relative to the logarithm of the surfactant concentration. Consequently, it is anticipated that the slope of a plot of surface tension versus the logarithm of the concentration should be linear. Figure 1.4 shows the anticipated relationship. The gradual change in slope as concentration increases reflects increasing adsorption at the air–water interface. The abrupt change in slope to a nearly horizontal line at the minimum surface tension identifies the concentration as the critical micelle concentration. The reason for the change in slope is the completion of one monolayer at the air–water interface, where the surface tension is determined by measuring the downward pull on a flat plate or ring. As the concentration exceeds the level needed for the completion of a monolayer at the air–water

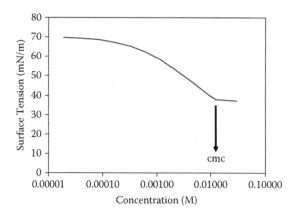

FIGURE 1.4 Plot of surface tension versus concentration for a typical surfactant

interface, additional surfactant has no room to fit in the layer and, therefore, forms micelles in solution or multilayers of surfactant under appropriate conditions. The surface tension does not decrease below the level associated with micelle formation. Consequently, the Gibbs surface excess is only applicable below the CMC.

1.6 CONTACT ANGLE

The adsorption of surfactant usually alters the underlying substrate's interaction with liquids. The interaction of a substrate's surface with a liquid is often characterized by the contact angle. The traditional three-phase contact model is shown in fig. 1.5. Based on the model, a force balance for the interacting surface tension components for each interacting phase combination (γ_{SL}, γ_{LO}, γ_{SO}) leads to Young's equation [2],

$$\gamma_{LO} \cos\theta = \gamma_{SO} - \gamma_{SL} \tag{1.3}$$

in which θ is the contact angle (degree) and γ is the interfacial tension (N/m), where the S subscript denotes the solid, the L subscript denotes the liquid, and the O subscript denotes the other phase, such as a gas or a second liquid.

1.7 ADHESION/COHESION

All molecules experience interaction forces when they encounter other molecules. As atoms approach each other, their orbiting electrons are continually changing positions. Consequently, charge-related interactions are constantly changing, and there is an induced dipole interaction between atoms in adjacent molecules that can provide a strong interaction force at close distances that is part of the van der Waals force. Other interaction forces include electrostatic or coulombic interactions between molecules with a net charge. Other forces that affect interactions are associated with solvent structuring around solute molecules.

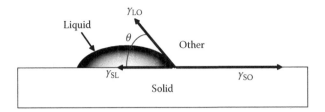

FIGURE 1.5 Typical view of the contact angle for a liquid on a solid substrate in the presence of another phase, such as a gas or a second liquid

Chemical interactions between atoms result in chemical bonding. Chemical bonding between atoms creates molecules. Chemical bonding between molecules such as adsorbate molecules and a substrate result in chemisorption. Chemisorbed molecules are tightly bound to surfaces and can provide a strong anchor for subsequent coatings or lubricant molecules.

Adhesion is one way of characterizing the sum of these interaction forces for different materials in contact. The ability of an adsorbed molecule to resist being forced off a surface by an impinging object with an applied load, which is critical to tribology, is related to the force with which it adheres to the surface. Adhesion is defined in terms of the energy or work per unit area needed to pull apart two different materials. Cohesion is the energy or work required to pull apart a homogeneous material. Thus, adhesion and cohesion are ways to characterize the net force of attraction between molecules in a way that is relevant to tribological applications.

1.8 CONCLUSIONS

Surfactants play an important role in tribology. The role of surfactants in tribology is related to the amphiphilic (dual) nature of surfactant molecules, which is both hydrophilic and hydrophobic. The amphiphilic nature of surfactant molecules leads to aggregation and adsorption, which are important to tribology. Surfactants have the ability to adsorb on metal surfaces to produce hydrophobic surfaces that are more receptive to lubricating oils than unaltered metal surfaces, which are naturally hydrophilic. Surfactants can also impart lubrication properties in the absence of oil. Thus, surfactant molecules play an important role in common tribological applications involving a variety of surfaces and lubricants.

REFERENCES

1. Hiemenz, P. C., and Rajagopalan, R. 1997. *Principles of colloid and surface chemistry.* 3rd ed. New York: Marcel Dekker.
2. Israelachvili, J. 1992. *Intermolecular and surface forces.* 2nd ed. London: Academic Press.
3. Rosen, M. J. 2004. *Surfactants and interfacial phenomena.* 3rd ed. New York: John-Wiley.

4. Adamson, A. W. 1997. *Physical chemistry of surfaces.* 6th ed. New York: Wiley-Interscience.
5. Lange, K. R. 1999. *Surfactants: A practical handbook.* Cincinnati: Hanser-Gardner.
6. Holmberg, K. 2002. *Surfactants and polymers in aqueous media.* 2nd ed. New York: Wiley-Interscience.
7. Myers, D. 1992. *Surfactant science and technology.* 3rd ed. Berlin: VCH.
8. Li, Z., Ina, K., Lefevre, P., Koshiyama, I., and Philipossian, A. 2005. *J. Electrochem. Soc.* 152: G299.
9. Peng, Y., Hu, Y., and Wang, H. 2007. *Tribol. Lett.* 25: 247.
10. Ratoi, M., and Spikes, H. A. 1999. *Tribol. Trans.* 42: 479.
11. Kimura, Y., and Okada, K. 1989. *Tribol. Trans.* 32: 524.
12. Yoshizawa, H., Chen, Y.-L., and Israelachvili, J. 1993. *Wear* 168: 161.
13. Yoshizawa, H., Chen, Y.-L., and Israelachvili, J. 1993. *J. Phys. Chem.* 97: 4128.
14. Vakarelski, I., Brown, S. C., Rabinovich, Y., and Moudgil, B. 2004. *Langmuir* 20: 1724.
15. Free, M. L. 2002. *Corrosion* 58: 1025.

Part II

Tribological Aspects
in Micro- and Nanodevices

2 Geometry and Chemical Effects on Friction and Adhesion under Negligible Loads

Yasuhisa Ando

2.1 INTRODUCTION

Friction force is proportional to the normal load as stated by Amontons–Coulomb's law. For microloads less than 1 mN, which are often found in mechanisms such as microelectromechanical systems (MEMS), Amontons–Coulomb's law is not valid due to the effect of the adhesion force between the contacting surfaces [1]. Studies show that, at such microloads, either the attraction force (adhesion force) caused by the surface tension of water condensed on the surface or the van der Waals force dominates the friction force, and that the friction force is proportional to the sum of the adhesion force (pull-off force) and the normal load [2, 3].

Some studies reported that even a slight roughness decreases the adhesion force significantly [4]. In magnetic storage devices, the technique of creating asperities, called texturing, is often used to prevent sticking at the head and disk interface [5]. The same technique has been used to reduce the sticking of microstructures [6]. Therefore, creating asperities was considered to be an effective way to reduce the adhesion force. However, when applying the friction reduction method by increasing surface roughness to MEMS, random asperities tend to prevent a smooth sliding because the contact area found in MEMS is small and on the order of square microns [7]. In contrast, a periodic array of submicrometer-scale asperities can reduce adhesion and friction forces unless preventing the smooth sliding. It is also very important to use the periodic arrays of asperities to clarify geometric effects on adhesion and friction.

In conventional-sized machines such as an engine system, lubricants are often used to reduce the friction force. Liquid lubricants, however, generate capillary force and often cause stiction in micromechanisms. Moreover, even a low-viscosity liquid tends to increase friction force because viscosity drastically increases when the spacing between solid surfaces becomes narrow and is on the order of nanometers [8]. Therefore, liquid lubricants could cause an increase of the friction force in MEMS. In micromechanisms, one solution for reducing the friction force is molecular boundary lubrication.

Techniques involving Langmuir–Blodgett (LB) films and self-assembled monolayers (SAMs) have been adopted as conventional methods to form organic monolayers on solid surfaces. Under a microload friction, where the normal load is negligible compared

13

with the adhesion force, organic monolayers are expected to reduce friction force in two ways. One is to *indirectly* reduce the friction force by reducing the adhesion force. Thus, the friction force is indirectly reduced because the friction force is proportional to the adhesion force (pull-off force) when the normal load is negligible [9]. The other way is to *directly* reduce the friction force. These two reduction effects (indirect and direct) might effectively reduce the friction force under microload conditions. Moreover, the friction force might be further reduced by the use of periodic asperity arrays. Therefore, these three means of reducing the friction force under microload conditions need to be clarified.

Surfactants are used for reducing friction. They are found in hair conditioner, fabric softener, etc. Determining the tribological characteristics of LB films and SAMs when the normal load is negligibly small is also important for understanding the friction-reducing effects of surfactants, because the molecular structures of some surfactants are very similar to those of the molecules composing LB films or SAMs.

In this study, we clarified these friction-reduction effects under microload conditions by measuring the friction and pull-off forces for two-dimensional asperity arrays on silicon plates. First, two-dimensional asperity arrays were created using a focused ion bean (FIB) system to mill patterns on single-crystal silicon plates. Each silicon plate had several different patterns of equally spaced asperities. Then, the friction and pull-off forces were measured using an atomic force microscope (AFM) that had a square, flat probe. This report describes the geometry effects of creating asperity arrays and the chemical effects of depositing LB films or SAMs on the friction and pull-off forces.

2.2 CREATION OF PERIODIC ASPERITY ARRAYS BY FOCUSED ION BEAM

2.2.1 FOCUSED ION BEAM SYSTEM AND PROCESSING METHODS

We created two-dimensional arrays of asperities having various heights on a silicon surface by milling with an ion-beam-focusing system, FIB 610 (FEI, Hillsboro, OR). The primary components of this FIB system are a liquid ion metal source (LIMS), ion-focusing column, specimen chamber, and electronics control console. The system extracts positively charged ions from the LIMS by applying a strong electric field, and then focuses the ions into a beam, which then scans the specimen using electrical lenses and deflectors. The nominal minimum diameter of the focused beam is 28 nm. The ion beam removes material from the specimen through a sputtering process and thus mills a narrow groove less than 100 nm wide on the specimen.

A groove can be made by scanning the beam continuously along the desired line on the specimen. The depth and width of the groove are determined by the radius of the beam and the processing time. By milling equally spaced grooves in two orthogonal directions, unmilled parts remained to form a two-dimensional array of "asperities." By varying the milling conditions, the groove depth, groove width, and the spacing between the grooves can be varied. To deposit a platinum film, a deposition gas (methylcyclopentadienylplatinum: $(CH_3)_3(CH_3C_5H_4)Pt$) is introduced, which reacts with the area where the ion beam strikes. By varying the processing time, the thickness of the film can be controlled.

2.2.2 Surface Modification Using Focused Ion Beam

A specimen on which the pattern is to be created must be metal or a semiconductor, because the electrostatic charge as the result of irradiation of ions makes the processing difficult. So, in this study, single-crystal silicon was used as the specimen because it has a very smooth surface, and it is much easier to mill a fine groove of about 100 nm width on it than on other materials such as polycrystalline metal.

Figures 2.1–2.4 show examples of periodic asperity arrays used in this experiment, which were measured by a conventional sharp probe that had a nominal radius of curvature of less than 40 nm. Figures 2.1 and 2.2 are two-dimensional

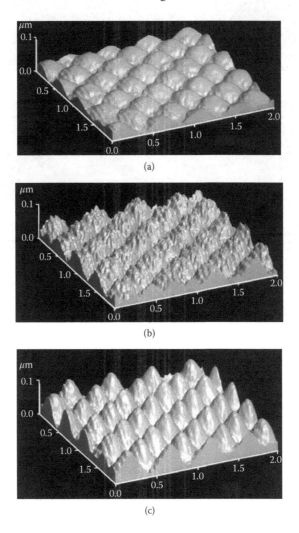

(a)

(b)

(c)

FIGURE 2.1 AFM images of the periodic asperity arrays. Asperity arrays were produced by using an FIB to mill grooves on a silicon surface. Groove depth was calculated from the AFM image. Each groove depth was determined from the topography measurements. (a) 21.2-nm groove depth, (b) 38.7-nm groove depth, (c) 49.2-nm groove depth.

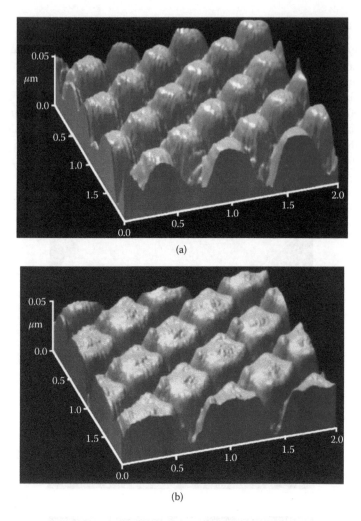

FIGURE 2.2 AFM images of periodic asperity arrays of various mound areas on a silicon plate. When fabricating the patterns, the number of scanning lines of the ion beam was varied to control the width of the groove, and the distance between the grooves was varied to control the area of each mound. The ratio of mound width to total width of mound and groove is 2/9 for (a) and 2/11 for (b), which were determined from the milling conditions. The groove depth is 23.1 nm for (a) and 19.7 nm for (b), which were determined from the topography measurements.

asperity arrays obtained by milling many grooves in orthogonal directions. With an increase in the processing time, the height of the asperities increased as shown in figs. 2.1a–2.1c. The microsurface roughness of asperities was extremely large for fig. 2.1b due to the different processing conditions. The distance between adjacent peaks was constant in fig. 2.1 and was about 240 nm for each asperity array, which was determined from the milling conditions. In fig. 2.2, the ratio of mound width to total width of mound and groove was varied by varying the spacing between the grooves. The depth of the grooves was relatively constant and was 20–27 nm. Figure 2.3

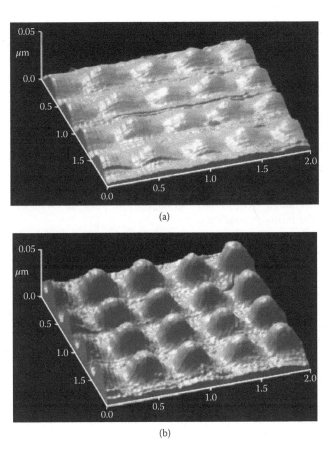

(a)

(b)

FIGURE 2.3 AFM images of periodic arrays of platinum asperities fabricated on a silicon plate. The asperity array was fabricated by depositing platinum mounds in two orthogonal directions at the same spacing. Each asperity height was determined from the topography measurements. (a) 11.1-nm asperity height, (b) 24.3-nm asperity height.

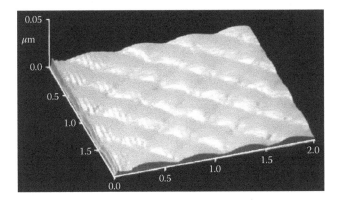

FIGURE 2.4 AFM image of asperity array fabricated on platinum layer that was deposited by FIB (6.5-nm groove depth).

shows an asperity array of platinum that was deposited by focusing small spots of ion beam equally spaced in two orthogonal directions while introducing the deposition gas. With an increase in the irradiation time, the height of the asperities increased, as shown in figs. 2.3a and 2.3b. Although fig. 2.4 also shows an asperity array of platinum, the pattern was obtained by milling many grooves in orthogonal directions in the same way as shown fig 2.1 after depositing a platinum layer in the wider area. The modified area created on the silicon plate was about 5×5 μm^2.

Table 2.1 summarizes the patterns used to determine the geometry effects and shows material of asperity, groove depth (asperity height), the distance between adjacent peaks (pitch), the ratio of mound width to total width of mound and groove, and the angle between the sliding direction and the direction of the array. The groove depths were obtained from the AFM images. The pitch and the ratio were determined from the milling conditions.

2.2.3 Curvature Radius Calculations

Some asperity arrays were selected and the curvature radii of the asperity peaks were calculated to examine the effects of contact geometry in a different perspective. The patterns on test specimen no. 3 in table 2.1 were selected because the curvature radii for the array patterns in this specimen were distributed widely.

The curvature radii of the asperity peaks were calculated from the AFM measurement data, as shown in fig. 2.1. For this calculation, first the AFM topography data were input into a personal computer. Next, we selected more than 10 asperity peaks and fitted each of them with a spherical surface curve. When fitting each peak, we included measured data within a 30–50-nm-radius area of each peak. Each fitted area contained about 80–200 digital data points. Using these data points, we calculated

TABLE 2.1

Patterns Used for Examining Geometry Effects

Specimen no.	Material	Groove depth (nm) [mound ratio[a]]	Pitch (nm)	Friction direction
1	Si	3.0–19.9 [2/7]	245	—
2	Si	17.2 [2/7]	240	0°
		19.3 [2/7]	240	45°
3	Si	6.2–49.2 [2/7]	240	45°
4	Si	21.2 [2/7]	240	45°
		19.5 [2/9]	310	
		20.5 [2/11]	380	
		27.2 [4/11]	380	
5	Pt	5.2–43.0	330	0°
6	Pt	2.5–26.2	330	0°
7	Si	3.7–15.5 [2/7]	240	45°
	Pt	3.7–7.9 [2/7]		

[a] Ratio of mound width to total width of mound and groove. Each width represents the number of scanning lines of the ion beam used for fabrication of patterns.

TABLE 2.2
Geometry of Asperity Array

Patterns used for determining geometry effects (specimen no. 3)

Chemical modification	Groove depth (nm)	Average curvature radius (nm) (σ: standard deviation)
None	6.2	790 (σ: 400)
	8.8	640 (σ: 130)
	18.0	330 (σ: 53)
	21.2	290 (σ: 38)
	38.7	170 (σ: 110)
	49.2	86 (σ: 5.5)

the curvature radius for each asperity by means of an approximation program that used a steepest-descent method. Then, we averaged the radius for each asperity array. Table 2.2 shows the geometry of the asperity arrays, i.e., groove depth, averaged curvature radius of the peaks, and standard deviation of the averaged curvature radius.

2.2.4 CHEMICAL MODIFICATIONS OF ASPERITY ARRAYS

In order to determine the effect of chemical modification on the friction and pull-off forces, self-assembled monolayers (SAMs) and Langmuir–Blodgett (LB) films were formed on the silicon surface. Films were formed on the asperity arrays having various groove depths, where the distance between the adjacent peaks was about 240 nm for each asperity array. The SAM and LB films were deposited as follows.

Each plate was cleaned and then coated with a SAM of alkylchlorosilane in a three-step process. First, to remove chemically or physically adsorbed contaminants, the plate was cleaned with a so-called piranha solution (3/7 v/v mixture H_2O_2/H_2SO_4) at about 70°C for 2 h and then with a 4/6 v/v mixture of benzene/ethanol for 48 h. After each cleaning process, the plate was rinsed with high-purity water. Second, to remove surface contaminants, the plate was placed in a UV (ultraviolet)/O_3 cleaner (UV output of 25 W) and exposed for 10 min. Finally, to form the SAM coating, the plate was immersed for about 5 s in a 0.5 mM solution of, for example, octadecyltrichlorosilane ($CH_3(CH_2)_{17}SiCl_3$:C_{18}) in hexane. We also prepared a control plate that contained an asperity array without a coating (C_0).

LB films were also formed on the silicon plate after the asperity array was processed by FIB. Before depositing the LB film, we cleaned the silicon wafers in a mixture of benzene and ethanol, rinsed them in pure water, and then exposed them to a UV–ozone atmosphere. Then the plate was immersed in ultrapure water where a monolayer of stearic acid ($C_{17}H_{35}COOH$:CH) or fluorocarboxylic acid ($C_6F_{13}C_{11}H_{22}COOH$:CFCH) was confined at a pressure of 30 mN/m. The monolayer on the water migrated onto the silicon surface and formed the LB film (CH-/CFCH-LB film) when the plate was removed from the ultrapure water. The temperature of the ultrapure water was 20°C. Table 2.3 shows the chemical modifications and the geometries of the asperity.

TABLE 2.3

Geometries and Chemical Modifications of Asperity Array

Patterns used for determining chemical effects

Sample name: Chemical modification	Average curvature radius (nm) (σ: standard deviation)
C_0(-SAM): (Uncoated)	250–1500 (σ: 41–1450)
C_6(-SAM): n-hexyltrichlorosilane ($CH_3(CH_2)_5SiCl_3$)	560–2330 (σ: 200–1350)
C_8(-SAM): n-octyltrichlorosilane ($CH_3(CH_2)_7SiCl_3$)	200–800 (σ: 44–83)
C_{14}(-SAM): n-tetradecyltrichlorosilane ($CH_3(CH_2)_{13}SiCl_3$)	1880, 1910 (σ: 680, 770)
C_{18}(-SAM): n-octadecyltrichlorosilane ($CH_3(CH_2)_{17}SiCl_3$)	330–2016 (σ: 85–1700)
CH-LB: stearic acid ($C_{17}H_{35}COOH$)	120–790 (σ: 10–150)
CFCH-LB: fluoroluorocarboxylic acid ($C_6F_{13}C_{11}H_{22}COOH$)	95–520 (σ: 6–66)

TABLE 2.4

Contact Angles and Surface Energies on SAM-Coated Plates

Sample name	Contact angle measured after deposition (degrees)		Surface energy (mJ/m^2)	Water contact angle measured after force measurement (degrees)
	Water	CH_2I_2		
C_0	4–5	39	73	34
C_6	105	79	24	94
C_8	105	79	25	—
C_{14}	111	72	21	103
C_{18}	112	70	20	—

The wettability of each SAM-coated plate was measured after the formation of SAM to confirm that the alkyl chain in alkylchlorosilane was exposed to the outside. The contact angles of water and diiodomethane (CH_2I_2) were measured by using a droplet on the flat area outside of the patterned area. These two liquids were used because the hydrogen bonding and dispersion force components manifest themselves in the surface energy with water and diiodomethane, respectively. The surface energies of the SAM-coated plates were determined from contact angle measurements by using the Fowkes equation [10]. Table 2.4 summarizes the contact angle measurements and surface energy of each plate.

2.3 MEASUREMENTS OF FRICTION AND PULL-OFF FORCES BY AN ATOMIC FORCE MICROSCOPE

2.3.1 Measuring Apparatus (Atomic Force Microscope)

Each modified area on the silicon surface was so small (5×5 μm²) that an optical lever type AFM system was used to measure the friction and pull-off forces. Figure 2.5 shows a schematic of the measurement method. The laser beam is deflected by the cantilever and strikes the quad-photodiode detector. When the cantilever probe is pressed against the specimen, the applied force by the cantilever is obtained from the output of the photodiode as $(a + b) - (c + d)$, where a, b, c, and d correspond to the outputs of sensor plates A, B, C, and D, respectively. The output for the height

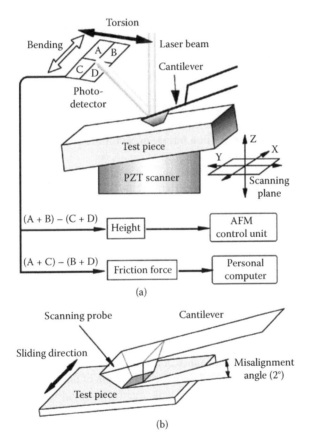

FIGURE 2.5 Schematic drawings of experimental setups (AFM) for measuring friction and pull-off forces. (a) The AFM probe scans along the x-direction, and the torsion and bending of the cantilever are detected by a quad-photodiode detector. (b) Arrangement of probe on a test specimen. The angle between the flat, square tip of the probe and the test piece was $2.5°-3°$.

is used for the conventional AFM system. The friction force between the cantilever probe and specimen is determined from $(a + c) - (b + d)$. The output for the friction force is analog-to-digital converted, and then the average friction force in a scanning is calculated using a personal computer.

To estimate the friction force from the output of $(a + b) - (c + d)$, first we measured the change in the value of the detector's output when tilting the cantilever. Then we calculated the torsional rigidity from the dimensions of the cantilever as 0.36×10^{-9} N·m/rad by using a shear modulus of 57.9 GPa, obtained from Young's modulus of 146 GPa [11] and a Poisson's ratio of 0.26 [12]. The two values (output change and torsional rigidity) were then used to calculate a conversion factor, which was used to convert the photodiode output to the friction force. This factor, however, changes as the position of the laser spot on the cantilever changes. This effect cannot be disregarded when exchanging a specimen or a cantilever. Therefore, we milled various patterns on the same silicon plate and examined the differences between the patterns on that plate.

Although an FIB can mill grooves less than 100 nm wide, this is still larger than the radius of curvature of a conventional AFM probe. The radius of curvature of the probe must be larger than one period of the pattern to examine the geometry effects. The tip of the scanning probe was a flat square, 0.7×0.7 μm². The SEM image of the tip of the scanning probe is shown in fig. 2.6. (This flat probe was used when measuring the pull-off and friction forces, except for the measurement shown in fig. 2.8 later in this chapter, in section 2.4.1.)

Both the installed cantilever and the flat square of the probe were inclined to the scanning plane of the PZT (lead zirconate titanate) scanner. To decrease this misalignment angle and to increase the contact area, the specimen was also inclined, almost parallel to the flat area of the probe (fig. 2.5). The misalignment angle was kept at 2°–3°, however, as the specimen and the square tip were parallel, the sensitivity of the AFM was significantly decreased.

2.3.2 METHOD FOR MEASURING FRICTION AND PULL-OFF FORCES

When measuring the pull-off force, we used a conventional AFM system that allowed us to use a force–curve mode that showed the force required to pull the scanning probe tip off the specimen. Figure 2.7 shows an example of how we measured the pull-off force from the force–distance curve. The photodiode output $(a + b) - (c + d)$ is plotted on the vertical axis, and the relative position of the specimen in the z-direction is plotted on the horizontal axis. This relative position is based on the applied voltage to the PZT scanner. In fig. 2.7, the probe jumped off the specimen just as the cantilever was pulled a distance H from its neutral position. The pull-off force was calculated using the distance H and the nominal spring constant of the cantilever (0.75 N/m). Because the pull-off force was originally calculated using the applied voltage to the PZT scanner, the measured value included an error caused by the hysteresis in the PZT.

Small particles can easily adhere to the square tip of an AFM probe and can influence the experimental results because the contact area between the flat probe and specimen was much larger than that of a conventional probe tip. Therefore, when

(a)

(b)

FIGURE 2.6 SEM images of the scanning probe. The flat square of the scanning probe is about $0.7 \times 0.7 \ \mu m^2$ from the image. (a) Side view, (b) flat square of the scanning probe.

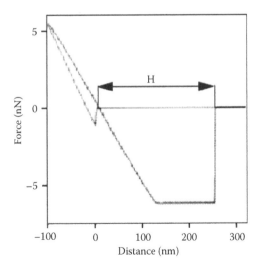

FIGURE 2.7 Force–curve technique used to determine the pull-off force. The force is calculated using the distance H and the spring constant of the cantilever.

TABLE 2.5

Experimental Conditions

Cantilever (Si$_3$N$_4$)	
Spring constant (N/m)	0.75/0.37
Torsional rigidity (N·m/rad)	$0.36 \times 10^{-9}/0.20 \times 10^{-9}$

Sliding conditions	
External normal load (nN)	~5 (0 to 10)
Raster scan area (μm^2)	2×2
Sliding speed ($\mu m/s$)	2
Relative humidity (%)	20–56

examining the difference in the pull-off force between the patterns, first we scanned an area of 20×20 μm^2 including some patterns. Second, these patterns were displayed together in the measurement window for the force–curve mode. Third, we selected more than six measurement points in each pattern, and then the pull-off forces were measured. Because a raster scanning was not performed between measurements, the possibility of any change in probe surface conditions caused by friction was very slim.

The friction force was measured as follows. First, an area of 2×2 μm^2 was selected in a pattern for friction-force measurement. Second, more than six measurement points were selected in that area, and the pull-off force was measured. Third, raster scanning was performed in the scanning direction of x, as shown in fig. 2.5a, and the photodiode output $(a + c) - (b + d)$ was registered on a personal computer. The personal computer calculated the average of the signal corresponding to the torsion angle of the cantilever over the measurement area. The friction force was obtained by multiplying the average by the conversion factor calculated in section 2.3.1. Finally, the pull-off force was measured again in the same way as before the friction-force measurement. The experimental conditions are summarized in table 2.5.

2.4 GEOMETRY EFFECTS ON FRICTION AND ADHESION [9, 13, 14]

2.4.1 FRICTION AND PULL-OFF FORCES ON SILICON PATTERNS

Figure 2.8 shows the relation between the pull-off force and groove depth. In this figure, solid squares show the pull-off force measured by a flat probe (fig. 2.6), and open triangles show the pull-off force measured by a conventional sharp probe. We did not carry out surface scanning between each measurement due to the possibility of altering the shape of the scanning probe either by a very small amount of wear or by contamination of the surface during friction measurements. Therefore, we assumed that the conditions of the probe were similar for all measurements. In fig. 2.8, the pull-off force measured with the flat probe clearly decreases with groove depth. By

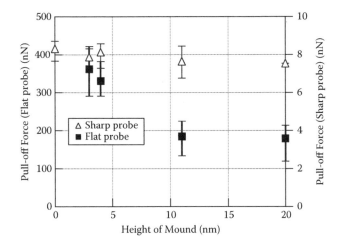

FIGURE 2.8 Relation between pull-off force and groove depth created on the silicon surface. The pull-off force was measured by using a flat probe as well as a normal sharp probe. Each probe was made of Si_3N_4.

contrast, the pull-off force with the conventional sharp probe remained constant, because the apparent contact area was so small that the contact conditions were not affected by the change in surface topography.

Figure 2.9 shows two kinds of friction forces obtained during surface scanning on the silicon asperity array. The abscissas in both figures represent lapse of time. The direction of the friction force changes when the sliding direction is reversed. Figures 2.9a and 2.9b show the friction force measured along the asperity array (i.e., parallel direction) and at a rotation of 45° to the array (i.e., 45° direction), respectively.

The average friction force during each reciprocating motion increases and decreases periodically, with the period depending on the sliding direction in fig. 2.9. The path of the sliding probe shifted during surface scanning. This shift caused changes in the contact conditions between the probe and the array, resulting in variations in the average friction force between reciprocating motions. When the average friction force is high, during the 3–10-s time period in fig. 2.9a, and 7–10-s and 15–18-s in fig. 2.9b, the probe seemed to be in contact with two rows of asperities simultaneously. This increase in the contact area caused the increase in the average friction force. The difference in the period was caused by the difference in the distances between the peaks along the direction of motion (d_p in fig. 2.9).

Spikes in the friction force are more prominent in fig. 2.9a compared with fig. 2.9b in the high average friction region. When the probe slid along a row of asperities and was in contact with two rows, the edge of the probe square contacted two asperities simultaneously; this did not occur when the probe slid at a 45° direction to the asperity array. The force needed for the probe edge to "climb" the asperities appears as a sharp increase in fig. 2.9a. Since this influence is different in different sliding directions (parallel direction or 45° direction), the sliding direction should be coincided when comparing the friction forces.

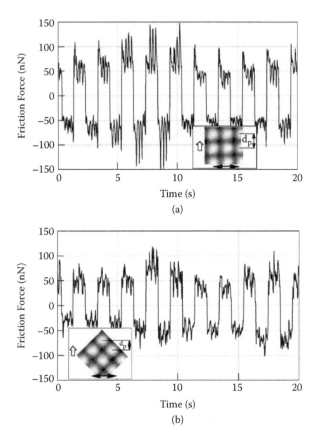

FIGURE 2.9 Friction forces obtained during surface scanning on a silicon asperity array when (a) the scanning direction is along the asperity array and (b) the scanning direction is inclined at 45° to the asperity array. Double arrow shows the primary scanning direction and open arrow shows the secondary scanning direction. The variable d_p shows the distance between the asperity peaks in the secondary scanning direction.

Figure 2.10a shows pull-off forces measured on the same pattern without surface scanning between each measurement (i.e., a no-scan pull-off force). Figure 2.10b shows the friction and pull-off forces measured for the arrays as a function of groove depth. The measured friction force was averaged over a scan area of about 2 to 4 μm². The pull-off forces in this figure were measured before and after each friction measurement at the same scan area, and were then averaged (i.e., a scan pull-off force).

Figure 2.10b shows that both friction and pull-off forces decrease with increasing groove depth. When measuring the friction force, a surface scan was carried out, and the probe was slid on the specimen not only to measure the friction force, but also to find the measurement position. Although the contact condition can change by this friction, the pull-off force shown in fig. 2.10b decreases with groove depth in the same way as that shown in fig. 2.10a. We also measured the pull-off and friction forces on a flat part of the specimen (i.e., unmilled) at the beginning and end of the series of measurements, and found that the differences were less than 1% and 6% for

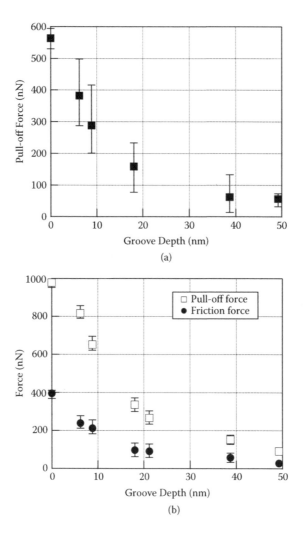

FIGURE 2.10 Friction and pull-off forces measured on a silicon periodic asperity array. (a) Pull-off forces were measured on each pattern without surface scanning between the measurements. (b) Pull-off forces were measured before and after each friction measurement at the same scanning area and were averaged.

the two forces, respectively. Therefore, changes in the friction force were not caused by changes in probe topography (due to wear or contamination) but by the geometry of the silicon asperity.

Figure 2.11 shows the friction and pull-off forces measured for various spacings and widths of periodic grooves using the patterns shown in fig. 2.2. The abscissa represents the ratio of mound width to total width of mound and groove, which was determined from the milling conditions. Both forces increase as this mound ratio increases.

From the topography data shown in fig. 2.1, the depth measured from the highest point in the image is divided into consecutive periods of 0.2 nm, and the surface areas included in all consecutive periods are added together over each period. Figure 2.12

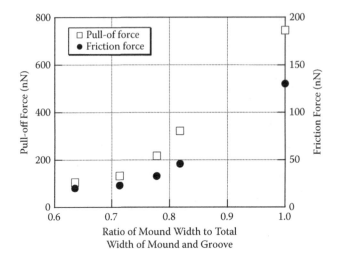

FIGURE 2.11 Friction and pull-off forces as a function of the ratio of mound width to total width of mound and groove.

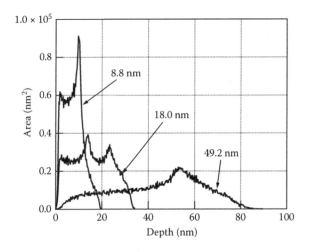

FIGURE 2.12 Depth distribution of silicon asperity arrays. The relations between the area and depth are calculated from topography data of a scanning range of 2×2 µm². The underlined dimension shows the groove depth of each pattern.

shows a distribution of the depth in the silicon patterns calculated as described above. Comparing the sum of the areas included from zero to 5 nm in depth, the included areas of the patterns with grooves of 8.8-, 18.0-, and 49.2-nm depth are calculated to be 1.11, 0.53, and 0.05 µm², respectively. Hence, a deeper groove not only increases the average depth, but also decreases the surface area close to the contact point. The adhesion force was probably strongly affected by this area.

Figure 2.13 shows the relation between the friction force and the pull-off force, from the data in figs. 2.10b and 2.11. Both friction forces are proportional to the

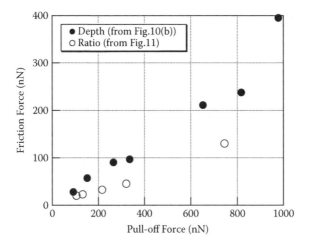

FIGURE 2.13 Relation between friction force and pull-off force measured on two kinds of asperity arrays. Friction forces are extracted from figs. 2.10b and 2.11 and are shown as solid circles and open circles, respectively.

pull-off force. The difference in the gradients of the two friction forces was caused by the difference in the sensitivities of the sensor, as described in section 2.3.1. The friction force is reportedly almost proportional to the sum of the normal load and pull-off force [2, 3]. In fig. 2.13, as the external applied load to the friction surfaces was less than 10 nN and thus was negligible compared with the pull-off force, the assumption that the friction force is proportional to the sum of the normal load and the pull-off force is true when the adhesion (pull-off) force varies.

2.4.2 FRICTION AND PULL-OFF FORCES ON PLATINUM PATTERNS

Figure 2.14a shows the relation between the no-scan pull-off force and the asperity height measured on a platinum pattern as shown in fig. 2.3. The pull-off force was measured without surface scanning between pull-off force measurements. The pull-off force scarcely changed as the asperity height increased. The pull-off force was measured on a silicon pattern by using a conventional sharp probe (fig. 2.8). The effect of groove depth could not be observed for the sharp probe because the contact area was too small to be affected by changes in surface topography. The relation between the pull-off force and asperity height for the flat probe on the platinum pattern (fig. 2.14a) was similar to that of the sharp probe on the silicon pattern. If wear debris were jammed between the probe and the substrate or formed a bump on the flat probe, the square tip of the probe could not make contact with platinum asperities. Thus, the pull-off force was not affected by the difference in asperity height.

Figure 2.14b shows the friction force measured on platinum patterns and the averaged scan pull-off force measured before and after each friction measurement for the same scanned area. In fig. 2.14b, both the friction and pull-off forces decreased as the asperity height increased, but not as rapidly as the forces with silicon groove depth (fig. 2.10b). The difference in contact conditions, as mentioned previously, possibly caused the differences in rates of decrease in the friction and pull-off forces.

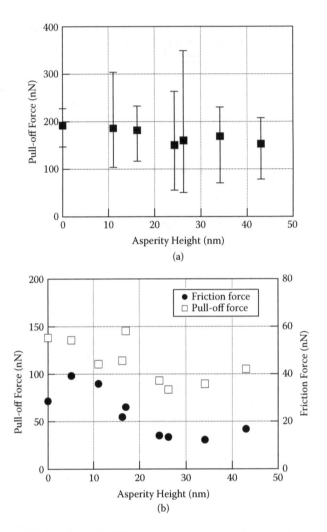

FIGURE 2.14 Friction and pull-off forces measured on platinum asperity arrays as shown in fig. 2.3. (a) Relation between pull-off force and groove depth and (b) effect of asperity height on friction and pull-off forces

If the reaction of platinum by the FIB was not sufficient, the bonding force between the platinum layer and substrate was probably weak, and a part of the deposited platinum easily turned into debris by rubbing. In order to remove the superficial layer that probably had a weak bond to the substrate, the specimen was placed in the FIB system again, and all the pattern areas were sputtered by FIB. Figure 2.15 shows AFM images of the sputtered pattern of the platinum asperities. The patterns shown in figs. 2.3a and 2.3b changed to figs. 2.15a and 2.15b, respectively. As the height of the asperity decreases, the small irregular ridges completely disappear in fig. 2.15.

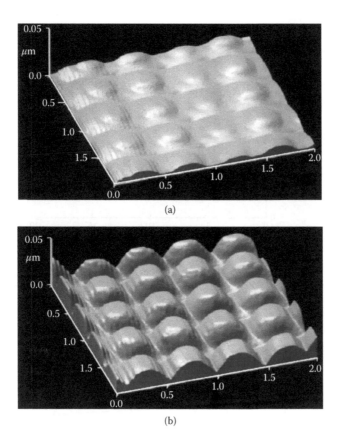

FIGURE 2.15 AFM images of platinum asperity arrays after sputtering. The shape of platinum asperity arrays shown in fig. 2.3 changed after removing the unreacted material. The height of the asperity decreased and the asperity peaks became flat. (a) 6.6-nm asperity height; (b) 17.2-nm asperity height.

Figure 2.16a shows the no-scan pull-off force on the sputtered platinum patterns (fig. 2.15). Figure 2.16b shows the friction and the scanned pull-off force measured on the same specimen. In fig. 2.16b, both the friction and pull-off forces clearly decreased as the asperity height increased. The superficial layer was probably removed by the sputtering process by FIB, because the fluctuation of the forces significantly decreased. In fig. 2.16a, the pull-off forces measured on the pattern of 6.6-nm asperity height showed wide scatter compared with the others because of the lower uniformity of the asperity shape.

Figure 2.17 shows the relation between the friction force and pull-off force, from the data in fig. 2.16b. The friction force was proportional to the pull-off force except for the point (open circle) measured on the silicon surface having no platinum asperity. This difference was caused by the difference in the material. Therefore, the friction force was proportional to the sum of the normal load and the pull-off force for the platinum patterns.

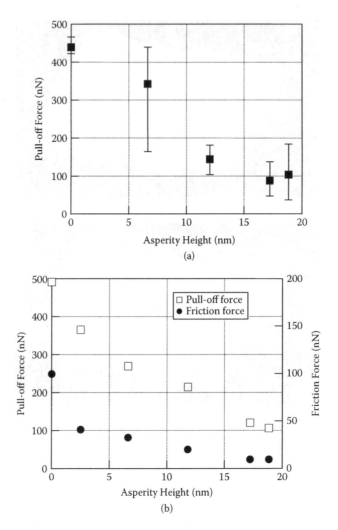

FIGURE 2.16 Friction and pull-off forces measured on platinum asperity arrays after sputtering. (a) Relation between pull-off force and asperity height and (b) effect of asperity height on the friction and pull-off forces.

2.4.3 COMPARISON OF FRICTION AND PULL-OFF FORCES OF SILICON AND PLATINUM

It was difficult to precisely compare the friction forces of different materials on different substrates (figs. 2.13 and 2.17) because the sensitivity of detecting the torsion angle was not always the same for each measurement. Platinum and silicon patterns were made on the same plate, and the friction and pull-off forces were measured to compare the friction coefficients calculated by dividing the friction forces by the pull-off forces for the different materials. Figure 2.18 shows the friction and scanned pull-off forces as a function of groove depth, and these forces were measured for

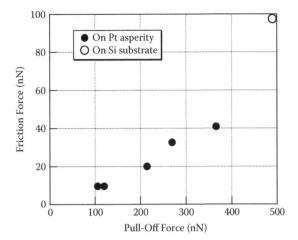

FIGURE 2.17   Relation between friction force and pull-off force measured on a platinum asperity array as shown in fig. 2.15. Friction and pull-off forces are extracted from fig. 2.16. The maximum friction force of 100 nN (shown as ◯) was measured at the asperity height of 0 nm in fig. 2.6, which means the friction force was measured on a bare silicon surface without platinum asperities.

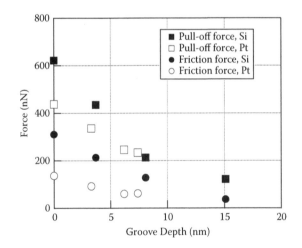

FIGURE 2.18   Comparison of friction and pull-off forces between silicon and platinum asperity arrays. The platinum asperity arrays were fabricated near the silicon asperity arrays on the same silicon plate, which makes it possible to directly compare the difference in the friction and pull-off forces between two different materials.

platinum and silicon asperity arrays on the same silicon plate. This figure shows that the friction force on the silicon asperity array was always greater than that on platinum, although the pull-off forces were rather comparable for both patterns.

Figure 2.19 shows the relation between the friction and pull-off forces using the data from fig. 2.18. The slope of each fitted line that passes through the origin (values

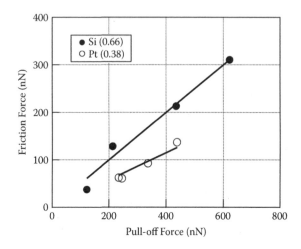

FIGURE 2.19 Comparison of friction force vs. pull-off force on platinum and silicon asperity arrays using data from fig. 2.18. The friction forces measured on each material were fitted with a line that passes through the origin. The gradient of each approximated line is shown in the parentheses in the inset box.

are shown in the inset) appears to correspond to the friction coefficient. This friction coefficient for silicon was about twice than that for platinum.

The pull-off forces shown in fig. 2.18 probably reflected the adhesion forces acting on the surfaces to some extent. If the friction forces were directly caused by the adhesion forces, then the same adhesion force must generate the same friction force. However, in these measurements, the friction coefficient calculated by dividing the friction force by the pull-off force depended on the material. This means that the mechanism causing the adhesion force was different from that causing the friction force, and that the adhesion force acted as a hidden normal load and thus indirectly generated the friction force.

2.4.4 Relation between Pull-Off Force and Curvature Radius

Johnson, Kendall, and Roberts (JKR) calculated the Hertzian contact area between two spherical surfaces when the adhesion energy could not be disregarded [15]. They verified their theory by their own experiments using the material combination of rubber and glass. In the JKR theory, it is assumed that adhesion energy is proportional to the contact area and that the attractive force acting on the outside of the contact area can be ignored. The JKR theory is outlined below.

Two spheres in fig. 2.20, with radii R_1 and R_2, respectively, have a Hertzian contact under normal load P_0. The apparent Hertzian load P_1 is given by the following equation:

$$P_1 = P_0 + 3\gamma\pi R + \sqrt{6\gamma\pi R P_0 + (3\gamma\pi R)^2} \qquad (2.1)$$

where γ is the adhesion energy for the two surfaces and R is the effective radius determined as $R = R_1 R_2/(R_1 + R_2)$. Considering P_1 as equivalent to the external force,

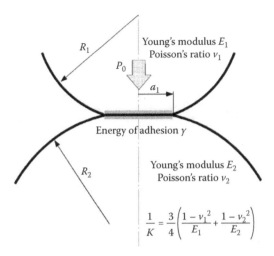

FIGURE 2.20 Contact geometry between two elastic solids showing outline of JKR theory. The two hemispherical surfaces having curvature radii of R_1 and R_2 are in contact under an external load of P_0. The radius of the contact circle is a_1. The energy of adhesion γ is generated between the two surfaces.

the radius a of the contact circle derived from the Hertz equation is given by the following equation:

$$a^3 = \frac{R}{K} \left\{ P_0 + 3\gamma\pi R + \sqrt{6\gamma\pi R P_0 + (3\gamma\pi R)^2} \right\} \tag{2.2}$$

where K is the elastic constant calculated from the Poisson's ratio and Young's modulus of each material. The condition in which both surfaces are in a stable contact is expressed as

$$P_0 \geq -\tfrac{3}{2}\gamma\pi R \tag{2.3}$$

From the above, the pull-off force, $P_{\text{pull-off}}$, which is measured when the surfaces are separated, is given by the following equation:

$$P_{\text{pull-off}} = -\tfrac{3}{2}\gamma\pi R \tag{2.4}$$

Equation (2.4) shows that the pull-off force is proportional to R. A similar relation would be obtained if a water capillary formed around the contact area and the surface tension of water was considered. When a hemispherical asperity whose radius of curvature is R_S contacts with a flat plane (fig 2.21), water condenses at the narrow spacing around the contact point. If the periphery of the capillary has a concave surface (i.e., a meniscus), the saturated water vapor pressure p_s on the meniscus surface is lower than that on a flat surface. Therefore, liquid water can exist when

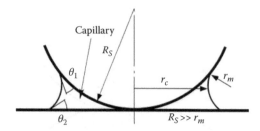

FIGURE 2.21 Capillary formed around contact area between hemispherical and flat surfaces. Size of the capillary can be geometrically determined based on the radius of curvature of the meniscus, contact angle of water, and the radius of curvature of the hemispherical surface.

relative humidity (p/p_s) is less than 100%, and thus r_m is given by Kelvin's equation [16] as

$$r_m = \frac{0.54}{\log(p/p_S)} \tag{2.5}$$

The area of the capillary πr_c^2 geometrically is

$$\pi r_c^2 = 2\pi R_S r_m (\cos\theta_1 + \cos\theta_2) \tag{2.6}$$

where θ_1 and θ_2 are the contact angles of water on the sphere and plane, respectively. The adhesion force F_W generated by the Laplace pressure is then given by

$$F_W = 2\pi R_S \gamma_L (\cos\theta_1 + \cos\theta_2) \tag{2.7}$$

Equation (2.7) shows that F_W is independent of r_m, indicating that the relative humidity does not affect the adhesion (pull-off) force, or F_W, between a hemispherical surface and a flat surface.

From the above discussion, the pull-off force is proportional to the curvature radius in both cases, i.e., whether the adhesion force acts in the contact area or the Laplace pressure acts in a capillary. Thus, the pull-off force must be proportional to the curvature radius of the hemispherical asperity that contacts with a flat plane. The pull-off forces shown in figs. 2.10a and 2.10b are then plotted in fig. 2.22 as a function of curvature radius for each pattern shown in table 2.2. The pull-off forces shown in figs. 2.10a and 2.10b are shown by solid and open squares, respectively. The lines in fig. 2.22 are approximate lines passing through the origin of the graph for the respective data sets. Both sets of pull-off forces were approximately proportional to the curvature radius of the asperity peak, and they agreed with eqs. (2.4) and (2.7). As shown in fig. 2.1b, the microsurface roughness of asperities was extremely large for the radius of curvature of 170 nm. This roughness, however, did not significantly reduce the pull-off force. This verifies that the pull-off force is only dependent on the radius of curvature of a sphere and not on microsurface roughness.

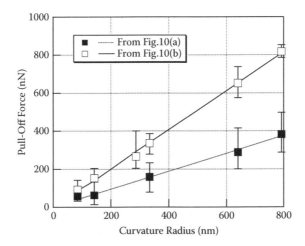

FIGURE 2.22 Relation between pull-off force and curvature radius of asperity peak measured on silicon asperity arrays of various groove depths. The pull-off forces were derived from fig. 2.10. The curvature radii were calculated from the AFM data by using a hemisphere approximation program.

In fig. 2.22, the gradients of the two lines are different. The higher gradient is almost twice the lower value. This is probably because the flat part of the probe was covered with slight contamination. If the probe tip was not flat but had a finite radius of curvature due to deposited contamination, the effective radius at the contact point became lower than the curvature radius of the asperity peak. Moreover, slight contamination affected both the adhesion energy in eq. (2.4) and the contact angles of water in eq. (2.7).

2.4.5 RELATION BETWEEN FRICTION FORCE AND CURVATURE RADIUS

When the friction force was measured, the applied load was very small (<10 nN) compared with the pull-off force. For the applied load P_0 at 0, the apparent Hertzian load P_1 is calculated from eq. (2.1) as

$$P_1 = 6\gamma\pi R \qquad (2.8)$$

Then the radius of the contact circle a_1 is calculated from eqs. (2.2) and (2.8) and is given by

$$a_1^{\ 3} = \frac{6\gamma\pi R^2}{K} \qquad (2.9)$$

Figure 2.23 shows the friction force as a function of R (the asperity curvature for silicon) and $R^{4/3}$ from fig. 2.10b. The friction force was proportional to R, but not to $R^{4/3}$. If the friction force was proportional to the total load that included adhesion force, then from eq. (2.8) the friction force was considered to be proportional to the apparent Hertzian load P_1 [17] as well as to R. If the contact circle between the AFM

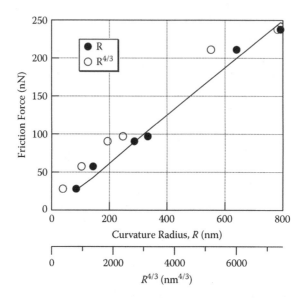

FIGURE 2.23 Relation between friction force and radius of an asperity curvature (R) and $R^{4/3}$.

probe and an asperity calculated from eq. (2.9) was equivalent to the real contact area, then the friction force should have been proportional to $R^{4/3}$. This discrepancy between theoretical and experimental results means that the contact circle given by eq. (2.9) was not a real contact area. This is also confirmed from our result that the friction force was independent of microsurface roughness.

The friction force was reportedly proportional to the contact area when the contact surface was an atomically flat surface, such as a cleavage plane of mica [18]. Under the conditions used in our experiments, the contact surface had a finite surface roughness. The area determined by eq. (2.9) represented only an apparent, not a real, contact area. Adhesion force due to the van der Waals force or capillary force acted on this apparent contact area. Because this adhesion force increased the real contact area as an equivalent normal load, the friction force was proportional to the adhesion force and, as a result, the friction force must be proportional to the radius of curvature.

2.4.6 Evaluation of Adhesion Energy

From eq. (2.4) and the data in fig. 2.22, we calculated that the adhesion energy γ was from 0.063 to 0.18 J/m². The van der Waals interaction energy per unit area, W, between two flat surfaces is given by [16]

$$W = \frac{-H_{12}}{12\pi d^2} \qquad (2.10)$$

where d is the distance between the surfaces, and H_{12} is the Hamaker constant between the two materials (in our case, silicon for the pattern and Si_3N_4 for the probe) expressed as

$$H_{12} = \sqrt{H_1 \cdot H_2} \qquad (2.11)$$

where H_1 is the Hamaker constant of silicon and H_2 is the Hamaker constant of Si_3N_4.

If the adhesion energy is determined by only the van der Waals energy, then γ is equivalent to W ($W = \gamma$). Then, for our case, the calculated average distance d was 0.30 to 0.18 nm when using a Hamaker constant of 2.56×10^{-19} J for silicon [19] and 1.8×10^{-19} J for Si_3N_4 [20]. Because this calculated d was equivalent to the lattice constant of silicon, contact at the atomic level appeared to be generated between the two surfaces over the entire contact area. However, fig. 2.22 shows that the pull-off force was proportional to only the asperity curvature radius and was independent of differences in the microroughness on the asperity (figs. 2.1a–2.1c). Figure 2.23 also shows that the contact area calculated from eq. (2.9) is not the real contact area. Therefore, the surface tension of the condensed water was dominant in the adhesion energy.

It is possible that a capillary condensed water bridge existed between the silicon asperity and the probe plane. When the Laplace pressure is considered, the adhesion force F_W caused by this type of meniscus between a sphere of radius R_S and a flat surface is proportional to the area covered by the capillary water and is expressed as eq. (2.7). When we measured the pull-off force for different asperity curvatures on the same plate, the attractive force F_W was determined only by the curvature, because we assumed that both contact angles were constant. Therefore, the contribution of the attractive force from the Laplace pressure gives a consistent interpretation of eq. (2.7) in agreement with the measured pull-off force shown in fig. 2.22.

The attractive force caused by the condensed water can be estimated using eq. (2.7). For a contact angle of 27° for Si_3N_4 [21] and 70° for silicon [22], the attractive force is, for example, 47 nN (for an R of 83 nm) and 430 nN (for an R of 760 nm). For a contact angle of 43° for oxidized silicon [22], this force is 62 nN (for an R of 83 nm) and 570 nN (for an R of 760 nm). Because these contact angles are easily altered by slight contamination or by a slight degree of oxidation, the difference in the pull-off force in fig. 2.22 was probably due to the difference in the Laplace pressure.

2.5 CHEMICAL EFFECTS ON FRICTION AND ADHESION [23, 24]

2.5.1 PULL-OFF FORCES ON ASPERITY ARRAYS COVERED WITH SAMs

The measured contact angle of water on C_{18} (table 2.4) is similar to that found in the literature, 112° ± 2° [25], which indicates that a SAM of alkylchlorosilane molecules was deposited on all substrates. Figure 2.24 shows the relation between the pull-off force and the curvature radius of the asperity peak covered with SAMs of alkylchlorosilanes. (Note that the x-axis represents the average curvature radius of the asperity peaks shown in table 2.3.) The number (#) of carbon atoms in each alkylchlorosilane is indicated as $C_\#$. C_0 represents an uncoated plate. The data for each plate (SAM) were fitted with a line passing through the origin. For the uncoated plate (C_0) and the SAM-coated plates except C_{14}, the pull-off force increased with increasing peak curvature radius. For C_6 and C_{18}, the SAM-coated plates, the pull-off force was roughly proportional to the curvature radius of the asperity peak. The slopes for

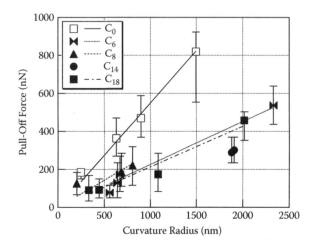

FIGURE 2.24 Relation between the pull-off force and the curvature radius of the asperity peaks for uncoated plate C_0 and SAM-coated plates C_6 to C_{18}. Measured data were fitted by a line that passes through the origin.

the other SAM-coated plates (C_6 to C_{18}) were half, or less, than that for the uncoated plate (C_0).

The experiment in the previous section revealed that the capillary force was predominant between the flat probe and the periodic asperity array. In fig. 2.21, the capillary geometry shows that a capillary can be formed even if one of the surfaces is hydrophobic, and eq. (2.7) shows that the adhesion force exists between the surfaces when $\theta_1 + \theta_2 < 180°$. Thus, the water capillary could form between the hydrophilic probe and SAM-coated asperity peaks, even though the SAM-coated surfaces are hydrophobic and the adhesion force F_W is given by eq. (2.7).

The contact angle of water on SAM-coated or on uncoated silicon (table 2.4) ranged from 4° to 113° after the SAM deposition and from 34° to 103° after the force measurements. For θ_1, we used the contact angles measured after the force measurements. For θ_2, we assumed it was 27° for Si_3N_4 [21]. Assuming these values for the contact angles of water, i.e., $\theta_1 = 34°$ to 103° and $\theta_2 = 27°$, then $\theta_1 + \theta_2 < 180°$, which means the condensed capillary could generate an adhesion force. Therefore, from eq. (2.7), the adhesion force is proportional to the radius of curvature of the fitted spherical surface.

If the contact angle for water on the probe is assumed to be 27°, then the adhesion force caused by a capillary on each surface can be calculated using the contact angles shown in table 2.4. Table 2.6 shows the slope of the fitted lines ($\partial P_{\text{pull-off}}/\partial R$) in fig. 2.24 and also shows $2\pi\gamma_L(\cos\theta_1 + \cos\theta_2)$ from eq. (2.7). The $\partial P_{\text{pull-off}}/\partial R$ from the measured data was only 30% to 45% less than $2\pi\gamma_L(\cos\theta_1 + \cos\theta_2)$. (The values from these two expressions would show a better agreement if a higher contact angle on Si_3N_4 were assumed.) Such agreement suggests that there was only one contact point between the asperity array and the probe.

TABLE 2.6

Comparison between $\partial P_{\text{pull-off}}/\partial R$ and Capillary Force on SAM-Coated Plates

Sample name	$\partial P\text{pull-off}/\partial R$ (N/m)	$2\pi\gamma_L(\cos\theta_1+\cos\theta_2)$ (N/m)
C_0	0.55	0.77
C_6	0.23	0.36
C_{14}	(0.16)	0.29

2.5.2 PULL-OFF FORCES ON ASPERITY ARRAYS COVERED WITH LB FILMS

Figure 2.25 shows the pull-off force measured for the asperity arrays covered with two kinds of LB films. The curvature radius on the x-axis was shown in table 2.3. The data for each plate with $CH(C_{17}H_{35}COOH)$-LB or $CFCH(C_6F_{13}C_{11}H_{22}COOH)$-LB film were fitted with a line passing through the origin. The pull-off force decreased with smaller curvature radius and was roughly proportional to the curvature radius. The pull-off force on the CH-LB film was about 1/5th of the pull-off force on the CFCH-LB film for the same curvature radius.

Figure 2.26 shows the pull-off forces on two kinds of asperity arrays as a function of the relative humidity. The average curvature radius of each asperity array was 150 and 440 nm for the CH-LB film and 95 and 370 nm for the CFCH-LB film. Each plot in fig. 2.26 is the average of 256 pull-off force measurements. The error bar shows the standard deviation for each data point. The average pull-off force clearly increased with higher relative humidity for the asperity array of 370-nm radius with the CFCH-LB film

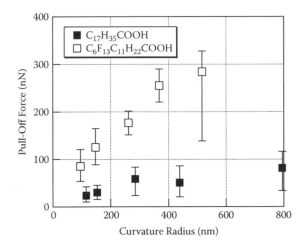

FIGURE 2.25 Relation between the pull-off force and the curvature radius of the asperity peaks for silicon plates coated with $CH(C_{17}H_{35}COOH)$-LB and $CFCH(C_6F_{13}C_{11}H_{22}COOH)$-LB films.

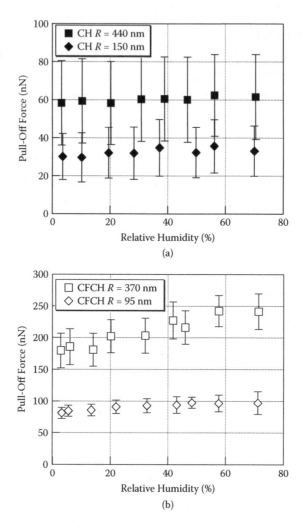

FIGURE 2.26   Relation between pull-off forces and relative humidity measured on asperity arrays. Each plot shows the average of pull-off forces from 256 measurements on (a) CH-LB film and on (b) CFCH-LB film. Two asperity arrays were selected from each plate, and their curvature radii are shown in the inset boxes.

(fig. 2.26b). For the asperity array with the CH-LB film, the pull-off force was nearly constant irrespective of changes in the relative humidity (fig. 2.26a).

In fig. 2.25, the pull-off force fluctuated, but it was nearly proportional to the curvature radius of the asperity peak with each LB film. Assuming that the water contact angle θ_2 on Si_3N_4 was 27°, the Laplace pressure would generate an adhesion force when the contact angle on the LB film was less than 153°, from eq. (2.7). In fig. 2.25, the pull-off force at a curvature radius of 400 nm ranged from about 50 nN for the CH-LB film to 350 nN for the CFCH-LB film. The adhesion force F_W as given by eq. (2.7) is 50 nN for $\theta_1 = 127°$ and 340 nN for $\theta_1 = 0°$, using $\gamma_L = 0.072$ N/m. The estimated contact angle has a moderate agreement with the value found in a report on the CH-LB film [27]. But, $\theta_1 = 0°$ for the CFCH-LB film is too small. Thus, we should refer to the possibility that average numbers of the contact points might be different for the CH-LB and CFCH-LB films. If the average contact points were *two* for CFCH-LB film, $\theta_1 = 85°$, and this would be a reasonable value. In fig. 2.26, comparing the standard deviation of the pull-off force against the mean value, the standard deviation for the CFCH-LB film (fig. 2.26b) is much smaller than that for the CH-LB film (fig. 2.26a). If the measured pull-off force were the average of two different curvature radii, fluctuation in the data would be suppressed. On the contrary, if the number of contacting points were only one, the pull-off force would be dominated by the curvature radius of one asperity, and thus varied by the contact position (contacting asperity) in the asperity pattern.

2.5.3 REDUCTION OF FRICTION BY ORGANIC MONOLAYER FILMS

Figure 2.27 shows the relation between the friction force and the peak curvature radius measured on SAM-coated plates. The data for each plate were fitted with a line. Similar to the pull-off force, the friction force for the SAM-coated plates increased with increasing peak curvature radius. The degree of reduction in the friction force due to the SAM coating was considerably larger than that in the pull-off force (fig. 2.24). For example, the slopes of the fitted lines for the SAM-coated plates C_6 to C_{18} are less than 1/5th of that for the uncoated plate C_0. Moreover, those for the SAM-coated plates C_{14} and C_{18} are particularly small.

Figure 2.28 shows the relation between the friction force in fig. 2.27 and the pull-off force in fig. 2.24. The data for each plate were fitted with a line passing through

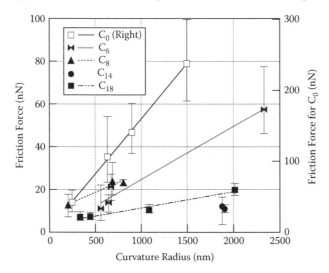

FIGURE 2.27 Relation between the friction force and the curvature radius of the asperity peaks for an uncoated plate C_0 and SAM-coated plates C_6, C_8, C_{14}, and C_{18}, where the force measured on the plate C_0 is on a different scale, shown on the right axis. Measured data were fitted with a line.

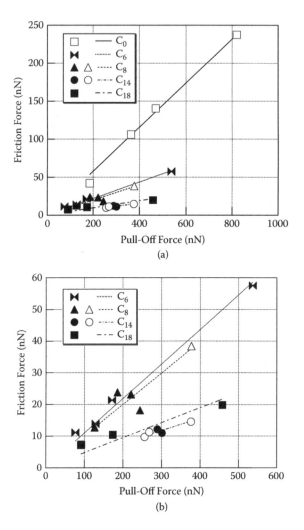

FIGURE 2.28 Relation between friction and pull-off forces for (a) uncoated plate C_0 and SAM-coated plates C_6 to C_{18} and for (b) SAM-coated plates C_6 to C_{18} on an expanded scale for clarity. Measured data were fitted with a line that passes through the origin. The data that are not shown in figs. 2.24 and 2.27 are added and are shown as the open triangles and circles.

the origin. On an expanded scale for clarity, fig. 2.28b shows the data for the SAM-coated plates shown in fig. 2.28a. In addition to the data from figs. 2.24 and 2.27, fig. 2.28 shows data for a patterned area that had an asperity array whose peaks could not be calculated because the height of the asperities was too low. (The curvature-radius-calculating program could not identify asperities whose heights were less than about 3 nm.) The friction force for all the plates is nearly proportional to the pull-off force. The slopes of the fitted lines for the SAM-coated plates are significantly lower than that for the uncoated plate C_0 (fig. 2.28a).

Figure 2.29 shows the friction force measured for the asperity arrays covered with LB films as a function of the curvature radius of the asperity peaks. The error

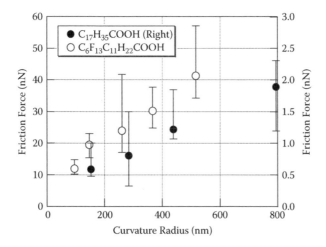

FIGURE 2.29 Relation between the friction force and the curvature radius of the asperity peaks for silicon plates coated with $CH(C_{17}H_{35}COOH)$-LB and $CFCH(C_6F_{13}C_{11}H_{22}COOH)$-LB films.

bars in fig. 2.29 show the maximum and minimum values of the pull-off and friction forces. The friction force decreased with smaller curvature radii and was roughly proportional to the curvature radius. The friction force on the CH-LB film was approximately 1/30th of the friction force on the CFCH-LB film for the same curvature radius, whereas the pull-off force on the CH-LB film was about 1/5th of that on the CFCH-LB film (fig. 2.25). The larger reduction rate of the friction force was magnified by the reduction of the pull-off force. Therefore, it is better to compare the gradients of the friction force (i.e., friction coefficients) to discuss the effect on friction.

Figure 2.30 shows the relationship between the friction force and the pull-off force taken from the data in figs. 2.25 and 2.29. The friction force was approximated with a straight line passing through the origin, and the gradient of the fitted line for each friction force (friction coefficient) on each LB film is shown in the parentheses in the plot legend. The friction coefficients for the CH-LB and CFCH-LB films were 0.021 and 0.14, respectively. The friction coefficient on the CH-LB film (0.021) was much lower than the friction coefficient (≈ 0.1) at a higher load measured using a tribology tester [27]. The friction coefficient on the CFCH-LB film (0.14) was comparable to the value (≈ 0.16) found in the same report. If we focus on the differences in the chemical properties between the two kinds of LB films, the surface of the CFCH-LB film exhibits a lower chemical interaction and thus should have a lower friction coefficient. But the CFCH-LB film exhibited a higher friction coefficient than the CH-LB film. If we focus on the mechanical properties, the CFCH-LB film probably had a lower stiffness because of its larger cross-sectional area (0.18 and 0.29 nm^2/molecule in the CH-LB and CFCH-LB films, respectively), and because of the lower interaction energy between molecules in the LB film.

Assuming that the CFCH-LB film had lower stiffness than the CH-LB film, the real contact area on the asperities with the CFCH-LB film was larger than that for the CH-LB film. Thus, the CFCH-LB film would exhibit a higher friction force than

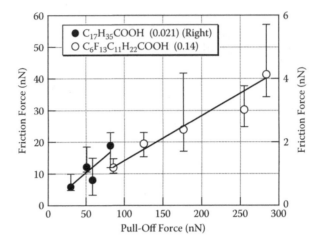

FIGURE 2.30 Friction force vs. pull-off force measured on asperity arrays on silicon plates coated with CH(C$_{17}$H$_{35}$COOH)-LB and CFCH(C$_6$F$_{13}$C$_{11}$H$_{22}$COOH)-LB films. Measured data were fitted with a line that passes through the origin. The gradient of each approximated line is shown in the parentheses in the inset box.

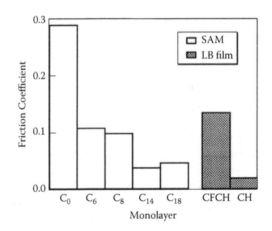

FIGURE 2.31 Friction coefficients for an uncoated plate (C$_0$) and SAM-coated plates (C$_6$–C$_{18}$) and for plates with LB films (CH-LB and CFCH-LB). Friction coefficients were calculated from the slopes of fitted lines in figs. 2.28 and 2.30.

the CH-LB film if the energy dispersion during sliding per unit area were comparable for each LB film. In our pull-off force measurements, slight differences between the heights of the adjacent asperity peaks could prevent the multipoint contact. If the LB film absorbs the height difference by deformation, capillaries would easily form at two or more asperity peaks, and a higher pull-off force would be obtained.

Figure 2.31 shows the friction coefficient for the uncoated plate (C$_0$) and the SAM-coated plates (C$_6$–C$_{18}$) and for the plates with LB films (CH-LB and CFCH-LB). These coefficients correspond to the gradients of the fitted lines shown in figs. 2.28 and 2.30. Comparing the SAM-coated plates (C$_6$–C$_{18}$) with the uncoated plate

(C_0), the friction coefficients for the SAM-coated plates were much lower than that for the uncoated plate. The molecular layer probably prevented a direct contact between solids. The friction coefficients for the SAM-coated plates C_{14} and C_{18} were about half that for the other SAM-coated plates. Some reports showed that the friction coefficient was inversely correlated with the alkyl-chain length of the SAM [28, 29]. Similar to the difference in the friction coefficients between CH-LB and CFCH-LB films, the difference in stiffness of the SAMs also might affect the friction force. The stiffness of a SAM is probably correlated with the alkyl-chain length because the van der Waals interaction between alkyl chains increases with longer chain length. The real contact area where the friction force operates is inversely correlated with the stiffness. Based on these assumptions, a longer alkyl chain tends to show a lower friction coefficient.

The length of the alkyl chain (number of carbon atoms) is the same for C_{18}-SAM and CH-LB films; therefore, the van der Waals interaction between the alkyl chains was probably the same for these two kinds of monolayers. The friction coefficient on the CH-LB film, however, was less than half that of the C_{18}-SAM. The difference was likely caused by the difference in the two-dimensional molecular density or in the degree of crystalline perfection. The CH-LB film has a more crystalline structure than the C_{18}-SAM because the structure of SAM on SiO_2 is complex [30]. The difference in the structure would result in differences in the stiffness and contact area. Thus, the CH-LB film showed a lower friction coefficient than C_{18}-SAM.

In this study, we showed that when LB films and SAMs are used to coat a silicon plate, the friction coefficient decreased to 1/3 to 1/12 (fig. 2.31). The friction force was further reduced by optimizing the geometry of the surface roughness. When spherical asperities are added to the surface, the adhesion force can be reduced by using asperities with a smaller curvature radius, thereby reducing the friction force. When this lubrication method involving a combination of an asperity array and a SAM coating is applied to sliding components in micromechanisms such as MEMS, an extremely low friction force can be achieved.

2.6 SUMMARY

Various patterns of two-dimensional asperity arrays were created by using FIB to deposit platinum asperities and to mill patterns on silicon plates and on a platinum layer deposited on the silicon plate. The pull-off and friction forces between the respective patterns and a flat scanning probe of an AFM were measured. Our findings are as follows:

1. The pull-off force decreased with increasing groove depth as well as with decreasing mound ratio, which suggested that the geometry of the asperity peaks dominated the adhesion force.
2. The pull-off force was proportional to the radii of curvature of the asperity peaks and was almost independent of the microsurface roughness of the asperities. The adhesion energy agreed well with the Laplace pressure due to capillary condensed water. These findings indicate that the Laplace pressure was a dominant factor in the adhesion force.

3. The friction force was proportional to the asperity curvature radius R, but not to $R^{4/3}$. This indicates that the friction force was not proportional to the contact area predicted by the JKR theory. This friction behavior was probably caused by the microroughness of the asperities.

4. The friction force was more proportional to the pull-off force than to the curvature radius. The friction coefficient (which was calculated by dividing the friction force by the pull-off force) for the silicon pattern was about twice that for the platinum pattern. These findings indicate that the adhesion force (pull-off force) did not directly affect the friction but, rather, indirectly affected friction, similarly to the effect of an external load.

5. The pull-off force decreased due to the SAM or LB film coatings on the asperity arrays. The magnitude of the pull-off force approximately corresponded to the capillary force calculated using the contact angle of water on the surface.

6. The degree of reduction of the friction force due to the SAM or LB film coatings was considerably larger than that of the pull-off force. While the friction force decreased to 1/10 to 1/30 for the same curvature radius, the pull-off force decreased to 1/2 to 1/5. The larger reduction of the friction force was magnified by the reduction of the pull-off force.

7. The differences in the pull-off and friction forces for the different kinds of LB films and SAM coatings might have been caused by differences in the stiffness of the molecules. The LB film of stearic acid showed the lowest friction coefficient of 0.021, which was probably due to its high stiffness.

ACKNOWLEDGMENT

The author would like to thank Jiro Ino, Yosuke Inoue, Takashi Igari, Shigeyuki Mori, and Kazuo Kakuta for their technical help and useful discussions.

REFERENCES

1. Etsion, I., and Amit, M. 1992. *J. Tribol. Trans. ASME* 115: 406–10.
2. Skinner, J., and Gane, N. 1972. *J. Phys. D: Appl. Phys.* 5: 2087–94.
3. Ando, Y., Ishikawa, Y., and Kitahara, T. 1995. *J. Tribol. Trans. ASME* 117: 569–74.
4. Fuller, K. N. G., and Tabor, D. 1975. *Proc. R. Soc. Lond. A* 345: 327.
5. Li, Y., and Menon, A. K. 1995. *J. Tribol. Trans. ASME* 117: 279.
6. Alley, R. L., Mai, P., Komvopoulos, K., and Howe, R. T. 1993. In *Proc. 7th Int. Conf. Solid-State Sensors and Actuators, and Eurosensors IX (Transducers '93)*, 288. Yokohama.
7. Gabriel, K. J., Behi, F., and Mahadevan, R. 1989. In *Proc. 5th Int. Conf. Solid-State Sensors and Actuators, and Eurosensors III (Transducers '89)*, 109. Montreaux.
8. Granick, S. 1991. *Science* 253: 1374–79.
9. Ando, Y., and Ino, J. 1997. *J. Tribol. Trans. ASME* 119: 781–87.
10. Sharma, P. K., and Rao, K. H. 2002. *Adv. Colloid Interface Sci.* 98: 341–63.
11. Petersen, K. E. 1982. *Proc. IEEE* 70: 420.
12. Dai, C., and Chang, Y. 2007. *Materials Lett.* 61: 3089–92.
13. Ando, Y., and Ino, J. 1996. *Sensors Actuators. A* 57 (2): 83–89.
14. Ando, Y., and Ino, J. 1998. *Wear* 216: 115–22.

15. Johnson, K. L., Kendall, K., and Roberts, A. D. 1971. *Proc. R. Soc. Lond. A* 324: 301.
16. Israelachvili, J. 1991. *Intermolecular and surface forces.* New York: Academic Press.
17. Ando, Y., Ishikawa, Y., and Kitahara, T. 1996. *J. Jpn. Soc. Tribologists* 41: 663–70.
18. Yoshizawa, H., Chen, Y., and Israelachvili, J. N. 1993. *J. Phys. Chem.* 97: 4128–40.
19. Okuyama, K., Masuda, H., Higashitani, K., Chikazawa, M., and Kanazawa, T. 1985. *Hun-tai Kogakkaishi* 22 (8): 451–75.
20. Bergström, L., Meurk, A., Arwin, H., and Rowcliffe, J. 1996. *J. Am. Ceram. Soc.* 79: 339–48.
21. Li, J., and Hattori, M. 1985. *Thermochimica Acta* 88: 267–72.
22. Hermansson, K., Lindberg, U., Hök, B., and Palmskog, G. 1991. *Proc. 6th Int. Conf. Solid-State Sensors and Actuators (Transducers '91),* 193–96. San Francisco.
23. Ando, Y., Inoue, Y., Kakuta, K., Igari, T., and Mori, S. 2007. *Tribol. Lett.* 27: 13–20.
24. Ando, Y., Igari, T., and Mori, S. 2007. *Tribol. Online* 2: 23–28. http://www.jstage.jst.go.jp/browse/trol/.
25. Calistri-Yeh, M., Kramer, E. J., Sharma, R., Zhao, W., Rafailovich, M. H., Sokolov, J., and Brock, J. D. 1996. *Langmuir* 12: 2747–55.
26. Ahn, S., Son, D., and Kim, K. 1994. *J. Mol. Struct.* 324: 223–31.
27. Igari, T., Oyamada, N., Nanao, H., and Mori, S. 2000. *J. Jpn. Soc. Tribologists* 45: 414–20.
28. Lio, A., Charych, D. H., and Salmeron, M. 1997. *J. Phys. Chem. B* 101: 3800–5.
29. Major, R. C., Kim, H. I., Houston, J. E., and Zhu, X. Y. 2003. *Tribol. Lett.* 14: 237–44.
30. Wirth, M. J., Fairbank, R. W. P., and Fatunmbi, H. O. 1997. *Science* 275: 44–47.

3 Effect of Harsh Environment on Surfactant-Coated MEMS

Joelle Frechette, Roya Maboudian, and Carlo Carraro

ABSTRACT

This chapter provides a concise review of the current status of the self-assembled monolayers as molecular lubricants for microelectromechanical systems operating in harsh environments, including elevated temperatures and in fluids. In particular, we focus on the similarities and differences in the structure–property relationships of two SAMs commonly employed to prevent in-use stiction in MEMS, namely octadecyltrichlorosilane (OTS, $CH_3(CH_2)_{17}SiCl_3$) and perfluorodecyl-trichlorosilane (FDTS, $CF_3(CF_2)_7(CH_2)_2SiCl_3$). We discuss the effect of harsh environments on these monolayers and how their degradation impacts their properties and the range of conditions for which these monolayers can be employed effectively for MEMS.

3.1 INTRODUCTION

Self-assembled monolayers (SAMs) are commonly used as molecular lubricants in microelectromechanical systems (MEMS) [1, 2]. They have been shown to reduce dramatically the adhesion (stiction) of free-standing or moving microstructures. The most common SAMs deposited on silicon are made from chlorosilane end-groups and alkyl or perfluorinated chains of variable length, with the chemical form RCl_3. The antistiction properties of a variety of monolayers have been reviewed extensively (see, for example, [3–5]). Relatively few investigations have dealt with the performance of molecular lubricants in harsh environments (such as elevated temperature, liquid and corrosive media, and high electric bias). It is of paramount practical importance to determine to what extent these films can be employed for antistiction in these conditions.

Packaging, biological, and sensing applications commonly involve conditions other than dry air, and it is unlikely that SAMs will have the same antiadhesive properties regardless of the operating conditions. Two important scenarios can occur: one in which the SAMs are briefly exposed to a harsh environment (such as high temperature during packaging), but after the brief exposure the device will be operated at conditions of low to moderate humidity and room temperature. A more demanding situation is one in which the device is constantly operating in a harsh environment and is required to maintain lasting antistiction properties.

The effect of a harsh environment on the structural and functional integrity of siloxane films is poorly understood. Harsh environments can damage the SAM or induce its desorption. A reduction of surface coverage can expose high-surface-energy groups on the silicon oxide surface and accelerate corrosion processes [6, 7]. In addition, for operation in solution, unwanted adsorption can cause failure, and surface treatments for bioMEMS should consider potential fouling issues after passivation. This chapter will cover the effects of elevated temperature and fluid environments on the effectiveness of various surface monolayers to prevent in-use stiction. We will not cover fouling and will focus mainly on two commonly employed antistiction monolayers, namely, octadecyltrichlorosilane (OTS, $CH_3(CH_2)_{17}SiCl_3$) and perfluorodecyltrichlorosilane (FDTS, $CF_3(CF_2)_7(CH_2)_2SiCl_3$).

3.2 EFFECT OF ELEVATED TEMPERATURE

The effect of annealing on the structural integrity of siloxane monolayers has been investigated at ambient and in ultrahigh vacuum (UHV) environments. Common techniques employed for the characterization of annealing effects on SAMs are contact angle measurements, x-ray photoelectron spectroscopy (XPS), and high-resolution electron energy loss spectroscopy (HREELS). This section reports on how thermal cycling affects surface coverage of monolayers deposited on smooth, single-crystal silicon wafers. It is important to note that MEMS involve rough surfaces, and, as will be discussed in the next section, the antistiction behavior of SAMs at elevated temperature is strongly affected by the presence of multi-asperity contacts.

Table 3.1 shows the reported temperature values for the onset of a significant, and irreversible, change in the structural integrity of OTS and FDTS films caused by annealing in different environments and for periods of time varying between 2 to 15 min. The large range in temperature values published in the literature is likely to be caused by the specific sensitivities of the various techniques employed to characterize the impact of thermal annealing on the surface films. As an illustration, for both OTS and FDTS, a 15%–20% loss of coverage is readily observed from XPS measurements but does not cause a significant change in static water contact angle values [8, 9].

The reported desorption mechanisms for OTS and FDTS are quite different. Thermal desorption of OTS has been shown to follow a "shaving" mechanism, i.e., the loss in the carbon content occurs via cleavage of the C-C backbones and successive

TABLE 3.1

Temperature for the Onset of Structural Change in the Siloxane Monolayer on Silicon

	OTS	FDTS
Air	125°C–240°C [9, 27]	100°C–300°C [12, 13]
N_2	400°C [1]	400°C [28]
UHV	475°C [10, 11]	100°C–300°C [12]

Note: The reported relative humidity for the measurements in air was 40%–50% [1, 9, 12].

shortening of the chain. The mechanism appears to govern the annealing both in air and in UHV [9–11]. Perfluorinated monolayers, such as FDTS, have been shown to desorb via the loss of a whole chain for temperatures as low as 100°C in vacuum. The desorption mechanism is believed to follow first-order kinetics, as it does not depend on the monolayer coverage. It is likely that the loss of coverage at low temperatures is caused by the desorption of physisorbed chains or chains that are poorly bonded to the silicon surface. Chain desorption is accompanied by a tilting of the monolayer, which allows the static water contact angle to maintain a value above 90° after annealing at 300°C and a loss of 20% of the fluorine content [8, 12]. The level of cross-linking for the monolayer can be varied by replacing one or two reactive Si-Cl groups by unreactive Si-CH$_3$ groups. Reducing the extent of cross-linking for perfluorinated monolayers has been shown to require a longer deposition time, often creating monolayers that are not fully packed, and to reduce the thermal stability [8, 13]. This effect is accentuated for annealing in air (compared with vacuum) for perfluorinated monolayers.

The stability of the monolayer at an elevated temperature depends on the duration of the annealing step. Using contact angle measurements, Zhuang et al. [13] have shown that FDTS can maintain a static water contact angle above 90° ($\theta > 90°$ is the minimum contact angle value necessary to avoid capillary forces) for up to 90 min at 400°C. On the other hand, the monolayer made out of a very similar tail, but not cross-linked, tridecafluoro-1,1,2,2,-tetrahydrooctyl ($CF_3(CF_2)_5(CH_2)_2(CH_3)_2SiCl$), can maintain a water contact angle above 90° for only 10 min.

3.3 IMPLICATION OF MONOLAYER DESORPTION ON STICTION

Intuitively, the capability of a monolayer to retain its structural integrity when exposed to an elevated temperature is a requirement for lasting antistiction properties. We have shown, however, that the relationship between the restructuring of a monolayer at elevated temperatures and its antistiction properties can be more complex. This is especially the case for highly hydrophobic surfaces, such as the ones produced with OTS or FDTS monolayers. As long as MEMS surfaces maintain their hydrophobicity, near-contact interactions are not dominated by capillary forces. Instead, they are governed by weaker forces, such as van der Waals interactions between surfaces that are quite rough. The complex surface topography and texture of the monolayer in the presence of defects create challenges to the modeling of surface interactions in real devices, especially when operating in hostile environments [14].

The behavior of FDTS monolayers under thermal stress offers an example of this complexity and of the counterintuitive results it can produce. We have shown that the monolayer can lose 25% of its fluorine content *and* display lower adhesion compared with when the surface is covered by a full monolayer, as can be seen from fig. 3.1. The lowering of the apparent work of adhesion (W_{adh}) at high temperatures is surprising, considering the fact that both water and hexadecane static contact angles decrease after the same annealing steps. Detailed XPS studies have shown that the low-temperature desorption of entire FDTS molecules is accompanied by a tilting of the remaining chains. It is also suspected that, due to the larger size of the fluorine molecule (compared with hydrogen), proper cross-linking of FDTS is difficult and that highly reactive Si-OH groups are left on the surface. Another indication that

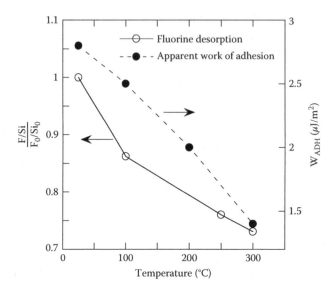

FIGURE 3.1 Temperature dependence of the fluorine loss and adhesion for a FDTS mono-layer. The fluorine content is measured from the F(1s)/Si(2p) ratio obtained before (indicated by subscript 0) and after annealing in the XPS chamber. The adhesion is obtained from cantilever beam arrays actuated at different temperatures. The adhesion is measured at the reported annealing temperature (and not after cooling to room temperature).

desorption occurs via the loss of whole, poorly cross-linked, chains comes from the similar losses in fluorine content for annealing in air and in UHV. This is not the case for alkylsiloxane monolayers that desorb via cleavage of C-C bonds, a reaction catalyzed by oxygen that displays accelerated desorption for annealing in air compared with UHV. Overall, it is suspected that the removal of some fluorine, chain tilting, and high temperature allow for the remaining Si-OH groups to react and cause the reduced adhesion upon annealing. The presence of buried Si-OH groups is unlikely to affect contact angle on a flat surface, but rough asperities involved in MEMS contact can pierce through the monolayer. These results illustrate both the importance of directly evaluating stiction at the device level and of understanding the desorption mechanisms of the film to be employed in harsh environments.

The effect of annealing on stiction of MEMS cantilevers covered with OTS has also been investigated by Ali et al. [15]. As shown in fig. 3.2, they first observed an increase in adhesion during annealing up to 200°C. Between 200°C and 300°C, they report a decrease in adhesion. Note that the reported values are for the sticking probabilities of cantilever beams with lengths between 480 and 540 µm; the detachment length or the apparent work of adhesion was not reported in their work. Also, it is important to note that the carbon desorption plotted in fig. 3.2 corresponds to the XPS data of Kim et al. [9]; therefore, correlation between the two curves is qualitative at best. Reported temperature values for "significant" structural damage caused by annealing in air of an OTS monolayer are between 125°C and 200°C. Similar to our work with FDTS, stiction appears to decrease in this temperature range in spite of the monolayer degradation. This reduction of adhesion can be caused by a loss of

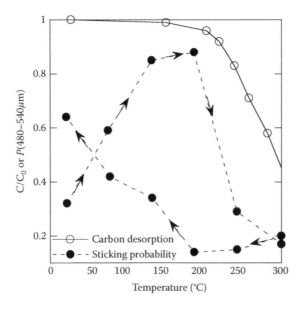

FIGURE 3.2 Desorption of OTS monolayers on Si(100), presented as C/C_0, where C_0 and C are carbon concentrations before and after annealing, respectively (data taken from Kim et al. [9]), and sticking probability for 480–540-μm cantilever beams (data taken from Ali et al. [15]). The loss in coverage is plotted as a function of annealing temperature. An increase in sticking probability corresponds to an increase of adhesion between the cantilever and the landing pad. The arrows indicate the hysteresis between heating and cooling.

water from the monolayer or via a process similar to the one described for FDTS. Interestingly, Ali et al. report XPS data showing 75% carbon loss after annealing to 300°C, while their detachment length appears to decrease from 710 to 590 μm. This change in detachment length corresponds to a significant increase in stiction, but the monolayer maintains a good degree of antistiction properties after sustaining serious damage. Further annealing cycles did not show more carbon loss or a further increase in stiction (as can be deduced from the sticking probability data).

Using atomic force microscopy (AFM), Kasai et al. [16] have also investigated how adhesion and friction between surfaces covered with FDTS and OTS respond to temperature variations. The temperature range investigated was 20°C–115°C. For both the OTS and FDTS monolayers, increasing the temperature was shown to decrease both the adhesion and friction between the tip and the surface. The decrease in both friction and adhesion was attributed to a loss in the water content within the monolayer. The adhesion and friction forces dropped by a factor of two over the temperature range investigated.

3.4 ADHESION IN FLUIDIC ENVIRONMENTS

Investigation of stiction in fluid environments is gaining importance due to the emergence of microfluidic devices and their technological applications for biological assays. There have been only a very few reports, however, on the investigation of stiction in fluid environments, especially for fluids at high or low pH. An additional issue is electrostatic actuation in a salt solution, as the establishment of the electrical double layer [17] prevents typical DC actuation. The problem can be circumvented by using high-frequency AC actuation [18]. Moreover, adhesion forces in a fluid environment are not the same as in air. Capillary forces are not present in a liquid environment, but double-layer repulsion, acid–base interaction, and hydrophobic forces can all play a major role [19].

Parker et al. [20] systematically investigated common antistiction SAMs (such as OTS and FDTS) for potential applications in various fluidic environments (such as water, isopropyl alcohol, iso-octane, and hexadecane). For operation in water, it was shown that bare silicon oxide was by far more effective in reducing stiction compared with any hydrophobic monolayers (three different monolayers and two different deposition conditions were investigated). Interestingly, adhesion with a bare silicon oxide surface was much lower than adhesion in air with surfaces covered with FDTS. The low adhesion between hydrophilic SiO_2 surfaces in solution can be explained by a reduced van der Waals interaction and double-layer repulsion (SiO_2 surfaces tend to have a negative surface potential of around -60 mV [21]). On the other hand, interactions between hydrophobic surfaces in solution can be very strong, especially in aerated solutions where air bubbles and cavitation can cause very strong adhesion [22]. A report by Parker et al. [20] on the use of hydrophobic SAMs (OTS and FDTS) as antistiction monolayers for operation in solution indicates that they are indeed detrimental, most likely because of hydrophobic interactions. Stiction did not appear to be an issue for any of the surface treatments (even bare SiO_2) for actuation in isopropyl alcohol. For operation in nonpolar solvents (hexadecane and iso-octane), stiction was greatly reduced with the use of a hydrophobic monolayer compared with a hydrophilic oxide surface.

An AFM image of an FDTS monolayer after it has been in water for a prolonged period of time shows a weblike structure, as shown in fig. 3.3, highlighting the pos-

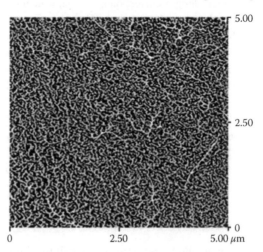

FIGURE 3.3 Tapping-mode AFM image of an FDTS film after immersion in water for 6 weeks. The z-scale is 10 nm. Figure taken from [12].

sibility of poor cross-linking (as also mentioned in section 3.2 about the effects of elevated temperature) and a result similar to hydrophobic surfactants adsorbed on mica after being in an aqueous solution [23, 24].

Geerken et al. [25] investigated the chemical and thermal stability of OTS and a perfluorinated equivalent (perfluorinated octyltrichlorosilane) in liquids as a function of time (up to 200 h), temperature (50°C–80°C), and pH (2–13). They evaluated the stability of the monolayers (not stiction) by measuring water and hexadecane contact angles after exposure to harsh conditions. They reported that perfluorinated monolayers were more stable than OTS in all the conditions they investigated.

3.5 CONCLUSIONS

Hydrophobic siloxane monolayers are effective at preventing capillary adhesion in MEMS operating in ambient air. However, the use of SAMs to prevent stiction in harsh environments (elevated temperature or fluids) has been investigated to a lesser extent. Packaging or specific application conditions can expose MEMS to elevated temperatures. Similarly, the advent of bioMEMS and microfluidic devices opens the door to a wide range of applications in liquid environments.

A brief review of the literature shows that, for applications at elevated temperatures, it is important to avoid testing solely the change in monolayer integrity (as measured with contact angle). It is shown that for both FDTS and OTS, a reduction of coverage does not necessarily directly lead to a significant increase in stiction. In particular, for an FDTS monolayer, we show that a small loss in fluorine content leads to improved antistiction behavior, justifying annealing the devices during, or right after, the monolayer deposition [26].

For MEMS operating in a fluid environment, a survey of the literature shows that the choice of surface treatment depends on the operating fluid. For operation in aqueous solutions, a hydrophilic surface is preferable to avoid strong hydrophobic interactions. A hydrophobic monolayer decreases adhesion for actuation in nonpolar fluids such as iso-octane and hexadecane.

REFERENCES

1. Srinivasan, U., Houston, M. R., Howe, R. T., and Maboudian, R. 1998. *J. Microelectromech. Syst.* 7: 252–60.
2. Tas, N., Sonnenberg, T., Jansen, H., Legtenberg, R., and Elwenspoek, M. 1996. *J. Micromech. Microeng.* 6: 385–97.
3. Ashurst, W. R. 2003. Surface engineering for MEMS reliability. Ph.D. thesis, UC Berkeley, Berkeley.
4. Ashurst, W. R., Carraro, C., and Maboudian, R. 2003. *IEEE Trans. Device Mater. Reliab.* 3: 173–78.
5. Maboudian, R., and Carraro, C. 2004. *Annu. Rev. Phys. Chem.* 55: 35–54.
6. Muhlstein, C. L., Stach, E. A., and Ritchie, R. O. 2002. *Acta Mater.* 50: 3579–95.
7. Muhlstein, C. L., Stach, E. A., and Ritchie, R. O. 2002. *Appl. Phys. Lett.* 80: 1532–34.
8. Frechette, J., Maboudian, R., and Carraro, C. 2006. *J. Microelectromech. Syst.* 15: 737–44.
9. Kim, H. K., Lee, J. P., Park, C. R., Kwak, H. T., and Sung, M. M. 2003. *J. Phys. Chem. B* 107: 4348–51.

10. Kluth, G. J., Sander, M., Sung, M. M., and Maboudian, R. 1998. *J. Vac. Sci. Technol. A* 16: 932–36.
11. Kluth, G. J., Sung, M. M., and Maboudian, R. 1997. *Langmuir* 13: 3775–80.
12. Frechette, J., Maboudian, R., and Carraro, C. 2006. *Langmuir* 22: 2726–30.
13. Zhuang, Y. X., Hansen, O., Knieling, T., Wang, C., Rombach, P., Lang, W., Benecke, W., Kehlenbeck, M., and Koblitz, J. 2006. *J. Micromech. Microeng.* 16: 2259–64.
14. Van Spengen, W. M. 2003. *Microelectronics Reliability* 43: 1049–60.
15. Ali, S. M., Jennings, J. M., and Phinney, L. M., *Sensors Actuators A* 113: 60–70.
16. Kasai, T., Bhushan, B., Kulik, G., Barbieri, L., and Hoffmann, P. 2005. *J. Vac. Sci. Technol. B* 23: 995–1003.
17. Russel, W. B., Saville, D. A., and Schowalter, W. R. 1989. *Colloidal dispersions.* New York: Cambridge University Press.
18. Sounart, T. L., Michalske, T. A., and Zavadil, K. R. 2005. *J. Microelectromech. Syst.* 14: 125–33.
19. Israelachvili, J. N. 1992. *Intermolecular and surface forces.* London: Academic Press.
20. Parker, E. E., Ashurst, W. R., Carraro, C., and Maboudian, R. 2005. *J. Microelectromech. Syst.* 14: 947–53.
21. Ducker, W. A., and Senden, T. 1992. *Langmuir* 8: 1831–36.
22. Christenson, H. K., and Claesson, P. M. 2001. *Adv. Colloid Interface Sci.* 91: 391–436.
23. Perkin, S., Kampf, N., and Klein, J. 2006. *Phys. Rev. Lett.* 96: 038301.
24. Meyer, E. E., Lin, Q., Hassenkam, T., Oroudjev, E., and Israelachvili, J. N. 2005. *Proc. Natl. Acad. Sci. U.S.A.* 102: 6839–42.
25. Geerken, M. J., Van Zonten, T. S., Lammertink, R. G. H., Borneman, Z., Nijdam, W., Van Rijn, C. J. M., and Wessling, M. 2004. *Adv. Eng. Mater.* 6: 749–54.
26. Bunker, B. C., Carpick, R. W., Assink, R. A., Thomas, M. L., Hankins, M. G., Voigt, J. A., Sipola, D., De Boer, M. P., and Gulley, G. L. 2000. *Langmuir* 16: 7742–51.
27. Calistriyeh, M., Kramer, E. J., Sharma, R., Zhao, W., Rafailovich, M. H., Sokolov, J., and Brock, J. D. 1996. *Langmuir* 12: 2747–55.
28. Srinivasan, U., Houston, M. R., Howe, R. T., and Maboudian, R. 1997. In *Proc. 9th IEEE Conf. Solid-State Sensors and Actuators (Transducers '97)*, 1399–402. Chicago.

4 Surfactants in Magnetic Recording Tribology

T. E. Karis and D. Pocker

ABSTRACT

Surfactants play key roles in the mechanical systems of magnetic recording disk drives. The most notable arena for surfactants is at the slider-disk interface, where the lubricant ensures reliable operation of a nanometer scale recording gap between the slider and the disk. The low surface energy and surface activity of functional end groups on the lubricant hold a monolayer in place for the lifetime of the disk drive. This chapter focuses on the novel development of a low surface energy coating of poly (1H,1H-pentadecafluorooctyl methacrylate) fluorohydrocarbon surfactant on the magnetic recording slider to improve tribological performance. Application of the film and methods for film characterization are detailed. Subambient pressure tribological test results show the ability of this film to reduce lubricant transfer and disk scratching as good as or better than several other low surface energy coatings.

4.1 INTRODUCTION

Even though magnetic recording is ubiquitous and pervasive in all aspects of modern society, one scarcely recognizes the miraculous harmony of physics and engineering embodied within these data-storage devices. A disk drive comprises magnetic recording disks and sliders in an enclosure. Each slider contains a magnetic recording head to read and write the data. The disks are rotated by a centrally located motor, and the heads are positioned over the magnetic data tracks by a servomechanism. The disks are typically rotating between 3,600 and 15,000 rpm, while the slider maintains the head within a few nanometers of the disk surface to detect the magnetic domain orientation. For reliable operation of the slider–disk interface, the disk is coated with approximately 1 nm of a perfluoropolyether lubricant with polar hydroxyl end groups [1]. The disk lubricant is an amphiphilic fluorosurfactant [2, 3].

The clearance between the slider and disk is extremely small, and an accumulation of lubricant on the slider can lead to wicking underneath the nanometer-thick spacing gap, causing data errors, scratching of the disk overcoat, and potential crashing of the slider. Accumulation of lubricant on the slider [4–6] must be avoided. Low-surface-energy coatings of fluorohydrocarbon surfactants [7] help to keep the slider clean. The focus of this chapter is the application and properties of slider surface coatings.

Various surface modifications can be done to the surface of the slider without significantly increasing the head–disk spacing gap. A low-surface-energy sputtered Teflon or chemical vapor-deposited fluorocarbon film can be applied at the slider-row level during slider fabrication. Another means to lower the surface energy of the slider is by solution-casting a thin film of fluoropolymer on the completed head gimbal assembly (HGA) by dip coating. For use in manufacturing, it is essential to have methods for precise and reproducible control of the film thickness and properties. This chapter describes x-ray photoelectron spectroscopy (XPS) and ellipsometry procedures developed to measure fluoropolymer film thickness. Ellipsometry was done to measure the film thickness on test samples consisting of slider rows, carbon- and non-carbon-overcoated silicon strips, and silicon wafers. XPS was needed to measure the fluoropolymer film thickness on the air bearing surface of sliders because of their small size. Thickness measured by XPS on the air bearing surface is compared with that measured by ellipsometry on strip test samples dip coated at the same time as the sliders. XPS is also employed to study the chemical composition of the fluorohydrocarbon surfactant thin films [8].

Since the slider coating film may be in contact with disk lubricant, it is possible that the lubricant may dissolve in and soften the film. This could potentially lead to some flow of the slider coating at the elevated temperatures inside the disk drive. To assess the possibility of film shear flow from the slider, or development of tackiness, the rheological properties of the fluorosurfactant and various concentrations of lubricant were measured. The fluorosurfactant coating is also compared with several other types of alternative coatings in an accelerated tribological test for the ability of the coating to inhibit lubricant transfer and abrasion of the disk by the slider.

4.2 EXPERIMENTAL

4.2.1 MATERIALS

The fluorohydrocarbon surfactant that is the focus of this chapter consists primarily of the poly (1H,1H-pentadecafluorooctyl) methacrylate chemical structure shown in fig. 4.1a and is referred to as PFOM. A small amount of fluorocarbon side group isomer is also detected by nuclear magnetic resonance (NMR). The monomer molecular weight is 468 g/mole. The degree of polymerization $n \approx 640$ was determined from the weight average molecular weight of 300,000 g/mol by light scattering. The glass transition temperature is $\approx 50°C$, and the index of refraction $n_f = 1.36$. The PFOM was obtained from commercial sources.

Several different types of substrates were employed to develop the film thickness measurement procedures. The first type of substrate was slider rows. Slider rows are strips (2.5 × 47 mm, 0.44 mm thick) of N58 (TiC/Al$_2$O$_3$ ceramic) overcoated with a nominally 12.5-nm-thick layer of sputtered carbon. Silicon wafers 0.4 mm thick (International Wafer Service, Santa Clara, CA, www.siwafer.com) were cut into strips having the same lengths and widths as the slider rows. Some of the silicon strips were overcoated with a nominally 12.5-nm-thick layer of sputtered carbon. Other silicon wafers 25.4 mm in diameter and 0.2 to 0.3 mm thick were used without cutting (Virginia Semiconductor, Inc., Fredericksburg, VA, www.virginiasemi.com). Film thickness was also measured on the air bearing surface of production-level sliders mounted on head gimbal assemblies (HGA).

To simulate exposure of the surfactant films to disk lubricant, PFOM films containing various amounts of perfluoropolyether Zdol 2000 were prepared for rheological

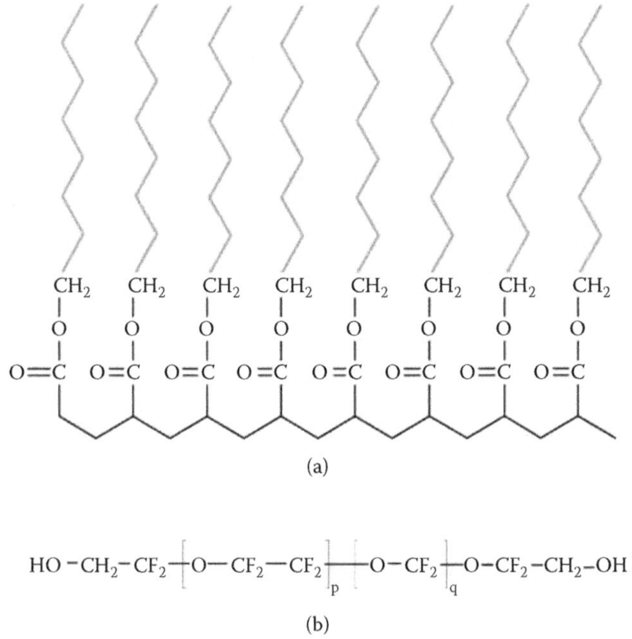

(a)

$$HO-CH_2-CF_2 \left[O-CF_2-CF_2 \right]_p \left[O-CF_2 \right]_q O-CF_2-CH_2-OH$$

(b)

FIGURE 4.1 **(See color insert following page 80.)** (a) poly (1H,1H-pentadecafluorooctyl) methacrylate (PFOM), molecular weight 300000 (g/mol); (b) Hydroxyl terminated perfluoropolyether Fomblin Zdol, molecular weight 2000 or 4000 (g/mol), p/q = 2/3.

measurements. The chemical structure of Zdol is shown in fig. 4.1b. Ztetraol has the same perfluoropolyether chain as Zdol, but the hydroxyl end groups have been reacted with glycidol, which doubles the number of hydroxyl groups at the chain ends. (Zdol and Ztetraol are products of Solvay Solexis, Inc., West Deptford, NJ.)

Sliders were coated with three other types of low-surface-energy coatings for tribological measurements in comparison with PFOM. Fluorinated carbon overcoat (FCOC) was prepared by plasma deposition of fluorinated monomers onto the slider [9]. ZNa, the sodium salt of Fomblin Zdiac [10], was deposited from dilute solution in 3M™ Novec™ Engineered Fluid HFE-7100 (nonafluorobutyl methyl ether) solvent. Zdiac has the same perfluoropolyether chain as Zdol, but with acid rather than hydroxyl end groups. Another coating tested was 3M Novec Electronic Coating EGC-1700, which was dip coated from solution in 3M Fluorinert™ Electronic Liquid FC-72 (perfluorohexane) solvent (www.3m.com/product).

4.2.2 METHODS

A motorized stage was employed to withdraw the test samples and HGAs from a PFOM solution tank at specified rates.

Ellipsometric measurements were done using a Gaertner model L-115B ellipsometer (HeNe laser, wavelength 632.8 nm, Gaertner Scientific, Chicago, IL). Six points were measured on the dip-coated slider rows and silicon strips. Three points were measured on the dip-coated silicon wafers. The PFOM thickness was calculated from the ellipsometric angles Δ and Ψ using the standard software package provided by Gaertner.

X-ray photoelectron spectroscopy (XPS) was done on some of the silicon test samples and on all the sliders. The narrow dimensions of the air bearing surface of the slider are difficult to measure using ellipsometry, while the spot size of the XPS is small enough to measure in these regions. The XPS measurements were done on Surface Science Labs SSX-100 spectrometers (Al K_α source) at resolution 3 with a 300-µm spot size. These conditions yield an Ag $3d_{5/2}$ line width of about 1.13 eV. The anode power was 50 W, and the irradiated area was about 300 × 500 µm because of the 35° angle of incidence. The XPS measurements were performed within 25 min to minimize film thickness erosion due to gradual ablation of the PFOM by the incident x-ray beam.

Tribological measurements, consisting of subambient pressure frictional hysteresis loop measurements, were carried out inside a sealed disk tester with a controlled leak (CETR Olympus, Center for Tribology, Campbell, CA). The disk rotation rate was 7200 rpm, and the slider suspension was mounted on a strain-gauge block to measure the friction force. Scratch measurements on the surface of the tested track were done with an optical surface analyzer.

Rheological measurements were performed in a stress rheometer fixture with a 2-cm cone and plate having a 1° cone angle and gap of 27 µm. Dynamic shear moduli were measured at 0.5% strain between 0.1 and 100 rad/s. Creep compliance was measured with a constant applied stress in the range of 0.1 to 5 kPa. Both measurements were performed over a series of temperatures to obtain data for time–temperature superposition.

Glass transition temperature measurements on the solutions of Zdol in PFOM were performed in a temperature-modulated differential scanning calorimeter.

The procedure for estimating the PFOM thickness from an XPS spectrum [11] requires calculating d/λ. This is the ratio of the PFOM film thickness d to the electron mean free path λ in the PFOM film. It is calculated from the XPS as described below. The electron emission intensity is corrected for the Scofield capture cross section of each element. The escape depths are corrected for kinetic energy E according to $E^{0.7}$ by the spectrometer analysis routine. Note that this kinetic energy correction is rigorously correct only for bulk samples and may cause some error when applied to thin films.

The C1s and O1s regions of the XPS spectrum on the air bearing surface of a slider coated with 2.5 nm of PFOM are shown in fig. 4.2. The binding energy of the peaks used in the calculation are indicated by the arrows in fig. 4.2. The peaks in the C1s spectrum, fig. 4.2a, are assigned to the chemical environments shown in table 4.1. The measured C1s, O1s, and F1s regions of the spectrum are peak fitted at each of the binding energies listed in table 4.1, and the total integrated area under the peaks is normalized to 100%. In the case of PFOM on non-carbon-overcoated silicon, the $Si2p$ peak area is also included. Part of the C1s spectrum is assigned to the PFOM, and the remaining portion of the spectrum is assigned to carbon in the substrate. The contribution of oxygen from the substrate is then determined from the known stoichiometry of the PFOM. The difference between the raw O1s percentage and the percentage assigned to the PFOM is due to the substrate. All the F1s signal is from the PFOM. Details of the calculation are given below.

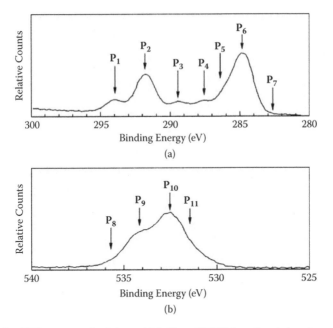

FIGURE 4.2 XPS spectra of a 2.4 nm thick film of PFOM on the air bearing surface of a slider. C1s spectrum (a), and O1s spectrum (b). The arrows indicate the binding energies of atoms with different chemical environments.

TABLE 4.1
Peak Assignments Used in Analysis of the XPS Spectra to Determine PFOM Thickness on the Air Bearing Surface of the Slider

Element	Label in fig. 4.1[a]	Binding energy (eV)	Chemical environment[b]
C1s	P_1	294	–CF3
	P_2	291.8	–CF2–
	P_3	289.5	–CF + –COOCH2–
	P_4	287.7	–CF2CH2O–
	P_5	286.3	CH3CCOO–
	P_6	284.8	–CH3 + –CH2– +carbon overcoat
	P_7[c]	282.3[c]	
O1s	P_8	535.9	—
	P_9	534.3	—
	P_{10}	532.7	—
	P_{11}	531.3	—
F1s	—	691.1[c]	—
	—	689	—
	—	686.5[c]	—

[a] The locations of peaks P_i are indicated by the arrows in the spectra in fig. 4.2.
[b] The underscore shows which carbon atom/chemical environment is emitting at the specified binding energy.
[c] Small peaks are due to non-Gaussian skirts arising from imperfections in the spectrometer transmission characteristics.

Among the carbons in the PFOM, fig. 4.1, there are seven carbons attached to fluorine ($-CF_3$ and $-CF_2-$), there is one ester carbon ($-COOCH_2-$), and there is one carbon in the ($-CF_2CH_2O-$) environment, i.e., the α carbon of the fluorocarbon side-chain. These nine carbons give rise to the four peaks with the higher binding energies in fig. 4.2a and table 4.1. The sum of these easily recognized parts of the PFOM in the C1s spectrum is given by

$$\sum_{i=1}^{4} P_i$$

In the calculations, P_i refers to the integrated area under peak i at the binding energy listed in table 4.1. The remaining C1s peaks at lower binding energies include both the PFOM and the carbon overcoat. The integrated area of the peaks at lower binding energies (table 4.1, P_{5-7}) is

$$\sum_{i=5}^{7} P_i$$

There are three PFOM carbons in this region of the C1s spectrum. The contribution of the three PFOM carbons is subtracted from the integrated area of the peaks at lower binding energies to get the contribution to the atomic% substrate from the carbon overcoat in the C1s spectrum,

$$\sum_{i=5}^{7} P_i - (3/9) \times \sum_{j=1}^{4} P_i$$

The next step is to determine the contribution to the atomic% substrate from the carbon overcoat in the O1s region of the spectrum. The oxygen-to-carbon ratio of the PFOM is (2/12), and the oxygen-to-fluorine ratio is (2/15). Thus, there are two ways to estimate the contribution of the PFOM to the O1s spectrum: (1) multiply the F1s percentage by (2/15), and (2) multiply the atomic% of carbon from the PFOM in the C1s spectrum by (2/12), i.e.,

$$(2/12) \times (1 + (3/9)) \times \sum_{i=1}^{4} P_i = (2/9) \times \sum_{i=1}^{4} P_i$$

The average of the two estimates from methods (1) and (2) above is then subtracted from the raw O1s percentage

$$\sum_{i=8}^{11} P_i$$

to estimate the contribution to the atomic% substrate from the O1s region of the spectrum.

The atomic% substrate in the spectrum is then the sum of the substrate portions of the C1s and O1s integrated areas (plus the integrated area of the Si2p spectrum in the case of a non-carbon-overcoated silicon substrate). The ratio of the PFOM film thickness to the electron mean free path d/λ measured by the XPS for determining PFOM thickness is given by

$$\frac{d}{\lambda} = \sin(\theta) \times \ln\left(\frac{100\%}{\text{atomic\% substrate}}\right) \qquad (4.1)$$

where θ is the electron take-off angle from the sample to the detector (35°). In this work, the electron mean free path λ in eq. (4.1) was empirically determined as follows. Ellipsometry was used to measure PFOM thickness d, XPS was used to measure d/λ on the same samples, and an experimental λ was estimated from these data. The substrate signal was detectable for all of the PFOM films measured by XPS in this study. (The PFOM film thickness can be measured as long as the signal from the substrate is detectable in the spectra.)

For the rheological measurements, the PFOM was dissolved at 2 wt% in FC-72 cosolvent with Zdol, and the solvent was evaporated, leaving behind the mixture. The mixtures were melted into the fixtures of the rheometer to measure creep compliance and dynamic moduli.

For the tribological measurements, the radial position of the slider during the tests was halfway between the inner and outer diameters of the disk, and the pivot-to-center distance was adjusted so that the slider skew angle was close to 0°. The disk rotation was started with the slider suspension on the load/unload ramp just off the edge of the disk. The slider was accessed to the test radius. The pressure control and measurement of the frictional displacement were performed with The National Instruments LabView software on a PC with interface cards. Average displacement and peak amplitude were recorded after each pressure decrement/increment (4000 samples at 20 kHz for 200 ms). The pressure inside the tester was varied linearly from ambient (100 kPa) to 15 kPa and back up to ambient over about 3 min in approximately 1-kPa increments. At the end of the test, the slider was translated back to the load/unload ramp before stopping the disk rotation.

4.3 RESULTS

4.3.1 MULTIPLE DIP TEST

A test series was carried out to determine the effects of multiple dip-coating cycles on the PFOM thickness on slider rows. It was necessary to measure the substrate optical constants n_s and k_s for each row before dip coating because of the significant variation from one row to another. These values for the uncoated slider rows are listed in table 4.2. The slider rows were dip coated and the PFOM thickness was measured. This procedure was repeated until some of the rows had been dipped up to four times (4×). The PFOM thickness measured after each dip-coating cycle is given in table 4.3.

4.3.2 DIP-BAKE-DIP TEST

A test was done to study the effects of multiple dip-coating cycles with baking in between on the PFOM thickness. Slider rows were dipped; the PFOM film thickness

TABLE 4.2

Substrate Optical Constants on Slider Rows, N58 Overcoated with 12.5 nm of Carbon

Average n	Standard deviation	Average k	Standard deviation
2.039	0.011	0.385	0.009

Note: Sample size = 20.

TABLE 4.3

PFOM Thickness as Measured by Ellipsometry on Carbon-Overcoated Slider Rows Showing the Effect of Multiple Dip Coatings

	No. of dips			
	1×	2×	3×	4×
Sample size	12	8	4	4
Average thickness (nm)	2.2	4.3	4.7	4.6
Standard deviation (nm)	0.6	0.4	0.5	0.4

Note: PFOM concentration 700 ppm, withdrawal rate 1.6 mm/s, 12.5-nm carbon overcoat on N58 ceramic.

was measured by ellipsometry; the rows were baked at 70°C for 15 h; and the PFOM thickness was measured by ellipsometry after the bake. The dip and bake process was repeated until the rows had been dipped four times. The results of this test are shown in table 4.4.

4.3.3 RELATIVE HUMIDITY EFFECTS

The effect of relative humidity (RH) on the apparent PFOM film thickness measured by ellipsometry was studied. The PFOM thickness on slider rows was measured after equilibration at 51% RH and 26% RH. These rows were then placed in a vacuum desiccator to be thoroughly dried, and the apparent thickness was measured. The thickness was remeasured following equilibration at 50% RH. The results of this test are shown in table 4.5.

4.3.4 CARBON-OVERCOATED SILICON STRIPS

Since the air bearing surface of the slider is carbon-overcoated, the same carbon overcoat was placed on some of the silicon strips to evaluate the PFOM film thickness and ellipsometric measurement procedure on carbon- and non-carbon-overcoated substrates. A nominally 12.5-nm-thick layer of sputtered carbon was deposited on silicon strips, and the strips were dip coated with PFOM. The ellipsometric angles Δ and Ψ were measured. The two-layer model (two films on an absorbing substrate) was used with the optical constants for the materials listed in table 4.6 in calculating the PFOM thickness from Δ and Ψ on carbon-overcoated silicon. The apparent

TABLE 4.4
PFOM Thickness as Measured by Ellipsometry on Carbon-Overcoated Slider Rows, Showing the Effect of Multiple Dip Coatings with Baking in Between

	Process step						
	Dip	Bake	Dip	Bake	Dip	Bake	Dip
Sample size	8	8	8	4	4	4	4
Average thickness (nm)	2.8	2.8	3.9	2.9	4.6	3.9	5.4
Standard deviation (nm)	0.4	0.5	0.6	0.2	0.8	0.6	0.9

Note: Bake for 15 h at 70°C, PFOM concentration 700 ppm, withdrawal rate 1.6 mm/s, and 12.5-nm carbon overcoat on N58 ceramic.

TABLE 4.5
PFOM Thickness as Measured by Ellipsometry on Carbon-Overcoated Slider Rows, Showing the Effect of Relative Humidity and Drying

Treatment	Thickness (nm)		
	Average	Standard deviation	Sample size
51% RH	2.9	0.2	3
26% RH	3.4	0.3	3
4 h in vacuum desiccator	3.6	0.4	3
18 h in vacuum desiccator	4.2	0.1	3
34 h in vacuum desiccator	4.2	0.3	3
71 h at 50% RH	3.0	0.1	2

Note: PFOM concentration, 700 ppm; withdrawal rate, 1.6 mm/s; 1× dip; and 12.5-nm carbon overcoat on N58 ceramic.

TABLE 4.6
Optical Constants from Ellipsometry and Ellipsometric Angles Δ and Ψ for the Carbon-Overcoated Silicon Strips

Material	n	k	Ψ	Δ
PFOM	1.36	0	—	—
Carbon overcoat	1.84	0	—	—
Si/SiOx	3.853	0.212	10.54	171.25

Note: Measurements taken at wavelength 632.8 nm.

PFOM thickness, assuming two different carbon-overcoat thicknesses in the calculation, is given in table 4.7. Two of the PFOM-coated strips were measured by XPS. The d/λ from XPS and the corresponding thicknesses are listed in table 4.7.

TABLE 4.7
PFOM Thickness on Carbon-Overcoated Silicon Strips as Measured by Ellipsometry and as Estimated from XPS d/λ

| Strip | Ellipsometry thickness (nm)[a] | | | | XPS[b] | |
	12.5-nm carbon	13.5-nm carbon	$\triangle$	Ψ	d/λ	Thickness (nm)
1	3.4	2.1	126.15	13.02	—	—
2	4.1	2.7	124.98	13.20	—	—
3	4.9	3.5	123.91	13.37	—	—
4	4.3	3.0	124.71	13.24	—	—
5	4.3	3.0	124.83	13.23	—	—
XPS 1st[c]	—	—	—	—	1.08	2.9
XPS 2nd[c]	—	—	—	—	1.10	2.9
Average	4.2	2.9	—	—	1.09	2.9

Note: Carbon-overcoat thickness, nominally 12.5 nm; PFOM concentration, 650 ppm; withdrawal rate, 1.6 mm/s; 1× dip.

[a] Indicates the carbon thickness used in the calculation of the PFOM thickness from ellipsometric angles Δ and Ψ.

[b] Estimate from XPS d/λ is based on the mean free path determined from PFOM on non-carbon-overcoated silicon in table 4.8, $\lambda = 2.66$ nm.

[c] XPS 1st and 2nd were measured on two separate strips that were coated at the same time as strips 1–5.

4.3.5 NON-CARBON-OVERCOATED SILICON STRIPS AND WAFERS

Since an additional ellipsometric measurement would be needed to determine the carbon-overcoat thickness, the ellipsometric measurement of PFOM thickness directly on non-carbon-overcoated silicon is more straightforward. Silicon strips and wafers were dip coated with PFOM. The PFOM thickness measured by ellipsometry and the d/λ from XPS are listed in table 4.8. The thickness measured by ellipsometry was divided by the d/λ from XPS for each sample (last two columns in table 4.8). The experimentally determined average electron mean free path for PFOM film is $\lambda = 2.66$ nm. Sliders were dip coated with PFOM at the same conditions as the silicon wafers and strips, and d/λ was measured on the air bearing surface of each slider by XPS. These d/λ were multiplied by $\lambda = 2.66$ nm, as determined above, to estimate the PFOM thickness on the air bearing surface. These results are listed in table 4.9. The concentration of the PFOM solution was 650 ppm, and the withdrawal rate was 1.6 mm/s.

4.3.6 CONCENTRATION AND WITHDRAWAL RATE

Tests were done to determine the effects of PFOM concentration and withdrawal rate on the PFOM film thickness deposited on the air bearing surface of sliders. The PFOM film thickness was estimated using XPS. The film thickness as a function of PFOM concentration is shown in fig. 4.3a. Run 1 was made in a prototype coating tank using a developmental procedure. Run 2 was made with the coating tank and

TABLE 4.8

PFOM Thickness as Measured by Ellipsometry on Silicon Strips and Wafers, and the Mean Free Path Determined from XPS Measurement on the Same Samples

Sample no.	Ellipsometry thickness (nm)		d/λ		λ (nm)	
	Si strip	Si wafer	Si strip	Si wafer	Si strip	Si wafer
1	2.9	2.3	1.08	0.923	2.69	2.49
2	2.6	2.4	0.989	0.95	2.63	2.53
3	2.8	2.4	0.948	0.897	2.95	2.67
Average	2.8	2.4	1.01	0.923	2.66	

Note: PFOM concentration, 650 ppm; withdrawal rate, 1.6 mm/s; 1× dip.

TABLE 4.9

PFOM Thickness as Estimated from XPS Measurements on the Air Bearing Surface (ABS) of Magnetic Recording Sliders

Slider ABS	XPS	
	d/λ	Thickness (nm)
1	1.14	3.0
2	1.32	3.5
3	1.23	3.3
4	1.01	2.7
5	1.19	3.2
6	1.05	2.8
Average	1.16	3.1

Note: Magnetic recording sliders were dip coated at the same time and conditions as the silicon strips and wafers in table 4.7 using the experimental average mean free path $\lambda = 2.66$ nm from table 4.8.

procedure intended for manufacturing. The 2× dip was done to determine the effect of a rework cycle on PFOM thickness. The PFOM thickness as a function of withdrawal rate between 0.2 and 1.6 mm/s is shown in fig. 4.3b.

4.3.7 CONTACT ANGLES AND SURFACE ENERGY

Three reference liquids with known polar and dispersion surface-energy components were employed to determine the surface energy of the surface coatings 2–3 nm thick on silicon wafers. The surface energy and its components for water, hexadecane, and Zdol and their contact angles measured on silicon wafers coated with PFOM and several other low-surface-energy coatings are listed in table 4.10. The Girifalco–Good–Fowkes–Young equation [12],

$$\cos\theta = -1 + \frac{2}{\gamma_1}\left[\left(\gamma_s^d\gamma_1^d\right)^{1/2} + \left(\gamma_s^p\gamma_1^p\right)^{1/2}\right] \tag{4.2}$$

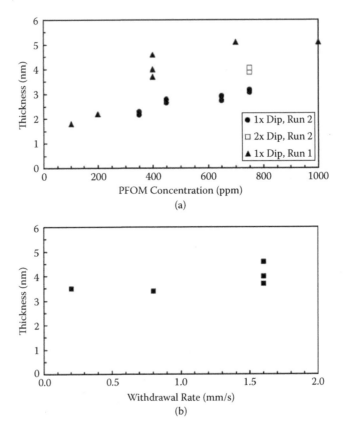

FIGURE 4.3 PFOM thickness estimated from XPS measurements on the air bearing surface of sliders as a function of PFOM concentration at a 1.6 mm/s withdrawal rate (a) and thickness as a function of withdrawal rate at a PFOM concentration of 400 ppm, 1× dip (b).

was used to determine the surface energies from the contact angles measured on the test samples. The superscripts d and p denote the dispersion and polar components of the surface energy; the subscripts s and l denote the solid (test sample) and reference liquid, respectively; and the total surface energy $\gamma = \gamma^d + \gamma^p$. The surface energies of the reference liquids are listed in the footnotes to table 4.10.

4.3.8 RHEOLOGICAL PROPERTIES

The storage and loss shear moduli, G' and G'', vs. oscillation frequency ω, and the creep compliance J vs. time t, measured at each concentration and temperature, were temperature shifted with respect to frequency or time. These temperature master curves at each concentration were then shifted to overlap one another along the frequency or time axis. The dynamic shear moduli master curves as a function of reduced frequency $\omega a_T a_C$ are shown in fig. 4.4, and the shear creep compliance master curves as a function of reduced time $t/a_T a_C$ are shown in fig. 4.5. Master curves

TABLE 4.10

Contact Angles and Surface Energies of Low-Surface-Energy Coatings on Silicon Wafers

Coating	Contact angle (degrees)			Surface energy (mJ/m²)					
				(water-hexadecane)[a]			(Zdol-hexadecane)[b]		
	water	hexadecane	Zdol 4000	γ^d	γ^p	γ	γ^d	γ^p	γ
PFOM	59.8	68.7	38.6	12.8	28.4	41.2	12.8	4.0	16.7
3M Novec EGC-1700	70.3	58.8	24.6	15.9	17.8	33.6	15.9	3.6	19.5
ZNa	33.8	62.0	16.5	14.8	46.5	61.3	14.8	5.4	20.2
FCOC	104.3	52.3	8.4	17.9	1.2	19.0	17.9	3.5	21.3

[a] Calculated from reference liquids water $\gamma^d = 22.1$ mJ/m², $\gamma^p = 50.7$ mJ/m², and hexadecane $\gamma^d = 27.47$ mJ/m², $\gamma^p = 0$, and contact angles.

[b] Calculated from reference liquids Zdol $\gamma^d = 15$ mJ/m², $\gamma^p = 6$ mJ/m², and hexadecane, and contact angles.

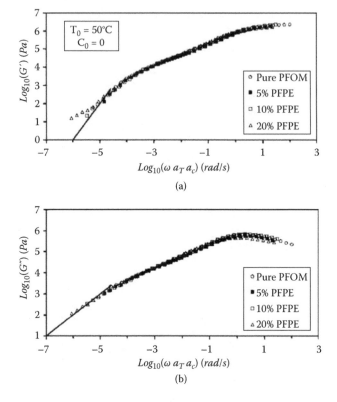

FIGURE 4.4 Dynamic shear moduli temperature-concentration shifted master curves for PFOM with dissolved PFPE Zdol: (a) storage modulus and (b) loss modulus.

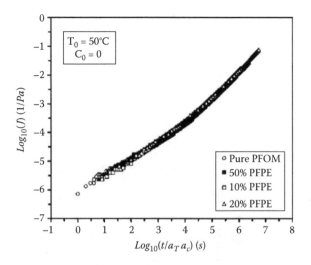

FIGURE 4.5 Shear creep compliance temperature-concentration shifted master curves for PFOM with dissolved PFPE Zdol.

are referenced to 50°C and pure PFOM ($a_T = a_C = 1$). The result is a set of temperature shift factors, a_T, and concentration shift factors, a_C. The shift factors are plotted in fig. 4.6 and are listed along with the glass transition temperatures in table 4.11.

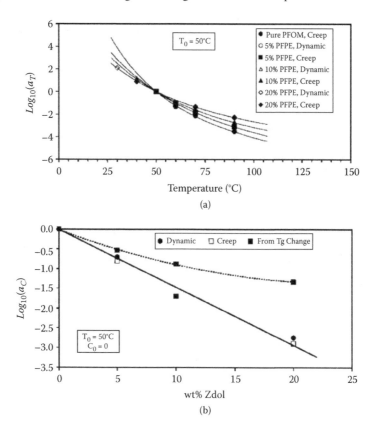

FIGURE 4.6 Shift factors from dynamic and creep rheological measurements on Zdol 2000 in PFOM: (a) temperature shift and (b) concentration shift.

TABLE 4.11
Concentration Shift Factor (a_C) from Dynamic and Creep Rheological Measurements, Glass Transition Temperature (T_g), WLF Coefficients (C_1 and C_2 [reference temperature $T_0 = 50$°C]), and the Expected Contribution of T_g Change with Concentration to the Concentration Shift Factor

| | $\log_{10}(a_C)$ | | | | | Log shift due to T_g change with |
PFPE (%)	Dynamic	Creep	T_g (°C)	C_1	C_2	concentration
0	0	0	45	9.84	70.77	0
5	−0.70	−0.80	40	10.21	91.50	−0.529
10	−1.70	−1.70	35	8.91	91.50	−0.878
20	−2.75	−2.90	25	7.43	91.50	−1.333

The WLF coefficients [13, 14] were calculated from the temperature shift factors a_T by nonlinear regression analysis using the functional form

$$\log\left(a_T\right) = \frac{-C_1\left(T - T_0\right)}{C_2 + \left(T - T_0\right)} \tag{4.3}$$

where the reference temperature $T_0 = 50°C$, and C_1 and C_2 are the WLF coefficients with respect to T_0.

4.3.9 TRIBOLOGICAL PERFORMANCE

Three metrics were employed to compare the performance of low-surface-energy coatings on sliders in the subambient pressure frictional hysteresis loop test: (1) frictional hysteresis, (2) lubricant accumulation on the slider during the hysteresis test, and (3) disk scratches. The typical friction force on the slider during loop tests, each carried out with a separate slider, is shown in fig. 4.7. Four test runs were done, each

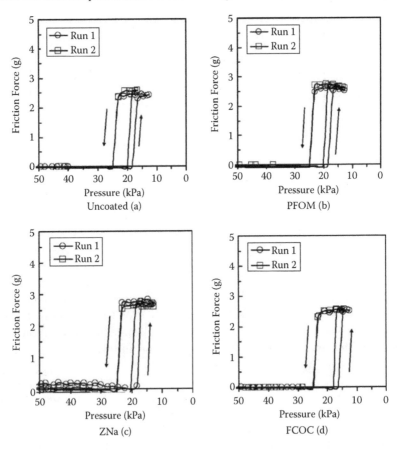

FIGURE 4.7 Results of subambient pressure frictional hysteresis loop tests on low surface energy slider coatings: (a) uncoated, (b) PFOM coated, (c) ZNa coated, and (d) FCOC coated.

with a different slider on a fresh disk test track near the middle diameter of the disk. The touch-down pressure (TDP), where the friction first increases as the air pressure is being decreased, was usually higher on the second run than on the first with a given slider/track. Representative optical micrographs of the slider taken within minutes after the end of the test on uncoated and coated sliders are shown in fig. 4.8. The splotchiness, which is more apparent on some of the sliders, is disk lubricant that was transferred to the slider. After 20 days, the splotchiness had mostly disappeared because the lubricant had spread out by surface diffusion. The increase in the number of scratches relative to the background before the test on the surface during the subambient pressure hysteresis loop tests is shown in fig. 4.9. Test tracks on two disk sides were measured. Most of the scratches were formed in the disk tests using the uncoated slider.

Overall, the tribological tests provide a ranking for the performance of the low-surface-energy slider coatings:

Lubricant transfer to the slider: uncoated > ZNa > PFOM > FCOC
Friction and hysteresis: no significant difference
Disk scratches: uncoated > FCOC > PFOM > ZNa

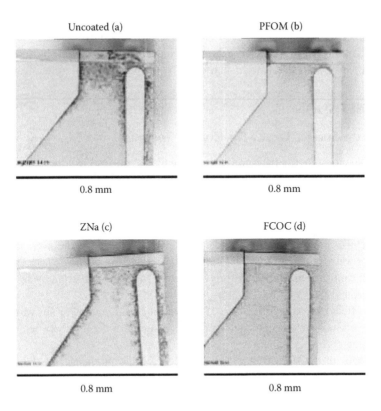

FIGURE 4.8 (**See color insert following page 80.**) Optical micrographs showing the relative level of lubricant accumulation on the slider during the subambient pressure frictional hysteresis loop tests on low surface energy slider coatings (a) uncoated, (b) PFOM coated, (c) ZNa coated, and (d) FCOC coated.

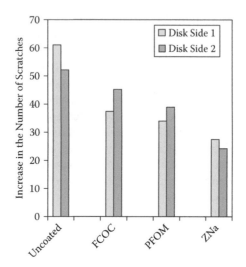

FIGURE 4.9 Scratches formed on the disk during the subambient pressure frictional hysteresis loop tests on low surface energy slider coatings.

None of the low-surface-energy coatings completely prevented disk scratching or lubricant transfer to the slider in the subambient pressure frictional hysteresis loop test. Friction and hysteresis were unaffected by the presence of the coating.

4.4 DISCUSSION

4.4.1 Low-Surface-Energy Dip-Coating Thickness

The initial ellipsometric measurements on the PFOM films were done using the TiC/ Al$_2$O$_3$ ceramic slider rows coated with 12.5 nm of carbon. Due to the variability of the ceramic optical properties and/or the carbon-overcoat thickness, n_s and k_s had to be measured for each row before coating with PFOM. A test was done to evaluate the effects of multiple dip-coating cycles (to simulate multiple dip-coating cycles during manufacturing rework) (see table 4.3). The largest increase in the average PFOM thickness was from 2.2 to 4.3 nm, following the second dip-coating cycle. There was no significant increase in PFOM thickness with further dip-coating cycles beyond 2× up through 4×.

The effect of baking at 70°C for 15 h in between dip-coating cycles was studied (table 4.4). The bake was intended to accelerate aging of the film between repeated dip-coating cycles.

Changes in the storage humidity produced significant changes in the apparent PFOM film thickness as measured by ellipsometry (table 4.5). A test was done to study the effect of extreme changes in humidity by subjecting coated slider rows to drying in a vacuum desiccator followed by equilibration at ambient 50% RH. A slider row without any PFOM coating was also subjected to the same drying cycle, and this produced no detectable changes in the optical properties of the substrate. The effect of the drying cycle on a 5-nm-thick PFOM film surface on silicon was measured by atomic force

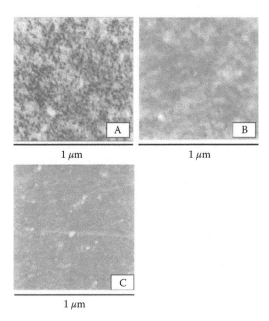

FIGURE 4.10 Atomic force images from $1 \times 1 \ \mu m^2$ scans on 5 nm thick PFOM film on silicon equilibrated at ambient 50% relative humidity, 0.26 nm rms roughness (a), the same PFOM film as in (a) following equilibration in a vacuum desiccator, 0.17 nm rms roughness (b), and an uncoated silicon strip, 0.09 nm rms roughness (c).

microscopy (AFM). Figure 4.10A shows an AFM micrograph of the PFOM film equilibrated at ambient 50% RH. Visible texture is apparent in this film, and the root mean square (rms) roughness was 0.26 nm. Figure 4.10B shows the same film after being dried in the vacuum desiccator. The micrograph of the dried PFOM film shows the absence of the texture observed in fig. 4.10A. The rms roughness of the dried PFOM film was 0.17 nm. The AFM micrograph of the uncoated silicon is shown in fig. 4.10C. The rms roughness of the silicon substrate was 0.09 nm. The PFOM film increased the surface roughness. The dried PFOM film was smoother than the film equilibrated at ambient 50% RH. However, these changes in the PFOM film texture upon drying cannot account for the observed variations in the apparent film thickness between ambient humidity and the dried film as measured by ellipsometry.

The increase in the apparent PFOM film thickness upon drying as measured by ellipsometry is attributed to polymer chain orientation within the film. For example, consider row 3 of table 4.5. In this row, the substrate $n_s = 2.052$ and $k_s = 0.393$. These were unaffected by drying in the vacuum desiccator. The ellipsometric angles measured at ambient humidity were $\Delta = 38.11°$ and $\Psi = 11.67°$, and those measured after drying were $\Delta = 40.15°$ and $\Psi = 11.64°$. The ellipsometric angle Ψ is related to attenuation during reflection, and was unaffected by drying. The ellipsometric angle Δ is the phase shift between the light polarized parallel and perpendicular to the plane of incidence. The value of Δ is determined by n_s, k_s, and the PFOM film thickness d and index of refraction n_f. The increase in the apparent thickness with drying is a result of the increase in Δ. Since the actual thickness d is unaffected by drying, and the AFM showed only a small change in film texture with drying, drying appears to change n_f. Since the n_f of a material depends primarily on its composition and density, n_f is often treated as a material property. However, anisotropy of stress or orientation causes small but significant differences between the values of n_f in the directions parallel and perpendicular to the plane of incidence. This effect is referred to as birefringence, or optical anisotropy. A molecular model of PFOM was studied to understand how molecular orientation could occur in the PFOM film. The monomer is shown in fig. 4.11a (energy minimized). The four-unit polymer in its energy-minimized free-space configuration is shown in fig. 4.11b. In this state, the side chains are randomly oriented, and the film is isotropic. Drying of the film removes water, increasing the interaction of the polar ester groups with the SiO_x surfaces. In fig. 4.11c, the side chains have been rotated about the C–O bonds (which have a low rotational energy barrier) to bring the ester groups closer to the surface. The effect of this configurational rearrangement is to produce a net orientation of the side chains perpendicular to the surface. In the oriented state, the value of n_f perpendicular to the surface > n_f parallel to the surface, leading to the apparent increase in thickness upon drying as measured by ellipsometry.

The time scale for the configurational rearrangement process is hours. When the sample was placed in the vacuum desiccator after equilibration at ambient humidity, more than 4 h were required for equilibration. Consequently, as long as the ellipsometric thickness measurements are done within about 4 h of dip coating, the PFOM film remains isotropic, and the ellipsometry provides an accurate value for the film thickness.

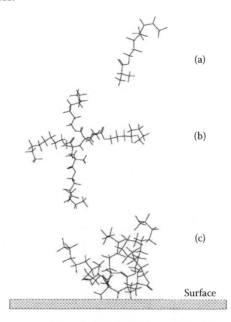

(a)

(b)

(c)

Surface

FIGURE 4.11   Wire frame image of PFOM monomer, energy minimized (a); polymer with 4 monomer units, energy minimized (b); and polymer with 4 monomer units, having side chains rotated about the C-O bonds, allowing the polar ester groups to more closely approach the surface (c).

In general, the deposition rate from solution during the dip-coating process is a function of the surface chemistry (surface tension and the heat of adsorption). A carbon overcoat was deposited on the silicon to match the surface properties of the carbon overcoat on the slider rails. However, when the single-layer model was used for analysis of ellipsometric measurements on the PFOM with the carbon-overcoated silicon strips, the calculated PFOM thickness was much too low. The ellipsometric model using two layers on a substrate was necessary. The two-layer model requires knowing the optical properties of the silicon, carbon overcoat, and PFOM (table 4.6), as well as the carbon-overcoat thickness, to calculate an unknown PFOM film thickness from the ellipsometric angles Δ and Ψ. The PFOM thickness calculated using the nominal carbon-overcoat thickness of 12.5 nm (table 4.7, column 2) seemed unusually high, so that the PFOM thickness was also calculated assuming a carbon-overcoat thickness of 13.5 nm (table 4.7, column 3). Two of the strips were also measured by XPS to obtain d/λ from eq. (4.1). The PFOM thickness estimated using the mean free path $\lambda = 2.66$ nm is in the last column of table 4.7. The PFOM thickness from XPS is closest to the PFOM thickness from ellipsometry calculated with a carbon-overcoat thickness of 13.5 nm. As a result of the tests with the carbon-overcoated silicon, it was concluded that measurement of the carbon-overcoat thickness would also be needed in conjunction with the more complicated two-layer model for the ellipsometry calculation if the carbon-overcoated silicon strips were to be used.

A test was done to evaluate the use of non-carbon-overcoated silicon wafers and strips as PFOM thickness monitors. The PFOM thickness on the silicon wafers and strips is shown in table 4.8 along with the d/λ from XPS measured on the same samples. The data in table 4.8 were employed to derive the experimental mean free path relating PFOM thickness from ellipsometry with d/λ from eq. (4.1) as

$$\text{thickness (nm)} = (2.66 \text{ nm}) \times (d/\lambda) \qquad (4.4)$$

The electron mean free path of $\lambda = 2.66$ nm is within the range of mean free paths reported for polymer thin films on surfaces [8]. Equation (4.4) was used to estimate the PFOM thickness on air bearing surfaces from d/λ. The values of d/λ and PFOM film thicknesses are given in table 4.9. The PFOM film was 0.5–0.7 nm thicker on the carbon-overcoated air bearing surfaces (table 4.9, column 3) and on the carbon-overcoated rows (table 4.7, columns 3 and 7) than on the silicon wafers (table 4.8, column 3). This is attributed to the difference between the surface chemistry of the SiO_x surface of the uncoated silicon and that of the carbon overcoat.

In conclusion, for the PFOM film thickness measurement, the non-carbon-overcoated silicon wafers can be used to monitor the thickness of PFOM being deposited on the air bearing surface as long as the presence of an offset in the deposited film thickness between the two types of substrates is taken into account.

The effects of PFOM concentration and withdrawal rate on the PFOM thickness deposited on slider rails during dip coating of HGAs were studied. The first PFOM concentration dependence run was done with a prototype coating tank and procedure (triangles in fig. 4.3a). The three triangles at 400 ppm are from three solutions separately formulated at the same concentration. Points above and below 400 ppm were from the same batch of formulation as the highest triangle at 400 ppm.

The points shown by the circles in fig. 4.3a were dip coated using an improved version of the tank and coating cycle designed for use in manufacturing. The squares are from a 2× dip using the latter conditions. The second dip, at 750-ppm PFOM concentration, increased the PFOM thickness by ≈0.8–1.0 nm. This is less than the average of ≈2.1-nm increase in thickness after 2× dips found on the slider rows with 700-ppm PFOM concentration (table 4.3). The second dip-coating cycle after baking increased the average PFOM thickness on slider rows by ≈1.0 nm (table 4.4).

For both of the PFOM concentration studies shown in fig. 4.3a, the PFOM thickness linearly extrapolated to zero PFOM concentration intercepts the vertical axis at about 1.5–2.0 nm. This thickness is nearly the same as the length of the monomer unit of the PFOM polymer shown in fig. 4.11a. Since the PFOM thickness-vs.-concentration curve must pass through zero, it seems that a layer of the PFOM forms on the surface at a very low concentration. The PFOM film thickness is also relatively independent of withdrawal rates between 0.2 to 1.6 mm/s, as shown in fig. 4.3b. The interpretation of the rapid layer formation, which is nearly one monomer unit thick, can be extended to suggest that the formation of a uniform layer may be possible at very low PFOM concentrations. An apparent nonzero intercept of the film thickness-vs.-concentration plot, and the film thickness being relatively independent of withdrawal rate, can also arise from rapid evaporation at the meniscus of a solvent that does not wet the substrate. It would be of interest to explore the region of even lower concentration to avoid excess PFOM deposit on surface topography features of the slider that accumulate a meniscus or pendant drops of solution.

4.4.2 Surface Energies of Slider Coatings

Hexadecane spreads out completely on bare carbon overcoat. To avoid this problem, the surface energy of the bare carbon overcoat can be evaluated with water and methylene iodide [15]. In this study, the dispersion surface-energy components of the low-surface-energy coatings were evaluated with either water–hexadecane or Zdol–hexadecane liquid pairs. The polar component varies considerably and is strongly dependent on the reference liquids used for measurement. For the fluorohydrocarbon surfactants and FCOC, the Zdol reports a lower polar surface energy component than water with hexadecane as the nonpolar reference liquid. This may be related to the hydrogen bonding density difference between water and Zdol, because Zdol only has hydrogen bonding at each end group, separated by the polymer chain in between. The ranking for surface energy of the coatings is based on the dispersion contribution, because this was most nearly independent of the reference liquids employed for measurement:

$$FCOC > ZNa > PFOM$$

The PFOM is probably the lowest because it has the most highly fluorinated interface arising from the fluoro-octanol ester chains being concentrated at the surface.

4.4.3 Rheology of PFOM and Mixtures with Zdol Disk Lubricant

PFOM mixtures containing less than 40 wt% Zdol remained single-phase solutions. The 40 wt% sample was initially a transparent viscoelastic liquid at room temperature. When

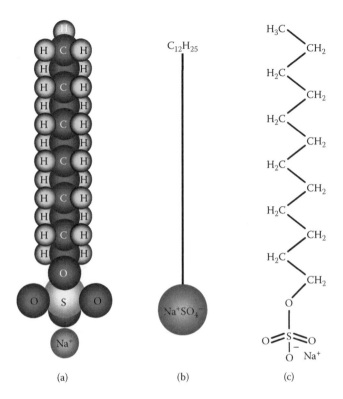

(a)   (b)   (c)

FIGURE 1.1 Comparison of (a) space-filling atomic, (b) simplified, and (c) simplified atomic views of sodium dodecyl sulfate. The letters represent the atomic symbols. These views represent the lowest energy state. However, due to molecular vibrations, the actual shape varies with time, and the average shape is not a straight chain as shown. Note that the angle for these views does not reveal the fact that the repeating carbon-hydrogen section of adsorbed sodium dodecyl sulfate is not perpendicular to the surface.

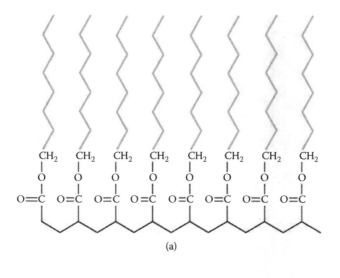

(a)

$$HO-CH_2-CF_2\left[O-CF_2-CF_2\right]_p\left[O-CF_2\right]_q O-CF_2-CH_2-OH$$

(b)

FIGURE 4.1 (a) poly (1H,1H-pentadecafluorooctyl) methacrylate (PFOM), molecular weight 300000 (g/mol); (b) Hydroxyl terminated perfluoropolyether Fomblin Zdol, molecular weight 2000 or 4000 (g/mol), p/q = 2/3.

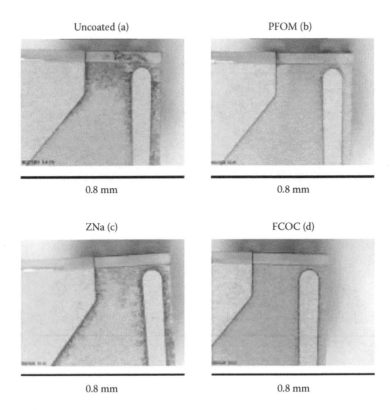

Uncoated (a)   PFOM (b)

0.8 mm   0.8 mm

ZNa (c)   FCOC (d)

0.8 mm   0.8 mm

FIGURE 4.8 Optical micrographs showing the relative level of lubricant accumulation on the slider during the subambient pressure frictional hysteresis loop tests on low surface energy slider coatings (a) uncoated, (b) PFOM coated, (c) ZNa coated, and (d) FCOC coated.

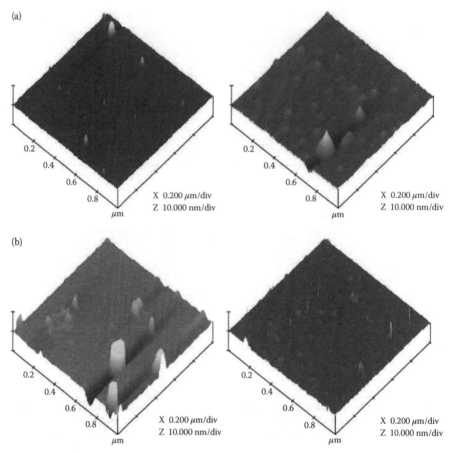

FIGURE 6.2 AFM images of (a) bare Si, (b) Si/OTS, (c) Si/APTMS, and (d) Si/GPTMS before (left image) and after (right image) coating with PFPE. The vertical scale is 10 nm in all images.

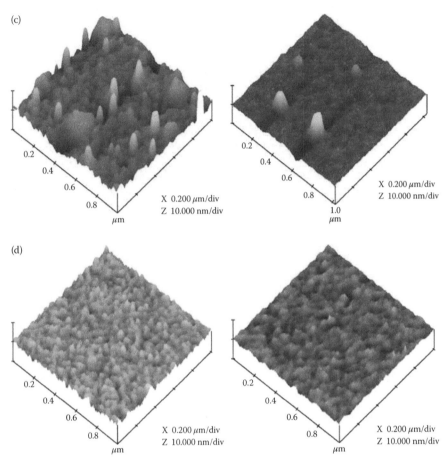

FIGURE 6.2 (CONTINUED) AFM images of (a) bare Si, (b) Si/OTS, (c) Si/APTMS, and (d) Si/GPTMS before (left image) and after (right image) coating with PFPE. The vertical scale is 10 nm in all images.

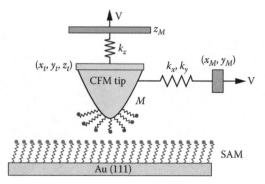

FIGURE 7.10 Mechanical model for the simulation system [80]. The tip was dragged by the support (z_M) through the spring (k_z) at a velocity v while the support (x_M, y_M) was fixed. (From Ghoniem, N. M., et al. 2003. *Philos. Mag.* 83: 3475. With permission.)

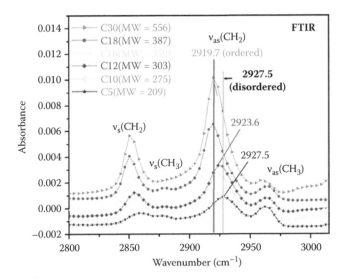

FIGURE 8.3 FTIR spectra in reflection mode of trichlorosilane SAMs with carbon chain lengths C_5–C_{30} on silicon. The asymmetric CH_2 band maxima are observed in the range of 2927–2912 cm^{-1}. Symmetric CH_2 stretches, observed at 2853–2846 cm^{-1}, are also indicated for various SAMs.

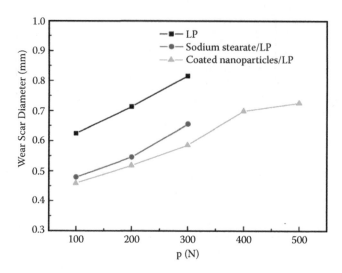

FIGURE 10.6 Effect of load on wear-scar diameter (four-ball test machine; additive concentration in liquid paraffin (LP): 0.1%; speed: 1450 rpm; test duration: 30 min).

heated to 90°C, the 40 wt% solution undergoes apparent crystallization within minutes. The crystallization occurs more slowly at 50°C. The 40 wt% sample turned from clear to cloudy when cooled to 10°C, apparently due to phase separation of the Zdol. This phase separation is reversible. The 40 wt% solution cleared upon heating to 20°C. The 60 wt% solution is a milky viscoelastic liquid; it cleared reversibly upon heating to 30°C. Apparent crystallization occurs in the 60 wt% solution over a day at room temperature.

The pure PFOM and its solutions with Zdol at and below 20 wt% Zdol concentration were thermorheologically simple fluids in that they superimposed on one another with a horizontal shift along the time or frequency axis. They also shifted to superimpose with respect to Zdol concentration (figs. 4.4 and 4.5). The temperature-shift factors are shown in fig. 4.6a. The smooth curves are fitted to the WLF equation, eq. (4.3). The concentration-shift factors plotted in fig. 4.6b show that both the dynamic and creep rheological measurements provided the same values. For typical glass-forming liquids, the shift factors follow the changes in the glass transition temperature induced by variations in molecular structure, hydrogen bonding, or molecular weight [16]. Specifically, the reference temperature in eq. (4.3) is expected to be offset by an amount equal to the change in the glass transition temperature with concentration. For the solutions of Zdol in PFOM, the concentration shift was more than expected from the change in the T_g alone. Within the context of the free-volume model, it may be that the Zdol not only mediates the intermolecular force between the PFOM chains, but also facilitates segmental translation. The temperature dependence of the shift factor in fig. 4.6a is consistent with a decrease in the flow activation energy with increasing Zdol concentration.

There was some concern that the PFOM on the slider exposed to Zdol disk lubricant might become runny and flow from the slider to the disk at the disk drive operating temperature. The rheological temperature–concentration master curves can be used to calculate the linear viscoelastic properties: zero shear viscosity η, equilibrium recoverable compliance J_e^0, and characteristic time $\tau_c = \eta J_e^0$. These are derived from the limiting low-frequency moduli according to $G''\{\omega a_T a_C \to 0\} = \eta \omega a_T a_C$ and $G'\{\omega a_T a_C \to 0\} = \eta^2 J_e^0 (\omega a_T a_C)^2$. The limiting low-frequency fit is shown by the solid lines in fig. 4.4. A hint of network formation, or phase separation, with 20 wt% PFPE in PFOM is suggested by the upward deviation of the low-frequency G' data in fig. 4.4a from the power-law slope [17]. The viscoelastic properties from the dynamic rheological data are given in table 4.12.

TABLE 4.12
Linear Viscoelastic Properties of PFOM and Its Solution with 20 wt% PFPE Zdol at 50°C

Property	Pure PFOM		20% PFPE Zdol	
	From dynamic	From creep	From dynamic	From creep
η (Pa·s)	1.6E+07	5.6E+07	2.4E+04	8.4E+04
J_e^0 (1/Pa)	1.0E–04	—	1.5E–07	—
τ_c (s)	1.6E+03	—	2.4	—

A steady viscosity is also provided by the steady-state creep compliance according to $J\left(\{t/a_T a_C\} \to \infty\right) \approx \left(t/a_T a_C\right)/\eta$. The power-law slope of the long-time creep compliance in fig. 4.5 is 0.984. The zero-shear viscosity from creep data is given in table 4.12. It is larger than the zero-shear viscosity value from the dynamic measurement. The difference is attributed to the slope of the long-time creep data being slightly less than 1. The viscoelastic properties are also calculated for 20 wt% PFPE in PFOM at 50°C with the average concentration log shift factor of −2.825, and these values are listed in table 4.12. For comparison, the viscosity of Zdol with molecular weight 2250 is 0.03 Pa·s, and the viscosity of another common disk lubricant Ztetraol 2000 is 0.6 Pa·s at 50°C. Thus, even if some lubricant becomes dissolved in the PFOM, a film of PFOM is not expected to flow from the slider.

Another possible concern, given the sensitivity of sliders to contamination, is that the combination of PFOM with disk lubricant could become tacky and collect particles. Over the range of temperature from 50°C to 90°C, the plot of the dynamic properties of PFOM and its solutions with Zdol (G′ vs. G″) transitions through the region of what is known as the viscoelastic window [18] for pressure-sensitive adhesives (PSAs). At high temperature, they should have the properties of a removable PSA, at intermediate temperature those of a general purpose PSA, and at low temperature those of a high-shear PSA. To test this assertion, PFOM solution was dip coated onto glass slides; glass particles with a size range from 5 to 50 μm were sprinkled onto the slides after heating to a series of different elevated temperatures; and the loose particles were blown off with an air gun. Micrographs showing the different amounts of glass particles remaining stuck in the PFOM film are shown in fig. 4.12. The number of residual particles clearly increased as the temperature was increased from 60°C to 70°C for the pure PFOM. With 20% Zdol in the PFOM, the transition to high adhesion for the glass particles was at a much lower temperature, between 20°C and 50°C. With the presence of PFOM coating on the slider, care must be taken to avoid particulate contamination in the disk drive to prevent particle accumulation on the slider.

4.4.4 Tribological Performance of Low-Energy Slider Coatings

All of the low-surface-energy coatings improved the tribological performance of the slider. The ranking of their relative surface energies and performance is summarized in table 4.13. The lowest-surface-energy coating PFOM had the next-to-the-least lubricant transfer and scratching. The FCOC had the least lubricant transfer.

In the absence of lubricant on the slider, the adhesion stress is taken to be the tensile strength of the interface between the slider carbon overcoat (material 1) and the disk lubricant (material 2). It is the force per unit area to separate the two materials. Only dispersion-force contributions to the tensile strength are included in the following, because the high velocity between the slider and disk asperities does not allow time for dipole orientation [19], and there are relatively few polar end groups per unit area of disk surface. An approximate expression for the tensile strength with atoms separated by distance r_{12} is derived as follows [20].

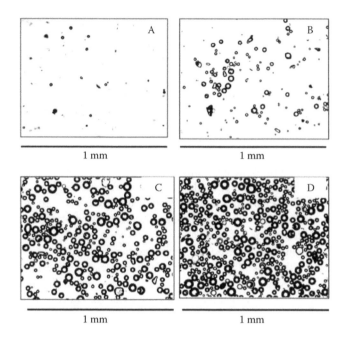

FIGURE 4.12 PFOM film dip coated on glass slide from 2 wt% solution in FC72. 5–50 μm glass microspheres dusted on dried polymer film at indicated temperature. Excess particles removed with 60 psi (414 kPa) air gun. Film thickness 50 nm, image area 1 × 1 mm. (a) 50°C, (b) 60°C, (c) 70°C, (d) 80°C.

TABLE 4.13

Ranking of Surface Energy, Lubricant Transfer, and Disk Scratches during the Subambient Pressure Loop Test

Ranking	Surface energy/ adhesion stress (MPa)	Transfer	Scratching
Highest	Uncoated/300	Uncoated	Uncoated
↓	FCOC/190	ZNa	FCOC
	ZNa/175	PFOM	PFOM
Lowest	PFOM/160	FCOC	ZNa

In this approximation, the dispersive interaction energy is

$$\sum_{12}^{d} \propto 1/r_{12}^{2}$$

and the interaction force is

$$d\sum_{12}^{d} \Big/ dr_{12} \propto -2 \sum_{12}^{d} \Big/ r_{12}$$

The interaction energy per surface is then taken to be

$$\sum_{12}^{d} \approx \sqrt{\gamma_1^d \gamma_2^d}$$

hence the adhesion stress is approximately given by

$$\sigma_{12} \approx \frac{4}{d_0} \sqrt{\gamma_1^d \gamma_2^d} \qquad (4.5)$$

where $d_0 = 0.317$ nm is the distance of closest approach between the surface atoms [21].

The thick-film limit of the lubricant dispersion surface energy is $\gamma_2^d \approx 13$ mJ / m^2 [22]. The dispersion component of the surface energy of the slider is γ_1^d. For the uncoated slider, $\gamma_1^d \approx 43$ mJ / m^2, and the coated slider values are given in table 4.10. The adhesion stress is listed in table 4.13. Further discussion of adhesion-controlled friction is given in the literature [23].

Lower adhesion stress is expected to reduce the tendency of the slider to pitch down and scratch the disk. The scratch ranking follows the adhesion stress except that the PFOM and ZNa are reversed. However, the adhesion stress values for both of these are nearly the same, so that the scratch count may be equivalent within the statistics of the measurement. Overall, the best coating for reduction of both scratches and lubricant transfer is PFOM.

4.5 SUMMARY

The development of an industrial procedure for applying a magnetic recording slider coating and for measuring the coating thickness was described. Both ellipsometry and XPS were employed to complement one another for thickness measurement and calibration. Ellipsometric measurement cannot be performed on slider rails, so it was done on slider rows or on silicon strips cut into the shape of slider rows. XPS was necessary to obtain the resolution needed for manufacturing process control.

The surface energy and tribological performance of poly (1H,1H-pentadeca-fluorooctyl methacrylate) fluorohydrocarbon surfactant were compared with several other types of slider coatings. The surface energy of the fluorinated acrylate polymer was the lowest, and it provides the best compromise for reduction of both lubricant transfer and scratches. The improvement is consistent with a reduction in the adhesion stress by the low-surface-energy coatings on the slider.

ACKNOWLEDGMENTS

The authors are grateful to P. Cotts at the IBM Almaden Research Center for light scattering measurements. ZNa was graciously provided by P. Kasai. The authors appreciate the optical surface analyzer measurements performed by A. Khurshudov, and thanks to C. Huang for assistance with the FCOC deposition.

REFERENCES

1. Karis, T. E., Kim, W. T., and Jhon, M. S. 2005. *Tribol. Lett.* 18 (1): 27–41.
2. Kasai, P. H., and Raman, V. 2004. *Tribol. Lett.* 16 (1–2): 29–36.
3. Karis, T. E., Marchon, B., Carter, M. D., Fitzpatrick, P. R., and Oberhauser, J. P. 2005. *IEEE Trans. Magn.* 41 (2): 593–98.
4. Li, Z. F., Chen, C.-Y., and Liu, J. J. 2003. *IEEE Trans. Magn.* 39: 2462.
5. Chiba, H., Musashi, T., Kasamatsu, Y., Watanabe, J., Watanabe, T., and Watanabe, K. 2005. *IEEE Trans. Magn.* 41 (10): 3049–51.
6. Ma, Y., and Liu, B. 2007. *Appl. Phys. Lett.* 90 (14): 143516-1–2.
7. Brown, C. A., Crowder, M. S., Gillis, D. R., Homola, A. M., Raman, V., and Tyndall, G. W. 1997. US Patent 5,661,618.
8. Clark, D. T., and Thomas, H. R. 1977. *J. Polym. Sci., Polym. Chem. Ed.* 15 (12): 2843–67.
9. Koishi, R., Yamamoto, T., and Shinohara, M. 1993. *Tribol. Trans.* 36 (1): 49–54.
10. Dai, Q., Kasai, P. H., and Tang, W. T. 2003. US Patent 6,638,622.
11. Kassis, C. M., Steehler, J. K., Betts, D. E., Guan, Z., Romack, T. J., DeSimone, J. M., and Linton, R. W. 1996. *Macromolecules* 29 (9): 3247–54.
12. Adamson, A. W. 1976. *Physical chemistry of surfaces.* New York: John Wiley.
13. Williams, M. L., Landel, R. F., and Ferry, J. D. 1955. *J. Am. Chem. Soc.* 77 (14): 3701–7.
14. Ferry, J. D. 1980. *Viscoelastic properties of polymers.* 3rd ed. New York: John Wiley.
15. Karis, T. E. 2000. *J. Colloid Interface Sci.* 225 (1): 196–203.
16. Kim, S.-J., and Karis, T. E. 1995. *J. Mater. Res.* 10 (8): 2128–36.
17. Roussel, F., Saidi, S., Guittard, F., and Geribaldi, S. 2002. *European Phys. J. E* 8 (3): 283–88.
18. Chang, E. P. 1991. *J. Adhesion* 34 (1–4): 189–200.
19. Karis, T. E. 2006. In *Synthetic, mineral oil, and bio-based lubricants: Chemistry and technology.* Ed. L. Rudnick. Boca Raton, FL: CRC Press.
20. Fowkes, F. M. 1965. In *Chemistry and physics of interfaces.* Washington, DC: American Chemical Society.
21. Marchon, B., and Karis, T. E. 2006. *Europhysics Lett.* 74 (2): 294–98.
22. Karis, T. E., and Guo, X.-C. 2007. *IEEE Trans. Magn.* 43 (6): 2232–34.
23. Gao, J., Leudtke, W. D., Gourdon, D., Ruths, M., Israelachvili, J. N., and Landman, U. 2004. *J. Phys. Chem.* 108 (11): 3410–25.

Part III

Self-Assembled Monolayers
and Ultrathin Films: Relevance
to Tribological Behavior

Part III

5 Frictional Properties of Organosilane Monolayers and High-Density Polymer Brushes

Motoyasu Kobayashi, Hideomi Ishida, Masataka Kaido, Atsushi Suzuki, and Atsushi Takahara

ABSTRACT

Frictional properties of organosilane monolayers and high-density polymer brushes on silicon wafer were investigated by lateral force microscopy (LFM) and ball-on-disk type tribotester. The ultrathin films of alkylsilane and fluoroalkylsilane monolayers with a series of chain lengths (C8-C18) were prepared by the chemical vapor adsorption method. LFM nanoscale measurements of tribological properties showed that the lateral force of alkylsilane monolayer decreased as the length of the alkyl chain increased. The macroscopic frictional properties were characterized by sliding a stainless steel ball ($\varphi 10$ mm) probe on its surface across a width of 20 mm at a rate of 90 mm/min under a load of 0.49 N in air at room temperature. The monolayer with longer alkyl chain and bilayer of organosilanes revealed excellent wear resistance. The microscopically line-patterned two-component organosilane monolayers were also fabricated and their macroscopic friction behavior was investigated. Even though the height difference between the two components was less than 1 nm, friction force anisotropy between the parallel and perpendicular directions against the line pattern was observed. Polymer brushes with high graft density were prepared by surface-initiated atom transfer radical polymerization of methyl methacrylate (MMA), (2,2-dimethyl-1,3-dioxolan-4-yl)methyl methacrylate (DMM), and 2-methacryloyloxyethyl phosphorylcholine (MPC) from silicon wafer immobilized with a 2-bromoisobutylate moiety. Poly(DMM) brush was further converted to the hydrophilic polymer brush consisting of 2,3-dihydroxypropyl methacrylate (DHMA) units, successively. Frictional properties of the polymer brushes were characterized by sliding a glass ball probe in air and various solvents under a normal load of 100 MPa. The PMMA brush was found to have a lower friction coefficient and much better wear resistance than the corresponding spin-coated PMMA film because of the strong anchoring of the chain ends in the brush. In addition, the friction coefficient of the polymer brush significantly decreased in response to soaking in toluene, and increased in response to immersion in cyclohexane. The hydrophilic poly(DHMA) and poly(MPC) brushes showed low dynamic friction coefficient in water. These results indicated the dependence of the tribological properties on the solvent.

5.1 INTRODUCTION

Macro- and nanotribological properties of ultrathin films as a boundary lubricant system have been the subject of great interest recently, particularly in medical devices, magnetic storage devices, and microelectromechanical systems (MEMS). The use of molecular films as boundary lubricants for solid surfaces has been known for a long time. A variety of organic molecular films have been studied as prospective boundary lubricants, including adsorbed organic layers [1], Langmuir–Blodgett (LB) monolayers [2], ultrathin films of end-functionalized fluoropolymers [3], grafted polymer brushes [4], and organic self-assembled monolayers (SAMs). A SAM is a good candidate for a boundary lubricant; therefore, many researchers have investigated how the frictional properties of the surfaces modified with SAMs depend on the molecular structure, the density of the monolayer, and the conformation of alkyl chains on a micro- and macroscopic level, as discussed below.

The number of studies on microscopic tribology started to mushroom, especially in the later 1990s, owing to development of lateral force microscopy (LFM), which can provide local surface information on friction, adhesion, modulus, and thermal properties for a wide variety of materials. Frictional properties of organosilane ultrathin films have been investigated by LFM from various points of view, including the dependence on the chain length of alkylsilanes [5], the effect of scan rate [6], the effect of different chain terminal structures of CF_3 and CH_3 [7], graft density [8], ordered structure of alkyl chains [9, 10], and so on. There is voluminous literature on the tribological properties of SAMs, such as alkanethiols on gold and alkylsilanes on silica, but alkylphosphonate on aluminum [11] and the combination of C-60 and fatty acids [12] have also been recently reported.

Early work on the macroscopic friction of SAM films was carried out by DePalma and others using a pin-on-disk type friction tester, and they observed a low friction coefficient [1, 13]; however, their frictional property containing a wear resistance was demonstrated under a limited condition. For example, the self-assembled octadecyltrichlorosilane monolayer revealed a good wear resistance at low load [14], but failed to lubricate at a normal load of 1.0 N due to wear [15]. To overcome the poor wear resistance of thin films, a dual-layer film was prepared by a combination of aminoalkylsilane and stearic acid, which improved the wear resistance of the SAM film [16]. In this work, the authors also proposed a defect-free double layer on a flat substrate by chemical adsorption of alkylsilane compound.

Recently, an anisotropic surface has been fabricated by micropatterned organic monolayers, which were prepared by the chemical vapor adsorption (CVA) method [17]. The CVA is one of the excellent and suitable processes for microfabrication based on photolithography of nanofilms because it produces a remarkably uniform surface without defects or aggregates [18, 19]. We reported previously that the line-patterned fluoroalkylsilane monolayers with 1–20-μm width showed an anisotropic wetting by a macroscopic water droplet with a 0.5–5-mm diameter [20]. Using this method, an anisotropic friction surface without microtrough or step structure can be achieved by organic monolayers aligned with two chemically different organic components. In this study, we have prepared line-patterned two-component organosilane monolayers to measure the friction force between the parallel and perpendicular directions against the line pattern.

The surface-grafted polymers have also attracted much attention to improve the physicochemical properties of solid surfaces. Long polymer chains tethered on a flat surface with a sufficiently high surface density are generally referred to as "polymer brushes" [21]. Klein, who is one of the pioneers in studying the tribology of polymer brushes, and his coworkers found a reduction in frictional forces between solid surfaces bearing tethered polymers using a surface force balance. They also reported that brushes of charged polymers (polyelectrolytes) could act as efficient lubricants between mica surfaces in an aqueous medium, even though the graft density was not as high [4, 22, 23]. Similarly, Osada and coworkers reported that the gel-terminated polyelectrolyte brushes, prepared by controlled radical polymerization using 2,2,6,6-tetramethyl-1-piperidinyloxy (TEMPO) radical, reduced the friction force between the hydrogels and a glass plate across water [24]. They also report that longer polyelectrolyte molecules exhibited even higher friction than that of a normal network gel. Sheth et al. measured the normal forces between a poly(ethylene glycol) brush against various surfaces [25]. They found that the brush did not adhere to a neutral lipid bilayer or to a bare mica. The lubrication properties of poly(L-lysine)-*graft*-poly(ethylene glycol) adsorbed onto silicon oxide and iron oxide surfaces have been studied by Spencer and coworkers using ultrathin-film interferometry, a minitraction machine, and pin-on-disk tribometry [26, 27]. They reported that the graft polymer formed a stable lubricant layer on the tribological interface in aqueous solution and reduced friction.

Over the last decades, high-density and well-defined polymer brushes have been readily synthesized owing to the progress in surface-initiated atom transfer radical polymerization (ATRP) techniques [28]. Tsujii and collaborators have measured topographic images and force–distance profiles of high-density polymethyl methacrylate (PMMA) brushes by scanning force microscopy using a microsilica sphere attached to a cantilever head [29]. At high graft density, the osmotic pressure extends the grafted polymer chains into the vertical direction, which gives rise to extremely strong resistance against compression. The steric repulsion between the polymers supporting high normal loads gives low frictional forces between brush-bearing surfaces. Extremely low friction coefficients were observed in nanotribological studies performed with atomic force microscopy using a colloidal probe immobilized with a high-density PMMA brush in toluene [30].

We have studied the macroscopic frictional properties of high-density polymer brushes prepared by surface-initiated ATRP of methyl methacrylate (MMA) [31] and hydrophilic methacrylates [32, 33] from silicon substrates. Friction tests were carried out using a stainless steel or glass ball as the sliding probe under a normal load of 100 MPa from the viewpoint of practical engineering applications. This chapter reviews the macroscopic frictional properties of polymer brushes under a high normal load, the dependence of solvent quality, the effect of humidity on hydrophilic brush, and wear resistance, and we compare these with alkylsilane monolayers.

5.2 EXPERIMENTAL

5.2.1 MATERIALS

Organosilane compounds, such as n-octadecyltrimethoxysilane (OTMS-C18, $CH_3(CH_2)_{17}Si(OCH_3)_3$, Gelest Inc., Morrisville, PA), n-hexadecyltrimethoxysilane (HTMS-C16, $CH_3(CH_2)_{15}Si(OCH_3)_3$, Gelest Inc., Morrisville, PA), n-dodecyltrimethoxysilane (DTMS-C12, $CH_3(CH_2)_{11}Si(OCH_3)_3$, Gelest Inc., Morrisville, PA), n-octyltrimethoxysilane (OCTMS-C8, $CH_3(CH_2)_7Si(OCH_3)_3$, Gelest Inc., Morrisville, PA), (perfluorooctyl)ethyltrimethoxysilane (FOETMS-C8F17, $CF_3(CF_2)_7CH_2CH_2Si(OCH_3)_3$, Shin-Etsu Chemical Co. Ltd., Tokyo), (perfluorohexyl)ethyltrimethoxysilane (FHETMS-C6F13, $CF_3(CF_2)_5CH_2CH_2Si(OCH_3)_3$, Shin-Etsu Chemical Co. Ltd., Tokyo), and nonadecenyltrichlorosilane (NTS, $CH_2=CH(CH_2)_{17}SiCl_3$, Shin-Etsu Chemical Co. Ltd., Tokyo), were used as received. Anisole was stirred with small pieces of sodium at 100°C for 6 h, followed by distillation from sodium under reduced pressure. Methanol was purified by refluxing over magnesium for 6 h followed by distillation under ambient pressure. Toluene for CVA was purified by shaking with cold concentrated sulfuric acid (10 mL of acid per 100 mL of toluene), once with water, once with saturated aqueous $NaHCO_3$, again with water, then drying successively with $MgSO_4$, and by final distillation from 1–3% n-butyllithium after refluxing for 6 h. Copper bromide (CuBr, Wako Pure Chemicals, Osaka) was purified by washing successively with acetic acid and ethanol and then drying under vacuum [34]. MMA (Wako Pure Chemicals, Osaka) was distilled under reduced pressure over CaH_2 before use. 4,4'-Di-n-heptyl-2,2'-bipyridine (Hbpy) was prepared by dilithiation of 4,4'-dimethyl-2,2'-bipyridine followed by coupling with 1-bromohexane according to the method of Matyjaszewski et al. [35]. Ethyl 2-bromoisobutylate (EB, Tokyo Chemical Inc., Tokyo) was distilled before use. 6-Triethoxysilylhexyl 2-bromoisobutylate (BHE) [36] was synthesized by hydrosilylation of 5-hexenyl 2-bromoisobutylate [37] treated with triethoxysilane using the Karstedt catalyst (platinum(0)-1,3-divinyl-1,1,3,3-tetramethyldisiloxane). MPC monomer (2-methacryloyloxyethyl phosphorylcholine) was prepared using the procedure reported previously [38]. Water for contact angle measurements and frictional tests was purified with the NanoPure water system (Millipore Inc., Tokyo).

5.2.2 LINE-PATTERNED MONOLAYER PREPARATION BY CVA [39]

The silicon (111) wafers (10 × 40-mm pieces) were cleaned by washing with piranha solution (H_2SO_4/H_2O_2 = 7/3, v/v) at 373 K for 1 h and by exposure to vacuum ultraviolet-ray (VUV) generated from an eximer lamp (UER20-172V, USHIO Inc., Tokyo, λ = 172 nm) for 10 min under reduced pressure (30 Pa). The substrates were placed together with a glass tube filled with 0.2 mL organosilane liquid into a 65-mL Teflon™ container. The container was sealed with a cap and was packed in a pressure-resistant stainless steel container. Then the container was placed in an oven maintained at 423 K (OTMS, HTMS, and DTMS monolayers) or 373 K (OCTMS, FOETMS, and FHETMS monolayers) for 2 h. In a similar process, the silicon wafers for surface-initiated ATRP and a glass vessel filled with 10% toluene solution of BHE were packed in a glass container purged with N_2 gas, and were placed in a heating oven at 373 K for 5 h. During heating at this temperature, alkylsilane vapors can adsorb on the surface of the wafers to form a uniform monolayer. The samples were rinsed with ethanol and dried in vacuo. The surface chemical composition was determined by x-ray photoelectron spectroscopy (XPS).

A microscopically line-patterned surface containing two organosilane compounds was also fabricated by VUV decomposition through a photomask and by a stepwise CVA [40, 41]. Figure 5.1 outlines the essential steps in fabricating two-component, microscopically line-patterned organosilane monolayers. A previously formed single-component organosilane monolayer on a silicon substrate was placed in an evacuated vacuum chamber and was covered with a photomask (20 × 20-mm square, 10-μm Cr pattern, 10-μm slit with 20-mm line length). Irradiation with VUV through the mask for 30 min leads to the cleavage of C-C bonds as well as the decomposition of the monolayers by the ozone generated from the trace amount of oxygen in the chamber, and thus to the formation of surface Si-OH groups [42]. After the patterned monolayer was sonicated for 10 min in ethanol and dried in vacuo to remove the decomposed residue, the second organosilane monolayer was then introduced onto the first irradiated area by a similar CVA process, to give a patterned monolayer consisting of two silane components.

5.2.3 GENERAL PROCEDURE FOR SURFACE-INITIATED ATRP

A typical surface-initiated ATRP of MMA was performed as follows. A few sheets of the initiator-immobilized silicon wafers, CuBr (0.025 mmol), Hbpy (0.050 mmol), a degassed anisole solution of MMA (50.0 mmol), and EB (0.0125 mmol) were introduced into a glass tube with a stopcock. The polymerization solution was degassed by a repeated freeze-and-thaw process to remove the oxygen. The polymerization reaction was conducted at 90°C for 2–24 h under argon to simultaneously generate a PMMA brush from the substrate and free PMMA from the EB. The reaction was stopped by opening the glass vessel to air, and the reaction mixture was poured into methanol to precipitate the free polymer. The silicon wafers were washed with toluene using a Soxhlet apparatus for 12 h to remove the free polymer adsorbed on their surfaces, and then dried under a reduced pressure at 373 K for 1 h. The free polymer in tetrahydrofuran (THF) solution was passed through a neutral alumina column to remove the catalyst. For DMM and MPC polymerization, (−)-sparteine and 4,4′-dimethyl-2,2′-bipyridyl were used instead of Hbpy, respectively. Polymerization conditions for DMM and MPC were previously reported [32, 33].

5.2.4 MEASUREMENTS

The thicknesses of monolayers and polymer brushes on the silicon substrates were evaluated as the height difference between the monolayer surface and bared Si-wafer surface by atomic force microscopy (AFM). The thickness of the OTMS monolayer was estimated to be ca. 2.0 nm. An ellipsometer (Imaging Ellipsometer, MORITEX Co., Tokyo) equipped with a YAG (yttrium aluminum garnet) laser (532.8 nm) was also used to determine the thickness of the polymer brushes.

The Fourier transform infrared (FT-IR) spectra were recorded on a Spectrum One KY (Perkin Elmer, Inc., Waltham, MA) system coupled with a mercury-cadmium-tellurium (MCT) detector. The incident angle of the p-polarized infrared

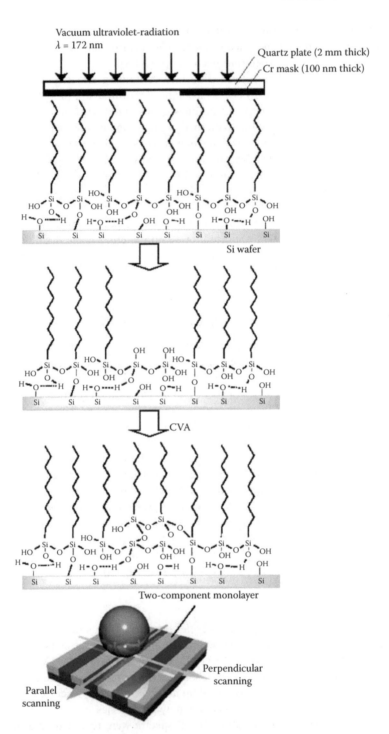

FIGURE 5.1 Preparation procedure for a two-component organosilane monolayer by photolithography and the CVA method.

beam was 73.7° (Brewster angle) in order to eliminate the multiple reflection of the infrared beam within the silicon wafer substrate [43, 44]. As substrates for the FT-IR measurements, Si wafers with both surfaces polished were used to reduce the influence of fringes in the spectra. For IR spectra, 1024 scans were accumulated at a resolution of 0.5 cm^{-1}.

The contact angles of water, hexadecane, and methylene iodide (droplet volume of 2 µL) were recorded with a drop shape analysis system (DSA10 Mk2, KRÜSS Inc., Hamburg, Germany) equipped with a video camera. XPS measurements were carried out on an XPS-APEX (ULVAC-PHI Inc., Chigasaki, Japan) using a monochromatic Al-K_{α} x-ray source at a takeoff angle of 45°.

Size-exclusion chromatography (SEC) analysis of PMMA and poly(DMM) was carried out with a JASCO LC system (JASCO Co., Tokyo) comprising three polystyrene gel columns of TSK-gel Super H6000, H4000, and H2500 (TOSOH Co., Tokyo), and a refractive index detector (JASCO RI-740) using THF as an eluent at a rate of 0.6 mL/min. The number-average molecular weight (M_n) and molecular weight distribution (MWD) were calculated from calibration against polystyrene standards. SEC chromatogram of water-soluble poly (MPC) was recorded on a JASCO instrument equipped with a JASCO 2031 plus refractive index (RI) detection, which runs through two directly connected poly(hydroxyethyl methacrylate) gel columns (Shodex OHpak SB-804$_{HQ}$, SHOWA DENKO K. K., Tokyo) using water containing 0.01 M LiBr as an eluent (0.5 mL/min) at 298 K. A calibration curve was made by a series of well-defined poly(MPC)s prepared by controlled polymerization based on a reversible fragmentation chain transfer system.

The nanotribological test was carried out with lateral force microscopy (LFM) in the air atmosphere. The LFM used in this study was an SPA400 with an SPI 3800N controller (SII NanoTechnology Inc., Tokyo), using a 20 × 20-µm scanner. A commercial rectangular 100-µm cantilever (Olympus Optical Co., Ltd., Tokyo), with a spring constant of 0.10 N·m^{-1} with a Si$_3$N$_4$ integrated tip, was used. Frictional curves were obtained for sliding velocities of 10–120 µm·s^{-1} with applied loads of 5 nN, and others were obtained for an applied normal load of 1–20 nN with sliding velocities of 40 µm·s^{-1}. The scan area was 20 × 20 µm. Each lateral force value reported in this chapter is the average of 10 data points obtained from different regions of each sample.

Macroscopic friction and wear tests were carried out on a conventional ball-on-disk–type reciprocating tribometer Tribostation Type32 (Shinto Scientific Co. Ltd., Tokyo) by sliding a stainless steel or glass ball (φ10 mm) on the substrates along a distance of 20 mm at a sliding velocity of 90–2400 mm/min under a load of 0.49–1.96 N at 298 K. The friction coefficient was determined by a strain gauge attached to the arm of the tester and was recorded as a function of time. Friction tests in various solvents were also carried out using the polymer brush substrates, which were immersed in the corresponding solvents for 24 h prior to the experiments. Humidity around the sample stage was controlled by blowing dry N$_2$ gas or by paving a wet paper towel.

5.3 RESULTS AND DISCUSSION

5.3.1 NANOTRIBOLOGICAL PROPERTIES OF ORGANOSILANE MONOLAYERS

The CVA of alkylsilanes afforded quite uniform thin monolayers on silicon substrates. Figure 5.2 shows the FT-IR spectra for the OTMS and DTMS monolayers in the range of 2800 to 3000 cm^{-1} measured at 293 K. The peaks observed at 2920.6 and 2851.0 cm^{-1} were assigned, respectively, to the antisymmetric CH$_2$ stretching band [v_a(CH$_2$)] and the symmetric stretching band [v_s(CH$_2$)] for the alkyl chain of the OTMS monolayer. On the other hand, the peaks assigned to the v_a(CH$_2$) and v_s(CH$_2$) bands for the alkyl chain of the DTMS monolayer were observed at 2925.7 and 2854.9 cm^{-1}, respectively. These results demonstrated that both OTMS and DTMS monolayers prepared by CVA were in an amorphous state at room temperature [17].

The magnitudes of lateral force for these monolayers were obtained as relative values against the FOETMS monolayer. Figure 5.3a exhibits an LFM image of the DTMS and FOETMS monolayers. The dark area was the DTMS monolayer, which had a lower magnitude of lateral force, while the bright area was the FOETMS monolayer, with a higher magnitude of lateral force. A representative frictional curve is shown in fig. 5.3b. The authors evaluated the magnitude of voltage between the trace and retrace lines. The x-axis corresponds to the lateral motion of the piezoelectric scanner, and the y-axis corresponds to the magnitude of lateral force observed as the torsion of the cantilever [45]. Figure 5.4a shows the dependence of the lateral force on the scan rate for alkylsilane monolayers having various chain lengths under a normal load of 5 nN. The alkylsilane monolayers gave much less lateral force than the silicon substrate surface. Moreover, the lateral force decreased as the alkyl chain length increased. The lateral force of a nonmodified Si substrate exhibited an almost constant value, regardless of the scan rate. On the other hand, the lateral force of the alkylsilane monolayers monotonically increased as the scan

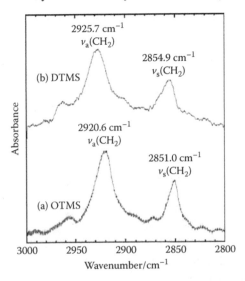

FIGURE 5.2 FT-IR spectra of (a) OTMS and (b) DTMS monolayers.

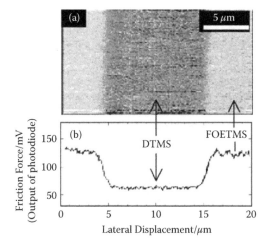

FIGURE 5.3   (a) LFM image and (b) frictional curve of DTMS and FOETMS monolayers.

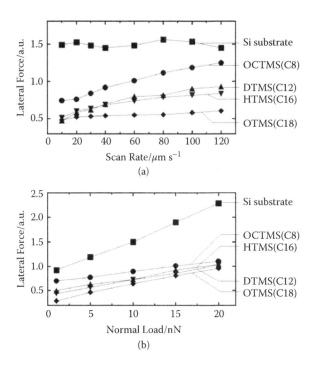

FIGURE 5.4   Lateral force measurements on organosilane monolayers: (a) scan rate dependence (normal load 5nN) and (b) normal load dependence (scan rate 40 µm·s⁻¹).

rate increased. This is because the monolayer behaves as a viscoelastic solid [46], since the organosilane monolayer is a polymer with a polysiloxane backbone. Similar behavior was observed for various self-assembled monolayers [47]. Figure 5.4b shows the normal load dependence of the lateral force with a scan rate of 40 µm·s⁻¹.

The lateral force of the Si substrate and all monolayers was proportional to the normal load. The lateral force decreased as the alkyl chain length increased, as was the case with scan rate dependence.

5.3.2 MACROTRIBOLOGY OF ORGANOSILANE MONOLAYERS

Figure 5.5 shows the normal load and scan rate dependences of the dynamic friction coefficients of various organosilane monolayers investigated by the friction tester. Under the conditions of normal load ranging from 20 to 200 g (0.2–1.96 N) at room temperature, the dynamic friction coefficient of the nonmodified Si substrate and all monolayers was proportional to the normal load. With the scan rate ranging from 1.0 to 40 mm·s^{-1} at room temperature, alkylsilane monolayers showed lower friction coefficients than the Si substrate. The dynamic friction coefficient of the nonmodified Si substrate and all monolayers exhibited almost constant values regardless of the scan rate. In contrast, the friction coefficient had no distinct chain-length dependence among the alkylsilane or fluoroalkylsilane monolayers. However, the organosilane monolayers with longer alkyl chains showed higher wear resistance in air than those with shorter chains. The bar graph in fig. 5.6 shows the lifetime of the friction-reduction effect as the number of sliding cycles at which the friction coefficient becomes 0.3 due to the peeling away of the monolayer. Since the intermolecular interaction among alkyl chains increases with chain length, the organosilane monolayers with long alkyl

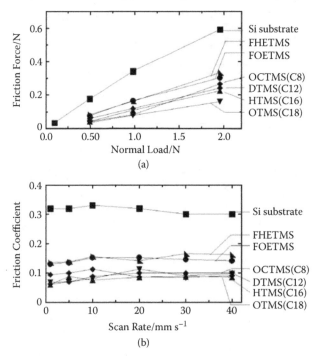

FIGURE 5.5 Macroscopic tribological properties of organosilane monolayers by ball-on-plate–type reciprocating tribometer: (a) normal load dependence (sliding velocity 80 mm/min) and (b) scan rate dependence (normal load 0.20 N).

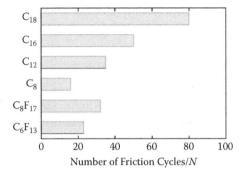

FIGURE 5.6 Wear resistance of organosilane monolayers in air: the lifetime of the friction-reduction effect as the number of cycles N, when the friction coefficient goes above 0.3 at 298 K.

chains showed strong wear resistance. The same experiment was performed with tetradecane as a lubricant. Compared with the friction tests in air, the friction coefficient was slightly lower with the tetradecane lubricant, and the frictional wear was much less than that in air. From these results, it can be concluded that organosilane monolayers exhibited excellent tribological performance in a lubricant.

5.3.3 FRICTION ANISOTROPY ON TWO-COMPONENT, MICROSCOPICALLY LINE-PATTERNED MONOLAYERS

Two-component, microscopically line-patterned organosilane monolayers of DTMS and FOETMS were prepared by CVA and the VUV photodecomposition process. The difference in height between the DTMS and FOETMS monolayers was less than 1 nm. The friction coefficients of the DTMS and FOETMS monolayers were 0.09 and 0.15, respectively. The spherical stainless steel probe was slid either parallel with or perpendicular to the direction of the line pattern of the two-component monolayer, as shown in fig. 5.1. The normal load and scan rate were 20 g and 80 mm·s^{-1}, respectively. The magnitude of the friction coefficient (0.14 ± 0.02) in perpendicular scanning against a line was higher than that (0.104 ± 0.005) in parallel scanning. In the case of parallel scanning, the contact component of the stainless steel probe and monolayer surfaces was always the same, hence there was no change in frictional behavior along the line. On the other hand, with perpendicular scanning, the contact component periodically varied as the friction probe advanced. Then, it was speculated that resistance arose at the interface between the DTMS and FOETMS monolayers. It is believed that in the case of the friction in the perpendicular direction, the stainless steel probe might be easier to move along the line direction. This was the first observation of friction force anisotropy originating from the surface chemical composition of a flat surface. This result was supported by wetting anisotropy. With the same line-patterned organosilane monolayer, wetting anisotropy was observed by sliding contact angle measurements for n-hexadecane. Sliding angle in the perpendicular direction to the line pattern was 80.0°, whereas that in the parallel direction was 28.4°. Even though the difference in the surface free energies between DTMS and FOETMS was small (16.4 mJ·m^{-2}), the wetting behavior was highly anisotropic.

5.3.4 FRICTIONAL PROPERTIES OF NTS BILAYER

To improve the wear resistance of organic monolayers, a bilayer thin film was prepared using NTS, as shown in fig. 5.7. After the first monolayer was prepared by dipping in bicyclohexane solution of NTS (5 mM), oxidation reactions were carried out by KMnO$_4$ (5 mM), NaIO$_4$ (195 mM), K$_2$CO$_3$ (18 mM), and water at 303 K for 24 h. The substrate surface was washed with water, with 0.3 M NaHSO$_3$, with water again, with 0.1 N HCl, with water, and with ethanol, and then dried under vacuum. Dipping in 5-mM bicyclohexyl solution of NTS for 24 h again produced an NTS bilayer. Water contact angles on the first NTS monolayer, the oxidized NTS monolayer, and the NTS bilayer were 103.4°, 52.9°, and 95.4°, respectively. Two times larger IR absorption due to asymmetric and symmetric vibrations of CH$_2$ was observed in the NTS bilayer compared with the monolayer. The carbonyl peak at 289.6 eV in the XPS spectra was observed in the oxidized NTS layer, but disappeared in the NTS bilayer. Figure 5.7A is an AFM image of a partially decomposed NTS bilayer film by VUV. The thickness of the NTS bilayer estimated from the line profile in fig. 5.7B

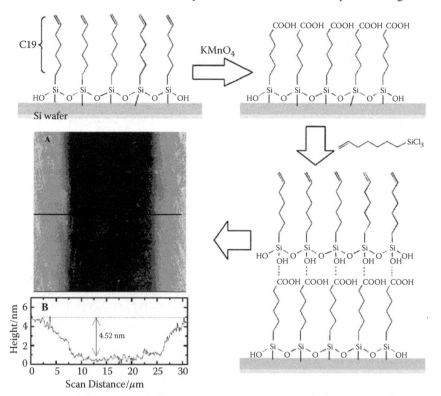

FIGURE 5.7 (A) AFM image and (B) depth profile of NTS bilayer partially decomposed by VUV.

was 4.52 nm, which is a reasonable value for the bilayer, considering the theoretical length (2.45 nm) of the alkyl chain in its extended all-*trans* conformation. These results support the formation of an NTS bilayer.

Macroscopic friction tests on the NTS monolayer and bilayer in air were performed by sliding a stainless steel ball along a distance of 20 mm at a sliding velocity of 90 mm/min under a load of 0.2 N. Both thin films showed similar dynamic friction coefficients of 0.080–0.083, while a large difference was observed in wear resistance. The friction coefficient of the NTS monolayer increased after the 150 friction cycles, as shown in fig. 5.8a. On the other hand, the bilayer demonstrated a stable friction coefficient until 2200 friction cycles (fig. 5.8b). A wear track was clearly observed on the surface of the monolayer after 300 friction cycles (fig. 5.8c), while only a slight wear track was present on the surface of the bilayer, even after 1000 slidings, as shown in fig. 5.8d. Generally, a wear track should be formed by direct contact of the stainless steel probe with the silicon substrate at the point where

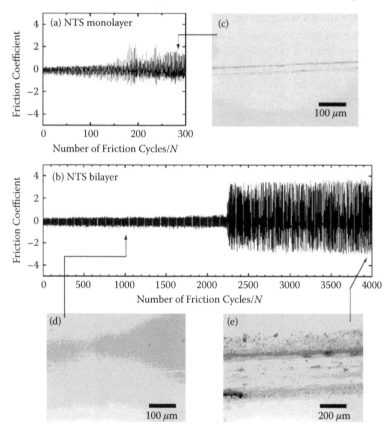

FIGURE 5.8 Evolution of the friction coefficient vs. the number of friction cycles N for silicon surface with (a) grafted NTS monolayer and (b) grafted NTS bilayer. The optical micrographs depict the wear tracks of (c) the NTS monolayer after 300-cycle tracking and (d) the NTS bilayer after 1000-cycle tracking and (e) after 4000-cycle slidings at a sliding velocity of 90 mm/min in air.

the organic layer was scratched out by local strong shear force due to the microscopic roughness on the surface. Apparently, the wear resistance of the organic film was improved by the NTS bilayer, probably because the thicker bilayer could restrict the direct contact of the probe to the substrate.

5.3.5 FRICTIONAL PROPERTIES OF POLYMER BRUSHES

Surface-initiated ATRP of MMA, DMM, and MPC from alkylbromide-immobilized silicon wafers was carried out in the presence of EB as a free initiator coupled with copper catalyst to produce polymer brushes. Figure 5.9 represents the schematic procedure for the preparation of polymer brushes. The AFM observation revealed that a homogeneous polymer layer was formed on the substrate, and the surface roughness was 0.8–1.5 nm in the dry state in the $5 \times 5 \ \mu m^2$ scan area. MWD values of the free polymers obtained were relatively low ($M_w/M_n = 1.1$–1.5), and the M_n values estimated by SEC were proportional to the monomer conversion. The M_n of surface-grafted polymer on a silicon wafer cannot be directly determined yet; however, it is widely known that a polymer brush should have the same molecular weight as the value for the corresponding free polymer [37, 48]. As shown in fig. 5.10, the thickness of the polymer brushes increased linearly with molecular weight because the polymer chain forms a fairly extended conformation due to high graft density and the excluded-volume effect. Therefore, the graft density σ was estimated by the relationship between the thickness L_d (nm) and M_n as follows:

$$\sigma = (d \cdot L_d \cdot N_A \cdot 10^{-21})/M_n$$

where d and N_A are the assumed density of bulk polymer at 293 K and Avogadro's number, respectively. The graft densities of PMMA, PDMM, and PMPC brushes were calculated to be 0.56, 0.36, and 0.22 chains/nm^2, depending on the cross-sectional area of the monomer. These values are comparatively high, taking into account the volume fraction of the polymer chain.

Macroscopic friction tests on PMMA brushes were carried out by sliding a stainless steel ball on the substrates at a rate of 90 mm/min in air under the normal load of 0.49 N at room temperature. In the case of a nonmodified silicon wafer under a normal

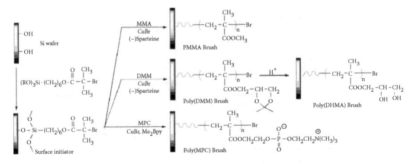

FIGURE 5.9 Preparation of ATRP initiator immobilized silicon wafer and polymer brushes by surface-initiated ATRP of methacrylates.

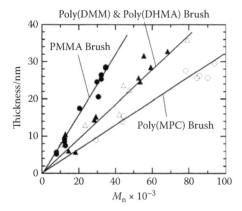

FIGURE 5.10 Relationship between polymer-brush thickness and M_n of the corresponding free polymers.

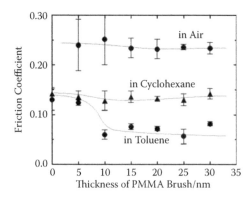

FIGURE 5.11 Friction coefficient of PMMA brush in air, cyclohexane, and toluene by sliding a stainless steel ball over a distance of 20 mm at a sliding velocity of 90 mm/min under a load of 0.49 N at 298 K.

load of 50 g (0.49 N), the theoretical contact area between the stainless steel probe and substrate can be calculated as 2.43×10^{-9} (m^2) by Hertz's contact mechanics theory,* and the average pressure on the contact area was estimated to be 201 MPa.

Figure 5.11 shows the dynamic friction coefficient of PMMA brushes of various thicknesses in air, cyclohexane, and toluene. Compared with the friction coefficient of the polymer brush in the dry state, the value decreased in both organic solvents

---

* If a circle with radius a (m) is regarded as the contact area between a stainless steel ball and substrate under a normal load P (0.49 N), Hertz's theory affords the following relationship using Young's modulus of stainless steel and silicon wafer, E_A (1.96×10^{11} Pa) and E_B (1.30×10^{11} Pa), and Poisson's ratio υ_A (0.30) and υ_B (0.28), respectively:

$$2/E = (1 - \upsilon_A^2)/E_A + (1 - \upsilon_B^2)/E_B$$
$$a = (3/4 \times 2/E \times P \times R_A)3/2$$

where R_A is the curvature radius (5.00×10^{-3} m) of a stainless steel ball. The contact area can be calculated by πa^2.

due to the fluid lubrication effect. The friction coefficient in toluene was lower than that in cyclohexane, and it decreased with increasing brush thickness. The friction coefficient of the PMMA brush with a 30-nm thickness in toluene was 0.06, which is half in magnitude compared with that of the nonmodified substrate, suggesting that the polymer in a good solvent performs as a good lubricant. In cyclohexane, however, the friction coefficient of the PMMA brush was 0.12, regardless of thickness. These findings suggest that the polymer brushes in poor solvents did not perform well as lubricants, i.e., only fluid lubrication occurred. The dynamic friction coefficient in toluene approached a value of 0.06 for a 25-nm-thick brush, while the friction coefficient for sliding friction in cyclohexane was higher than in toluene, maintaining a constant value around 0.14, regardless of the brush thickness. These results suggest that the interaction between the brush surface and the stainless steel ball (friction probe) was moderated because toluene is a good solvent for PMMA. On the other hand, the PMMA brush surface would be unwilling to be in contact with poor solvents such as hexane and cyclohexane, and would prefer to adsorb on the stainless steel probe, thus giving a higher friction coefficient.

To evaluate the wear resistance of the PMMA brush, a continuous friction test for 600 s was performed by sliding a stainless steel ball on substrates covered with 20 nm of thick polymer brush or cast film under a load of 0.49 N at a sliding rate of 90 mm/min (fig. 5.12). In the early stage of the friction test, the friction coefficient of the brush surface increased to 0.5 from 0.25, but there was no further increase during continuous friction, as shown in fig. 5.12a. On the other hand, the friction coefficient of the cast film continuously increased with friction cycles (fig. 5.12b), and the surface polymer gradually peeled away to form debris ("wear elements"), some of which was adsorbed on the surface of the probe ("transfer particles") or left in the wear track, interfering with the smooth sliding of the probe. Finally, the PMMA layer in the wear track had completely peeled off.

Solvent dependency was also observed in poly(DMM) and poly(DHMA) brushes. Hydrophilic poly(DHMA) was prepared by hydrolysis of poly(DMM) brushes with an acidic solution consisting of MeOH/2N HCl (250/1, v/v) for 6 h at room temperature. As shown in fig. 5.13, the friction coefficients of the poly(DHMA) brush with a sliding glass probe in air was found to be higher than that of poly(DMM) because hydrophilic poly(DHMA) does not prefer to contact with air, thus interacting with the glass ball to increase adhesion. In the case of friction in water (fig. 5.13), the friction coefficients of both poly(DMM) and poly(DHMA) brushes were reduced due to the fluid lubrication effect of water. However, it is noteworthy that the friction coefficient of the poly(DHMA) brush was lower than that of poly(DMM) under aqueous conditions. As water is a good solvent for poly(DHMA), the interaction between poly(DHMA) and the glass probe would be moderated, resulting in a low-friction surface.

The poly(MPC) brush gave a quite wettable surface. The contact angle of the water droplet (2.0 μL) on poly(MPC) brush surface was very low (below 5°). The contact angles of methylene iodide and hexadecane were 45° and <5°, respectively. Using the Owens–Wendt equation [49], the surface free energy of the poly(MPC) brush surface was estimated to be 73 mJ/m^2, which is quite similar to that of water. Therefore, water plays the role of a good solvent, resulting in low friction of the poly(MPC) brush. Ho et al. [50] prepared a low-friction surface on polyurethane

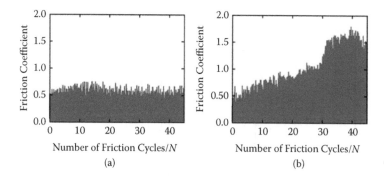

FIGURE 5.12 Evolution of the friction coefficient vs. the number of friction cycles N for the surface of (a) PMMA cast film and (b) PMMA brush under a load of 0.49 N at a sliding velocity of 90 mm/min in air (M_n of PMMA = 26,000; thickness of brush and cast film = 20 nm).

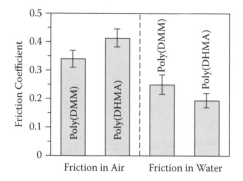

FIGURE 5.13 Friction coefficient of poly(DMM) and poly(DHMA) brushes by sliding a glass probe in air (left) and in water (right).

catheter material by coating it with poly(MPC) copolymer. They suggested that the water molecules bound to the surface of poly(MPC) due to hydration and thus provided effective and enhanced boundary lubrication.

Figure 5.14 shows the friction coefficient of a poly(MPC) brush surface measured by sliding a glass ball in air and water at 298 K. The M_n of the poly(MPC) brush for the friction test was approximately 80,000, which was estimated by SEC determination on the corresponding free polymer using water as an eluent. Under the dry N_2 gas condition, a high-friction coefficient was observed, probably due to the strong adhesion interaction between the poly(MPC) brush and the glass probe. In contrast, the brush surface afforded a lower friction coefficient with increasing humidity. It is supposed that water molecules adsorbed on the surface of the highly hydrophilic poly(MPC) brush and worked as a lubricant to reduce the interaction between the brush and the probe. In addition, a water-swollen poly(MPC) brush compressed by the probe sphere would produce a repulsive force against a high normal load due to the osmotic pressure originating from the steric repulsion of the high-density brushes [51]. These water-lubrication systems restricted a direct contact of the probe with the substrate to reduce the friction force. Contrary to our expectation, the friction coefficient in water

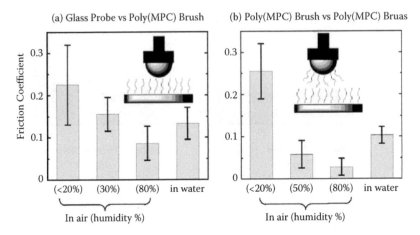

FIGURE 5.14 (a) Friction coefficient of poly(MPC) brush in dry air, humid air, and water by sliding a glass ball over a distance of 20 mm at a sliding velocity of 90 mm/min under a load of 0.49 N at 298 K. (b) A glass ball immobilized with poly(MPC) brush was also used as a sliding probe.

was higher than that in a humid air condition. We suppose that swollen and extended poly(MPC) brush chains in water should have a larger actual contact area between the brush and the probe, resulting in a higher friction coefficient compared with that obtained in the case of humid air.

A similar lubrication tendency was observed when a brush-immobilized glass ball was used as a sliding probe, as shown in fig. 5.14b. The friction coefficient under the highly humidified air condition was significantly reduced to 0.02, which is lower than that in aqueous conditions. It is supposed that hydrated MPC units effectively reduced the friction force, creating a good water-lubrication system. Basically, a mutual inter-penetration of brushes in water is supposed to be limited by the excluded-volume effects of dense grafting chains. However, a polymer brush has a distribution in chain length, as mentioned above, and the density of chains at the outer surface of the brush must be lower than that within the brush layer. The interdigitation of brush chains can occur in the limited outer surface of the brush layer, in particular at the thin layer of the brush–brush interface in water. Fairly extended poly(MPC) chains in water are expected to have a larger interdigitated region than the swollen brush in humidified air. This is the reason why the poly(MPC) brush in water showed a higher friction coefficient than in humid air. In addition, water molecules adsorbed on the surface of the hydrophilic poly(MPC) brush would form a thin boundary layer at the brush–brush interface under the humid air condition, giving a fluid lubrication and a low fric-tion. On the other hand, interdigitation of brushes in water would afford a continuous layer between the brush on the substrate and the brush on the probe to give a larger friction resistance, although additional experimental evidence is needed to confirm this hypothesis. Klein and coworkers [52] pointed out that highly compressed polymer brush layers between surfaces behaved in a solidlike manner. For a further discussion on the frictional mechanism of poly(MPC) brushes, the viscoelastic properties of the

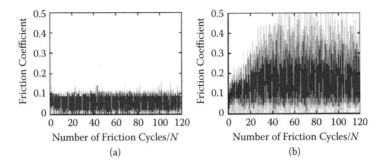

FIGURE 5.15 Evolution of the friction coefficient vs. the number of friction cycles N for the surface of (a) a poly(MPC) cast film and (b) a poly(MPC) brush at a sliding velocity of 90 mm/min under a load of 0.098 N in air (M_n of poly(MPC) = 80,800; (a) humidity = 50%, (b) humidity >75%).

brush, the effect of squeezed hydrodynamic lubrication [53], and adhesion force measurements on the brush surface would be required.

Figure 5.15a shows friction test results on a poly(MPC) brush surface under a load of 0.098 N in humid air (humidity >75%). A low-friction coefficient of poly(MPC) brush was continuously observed despite a higher normal load, implying a good wear resistance. Actually, the friction coefficient slightly increased with sliding cycles; however, it was around 0.1 even after 100 friction cycles. As a control experiment, the cast film was prepared by spin casting from a methanol solution of poly(MPC). Figure 5.15b shows the evolution of the friction coefficient of cast film vs. the number of friction cycles by a glass ball under a normal load of 0.098 N in air (humidity = 50%). The friction coefficient of the cast film began to increase in the early stage of the friction test and attained a magnitude of 0.2 within 20 tracking cycles. This result indicates that the surface polymer was easily scratched away by the sliding glass probe. In air with over 75% humidity, the cast film adsorbs moisture from the air to become a swollen layer, which is readily swept out from the silicon wafer by the sliding probe. Wear resistance was improved due to the anchoring effect of the tethered polymer chain end by covalent bonds on the substrate.

5.4 CONCLUSION

In this study, both microscopic and macroscopic frictional properties of oraganosilane monolayers, bilayers, and polymer brushes were investigated by LFM and a ball-on-disk–type friction tester. Organosilane monolayers with longer alkyl chains, which were prepared by CVA, showed lower lateral force in LFM and greater wear resistance in the macroscopic friction test than those with shorter chains. An NTS bilayer showed excellent tribological properties. Stable friction was observed for the NTS bilayer even after sliding a stainless steel ball 2000 cycles at a sliding velocity of 80 mm/min under a load of 0.2 N. From these results, it can be concluded that organosilane ultrathin films exhibited good tribological properties in a lubricant. In addition, friction-force anisotropy of organosilane monolayers was observed between parallel and perpendicular directions against the line pattern.

In the case of high-density polymer brushes prepared by surface-initiated ATRP, the friction coefficient depends on solvent quality and humidity. For instance, the friction coefficient of the polymer brush decreased in response to soaking in a good solvent, whereas it increased in response to immersion in a poor solvent. The macroscopic frictional properties of the poly(MPC) brush surface were strongly dependent on the humidity. An extremely low friction coefficient was observed under humid air conditions by sliding a brush-bearing glass probe under a load of 0.49 N. We suppose that boundary lubrication was achieved through the superhydrophilic poly(MPC) brush even in humid air, thus significantly decreasing the friction coefficient of the surface. The organosilane thin films, such as the NTS bilayer, show greater wear resistance than polymer brushes, even though the poly(MPC) brush under humid conditions demonstrated a stable friction coefficient, even after sliding over 100 friction cycles. Macroscopic friction tests at longer sliding times and further research into the relation between adhesion and lateral lubrication forces between brush surfaces are currently in progress.

ACKNOWLEDGMENTS

We are grateful to Prof. K. Ishihara, of the University of Tokyo, for donation of the MPC monomer and helpful discussions. This research was supported, in part, by a grant-in-aid for the 21st Century COE program "Functional Innovation of Molecular Informatics" and a grant-in-aid for "Joint Project of Chemical Synthesis Core Research Institutions" from the Ministry of Education Culture, Science, Sports and Technology of Japan.

REFERENCES

1. DePalma, V., and Tillman, N. 1989. *Langmuir* 5: 868–72.
2. Ulman, A. 1991. *An introduction to ultrathin organic films: From Langmuir–Blodgett to self-assembly.* San Diego: Academic Press.
3. Suzuki, M., Saotome, Y., and Yanagisawa, M. 1988. *Thin Solid Films* 160: 453–62.
4. Klein, J., Kumacheva, E., Mahalu, D., Perahia, D., and Fetters, L. J. 1994. *Nature* 370: 634–36.
5. Xiao, X., Hu, J., Charych, D. H., and Salmeron, M. 1996. *Langmuir* 12: 235–37.
6. Liu, Y., Evans, D. F., Song, Q., and Grainger, D. W. 1996. *Langmuir* 12: 1235–44.
7. Kim, H. I., Koini, T., Lee, T. R., and Perry, S. S. 1997. *Langmuir* 13: 7192–96.
8. Clear, S. C., and Nealey, P. F. 2001. *J. Chem. Phys.* 114: 2802–11.
9. Kim, H. I., Voiadjiev, V., Houston, J. E., Zhu, X.-Y., and Kiely, J. D. 2001. *Tribol. Lett.* 10: 97–101.
10. Sugimura, H., Ushiyama, K., Hozumi, A., and Takai, O. 2002. *J. Vac. Sci. Technol. B* 20: 393–95.
11. Hoque, E., DeRose, J. A., Hoffmann, P., Mathieu, H., Bhushan, J. B., and Cichomski, M. 2006. *J. Chem. Phys.* 124: 174710-1–174710-6.
12. Zhang, P., Lu, J., Xue, Q., and Liu, W. 2001. *Langmuir* 17: 2143–45.
13. Cléchet, P., Martelet, C., Belin, M., Zarrad, H., Renault, N.-J., and Fayeulle, S. 1994. *Sensors and Actuators A* 44: 77–81.
14. Rühe, J., Novotny, V. J., Kanazawa, K. K., Clarke, T., and Street, G. B. 1993. *Langmuir* 9: 2383–88.

15. Ren, S., Yang, S., Zhao, Y., Zhou, J., Xu, T., and Liu., W. 2002. *Tribol. Lett.* 13: 233–39.
16. Ren, S., Yang, S., and Zhao, Y. 2003. *Langmuir* 19: 2763–67.
17. Koga, T., Morita, M., Ishida, H., Yakabe, H., Sono, S., Sakata, O., Otsuka, H., and Takahara, A. 2005. *Langmuir* 21: 905–10.
18. Hozumi, A., Ushiyama, K., Sugimura, H., and Takai, O. 1999. *Langmuir* 15: 7600–4.
19. Hozumi, A., Yokogawa, Y., Kameyama, T., Sugimura, H., Hayashi, K., Shirayama, H., and Takai, O. 2001. *J. Vac. Sci. Technol. A* 19: 1812–16.
20. Morita, M., Koga, T., Otsuka, H., and Takahara, A. 2005. *Langmuir* 21: 911–18.
21. Rühe, J. 2004. In *Polymer brushes*. Ed. R. C. Advincula, W. J. Brittain, K. C. Caster, and J. Rühe. Weinheim: Wiley-VCH.
22. Kampf, N., Gohy, J.-F., Jérôme, R., and Klein, J. 2005. *J. Polym. Sci. Part B Polym. Phys.* 43: 193–204.
23. Raviv, U., Giasson, S., Kamph, N., Gohy, J.-F., Jérôme, R., and Klein, J. 2003. *Nature* 425: 163–65.
24. Ohsedo, Y., Takashina, R., Gong, J. P., and Osada, Y. 2004. *Langmuir* 20: 6549–55.
25. Sheth, S. R., Efremova, N., and Leckband, D. E. 2000. *J. Phys. Chem. B* 104: 7652–62.
26. Lee, S., Müller, M., Ratoi-Salagean, M., Vörös, J., Pasche, S., De Paul, S. M., Spikes, H. A., Textor, M., and Spencer, N. D. 2003. *Tribol. Lett.* 15: 231–39.
27. Müller, M., Lee, S., Spikes, H. A., and Spencer, N. D. 2003. *Tribol. Lett.* 15: 395–405.
28. Tsujii, Y., Ohno, K., Yamamoto, S., Goto, A., and Fukuda, T. 2006. *Adv. Polym. Sci.* 197: 1–45.
29. Yamamoto, S., Ejaz, M., Tsujii, Y., and Fukuda, T. 2000. *Macromolecules* 33: 5602–8.
30. Tsujii, Y., Okayasu, K., Ohno, K., and Fukuda, T. 2005. *Polym. Prepr. (Am. Chem. Soc., Div. Polym. Chem.)* 230: U4189.
31. Sakata, Y., Kobayashi, M., Otsuka, H., and Takahara, A. 2005. *Polym. J.* 37: 767–75.
32. Kobayashi, M., and Takahara, A. 2005. *Chem. Lett.* 34: 1582–83.
33. Kobayashi, M., Terayama, Y., Hosaka, N., Kaido, M., Suzuki, A., Yamada, N., Torikai, N., Ishihara, K., and Takahara, A. 2007. *Soft Matter* 3: 740–46.
34. Grimaud, T., and Matyjaszewski, K. 1997. *Macromolecules* 30: 2216–18.
35. Matyjaszewski, K., Patten, T. E., and Xia, J. 2005. *J. Am. Chem. Soc.* 119: 674–80.
36. Ohno, K., Morinaga, T., Koh, K., Tsujii, Y., and Fukuda, T. 2005. *Macromolecules* 38: 2137–42.
37. Husseman, M., Malmstrom, E. E., McNamura, M., Mate, M., Mecerreyes, D., Benoit, D. G., Hedrick, J. L., Mansky, P., Huang, E., Russell, T. P., and Hawker, C. J. 1999. *Macromolecules* 32: 1424–31.
38. Ishihara, K., Ueda, T., and Nakabayashi, N. 1990. *Polym. J.* 22: 355–60.
39. Tada, H., and Nagatama, H. 1994. *Langmuir* 10: 1472–76.
40. Takahara, A., Sakata, H., Morita, M., Koga, T., and Otsuka, H. 2003. *Composite Interfaces* 10: 489–504.
41. Koga, T., Otsuka, H., and Takahara, A. 2002. *Chem. Lett.* 31: 1196–97.
42. Saito, N., Wu, Y., Hayashi, K., Sugimura, H., and Takai, O. 2003. *J. Phys. Chem. B* 107: 664–67.
43. Kojio, K., Ge, S.-R., Takahara, A., and Kajiyama, T. 1998. *Langmuir* 14: 971–74.
44. Maoz, R., Sagiv, J., Degenhardt, D., Möhwald, H., and Quint, P. 1995. *Supramol. Sci.* 2: 9–24.
45. Clear, S. C., and Nealey, P. F. 2001. *Langmuir* 17: 720–32.
46. Zhang, Q., and Archer, L. A. 2005. *Langmuir* 21: 5405–13.
47. Brewer, N. J., Beake, B. D., and Leggett, G. J. 2001. *Langmuir* 17: 1970–74.
48. Matsuno, R., Yamamoto, K., Otsuka, H., and Takahara, A. 2004. *Macromolecules* 37: 2203–9.

49. Owens, D. K., and Wendt, R. C. 1969. *J. Appl. Polym. Sci.* 13: 1741–47.
50. Ho, S. P., Nakabayashi, N., Iwasaki, Y., Boland, T., and LaBerge, M. 2003. *Biomaterials* 24: 5121–29.
51. Yamamoto, S., Ejaz, M., Tsujii, Y., Matsumoto, M., and Fukuda, T. 2000. *Macromolecules* 33: 5608–12.
52. Klein, J., Kamiyama, Y., Yoshizawa, H., Israelachivili, J. N., Fredrickson, G. H., Pincus, P., and Fetters, L. J. 1993. *Macromolecules* 26: 5552–60.
53. Fredrickson, G. H., and Pincus, P. 1991. *Langmuir* 7: 786–95.

6 Tribology of Ultrathin Self-Assembled Films on Si

The Role of PFPE as a Top Mobile Layer

N. Satyanarayana, S. K. Sinha, and M. P. Srinivasan

ABSTRACT

The search for a truly ultra-thin (only a few to tens of nm thick) coating on solid substrates has gained importance in recent times because of the growth in microsystems technologies. Historically, Si has been the mainstay material for many microsystems where interfacial cyclic contacts and sliding are encountered, for example in microelectromechanical systems (MEMS). Si is known for poor tribological performance as it wears out easily when in dynamic contact with itself or with another similar material due to its strong hydrophilic nature and brittle mechanical response to contact stress. Hence, tribology of Si surface is a technologically very important area of research. In this chapter, a review of the recent developments in the area of tribology (friction, adhesion and wear properties) of self-assembled monolayers (SAMs) on Si is presented. Further, examples of the studies conducted on SAMs overcoated with perfluoropolyether (PFPE) are demonstrated to elucidate the importance of a liquid-like film on SAMs for enhancing the wear life of Si surfaces. PFPE overcoated-SAMs on Si have reduced friction and adhesion vis-à-vis only SAMs or bare Si surfaces, and have shown greater improvement in wear durability. These ultra-thin films can find applications in a variety of Si made components and devices.

6.1 INTRODUCTION

MEMS (microelectromechanical systems) are devices made by the integration of mechanical elements, sensors, actuators, and electronics on a common Si substrate through microfabrication technology [1]. MEMS find applications in areas such as sensors, actuators, power-producing devices, telecommunications, chemical reactors, biomedical devices, etc. [2, 3]. In MEMS components, the electrostatic and other surface forces become predominant when compared with inertial and gravitational forces because of the large ratios of surface area to volume, and hence the performance of MEMS components is limited by the nature of the surface. Therefore, contact-related tribological properties such as adhesion, friction, and wear phenomena become important in MEMS components [4, 5]. Moreover, Si, which is the common material used for producing MEMS components, is a poor tribological material and shows high friction, stiction, and wear without suitable modification of the surface [6, 7]. Therefore, the tribological properties of Si need to be improved considerably to realize the true potential of MEMS.

Conventional oil-based lubricants are not suitable for application in MEMS, as the viscous forces can be quite large when compared with the forces involved in operating these components because the size of the oil molecules is of the same order as those of MEMS components [5]. Therefore, many researchers have proposed ultrathin organic molecular layers as the lubricants for Si-based MEMS systems [3, 5, 8]. These ultrathin molecular lubricant layers are generally formed by two methods: (a) the Langmuir–Blodgett (LB) method and (b) the self-assembly method [9]. The LB monolayers have two main difficulties for their use as the lubricant layers for MEMS components: (a) LB films are less wear resistant, as the interaction forces between LB films and substrate (van der Waals) are weaker [10], and (b) application of the LB method is restricted to flat surfaces only, and it is not practical to coat three-dimensional surfaces of a structure [11–14].

Since LB films do not provide high wear resistance, SAMs have been proposed as prospective candidates for robust molecular lubricants [8, 15–18]. These nanometer-thick monolayers were introduced in the 1980s for surface modification through chemical adsorption of functional organic molecules, with the formation of chemical bonds between end reactive groups and surfaces [9]. Much literature has been reported on the SAMs. The two important SAMs that have been extensively studied are alkylsilane SAMs on Si [4, 15, 16, 19–22] and alkylthiol SAMs on Au [23–25]. Because Si is the base material for MEMS components, alkylsilane SAMs have become an obvious candidate for the lubrication of MEMS components. These SAMs have shown greater wear durability and lower friction coefficients than the classical physisorbed monolayers (LB films). One of the important technological advantages of these SAMs is that self-assembly processes are compatible with the wet-chemistry processes used in microfabrication technology. Therefore, much attention has been paid recently to SAMs because of their excellent properties, including easy preparation and deposition, low thickness, stable chemical and physical properties, and strong covalent bonding with the substrate. Moreover, SAMs are very attractive because of the great flexibility in varying their properties by changing the chain length [16], chemical nature [24, 25], terminal end-group chemistry [26, 27], and degree of cross-linking [24] within the layer. SAMs belong to the category of surfactants, as the molecules of SAMs contain chemical groups at both ends, out of which at least one group is reactive/hydrophilic, a property that is utilized to chemisorb them onto a substrate [9].

Much has been reported on the tribological properties of SAMs, especially of silane SAMs, which can chemically bond with the Si surface. Most of the work has been carried out using atomic force microscopy/frictional force microscopy (AFM/FFM), which simulates single asperity surface interactions [28], but only a few research results are available on the sliding experiments [15, 16, 19, 20, 29, 30], which simulate the actual rotating parts of MEMS, often moving at high velocities. An extensive review of the literature on AFM/FFM studies of SAMs can be found in the review articles by Bhushan [31], Tsukruk [32], and Zhang and Lan [33] (and references therein).

By preparing silane SAMs on a silicon surface, the coefficient of friction (COF) is reduced from 0.5–0.6 to 0.1, as reported by many groups [29, 34] using a reciprocating tribometer, and from 0.15 to 0.018, as reported by Tsukruk et al. [35] using FFM.

The effect of increasing the chain length of silane SAMs on tribological properties such as friction and wear has also been extensively studied [16, 23, 36, 37]. Ruhe et al. [16] found that the initial dynamic coefficient of friction was independent of the length of the alkylsilane chains, while the wear durability increased as the chain length of the SAMs increased. They relate this behavior to (a) the higher flexibility of the longer chains, which can dissipate more mechanical energy during shearing processes than the short-chain molecules, and (b) the stronger intermolecular interactions [16]. Similar findings have been reported for methyl-terminated thiols on gold surfaces [36]. Nanotribological investigations involving the study of the effect of chain length of silane SAMs on friction have shown that friction is higher with shorter chains compared with longer chains [37]. The reason for this behavior is the increase in the energy dissipation modes due to the increased disorder with shorter chains [23]. Kim et al. [26, 27] have systematically studied the effect of the surface terminal group chemistry on frictional properties and found that the introduction of a small percentage of bulky groups, such as CF_3 and $CH_2(CH_3)_2$, in place of CH_3 dramatically increased the frictional forces. They attributed this behavior to the increased amount of resistance to shearing offered by the bulky terminal groups.

Silane SAMs have been deposited onto MEMS components and micromotors, and their effect on the performance of the components has also been evaluated [38, 39]. They improved the performance of the components by reducing the stiction in micromachines [4].

Among the alkylsilane SAMs, octadecyltrichlorosilane (OTS) SAM has been the most widely studied and used in industry as an antistiction and friction-reducing organic modifier [28, 40]. OTS SAM has shown higher hydrophobicity, lower coefficient of friction, and reasonable wear durability compared with the other SAMs with different terminal groups, such as azide (N_3) and CF_3 [15, 41].

All of the studies discussed above have shown that some silane SAMs are efficient in reducing the coefficient of friction, the work of adhesion, and stiction properties; however, their wear resistance is not sufficient to provide high durability to the MEMS components [42]. One possible reason for the low wear durability of SAMs is the lack of a mobile portion in the lubricant. Hence, there is no replenishment in these layers as molecules are continuously removed from the contact area during the wear process. Moreover, the worn particles generated as a result of material removal act as a third body and further accelerate the wear of the film. Therefore, we proposed a lubrication concept of overcoating SAMs (bonded) with an ultrathin layer of perfluoropolyether (PFPE) (bound + mobile) to improve the wear durability of SAMs and hence that of the Si substrate (fig. 6.1) [43, 44]. The mobile PFPE is expected to lubricate and replenish the worn regions and hence enhance the wear durability.

The present concept of overcoating SAMs with PFPE to improve the wear durability of SAMs of different functional groups has not been tried previously from the view of MEMS lubrication. The concept of overcoating PFPE onto epoxy nanocomposite bilayers with the purpose of improving the wear durability has been proposed by Julthongpiput [45]; however, the results are not available. A similar concept of overcoating a mobile hydrocarbon-based lubricant onto hydrocarbon layers chemically bound onto

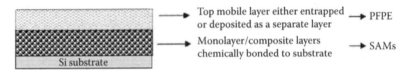

FIGURE 6.1 Representation of the lubrication scheme of PFPE overcoated onto SAM surfaces. (The size and the pattern of the rectangular boxes have no significance; diagram not to scale.) (Reprinted from Satyanarayana et al. 2007. *Phil. Mag.* 87 (22): 3209–3227. With permission from Taylor & Francis.)

Si-based MEMS components has been used by Eapen et al. [46] to enhance the wear durability. We do not compare our results with those of Eapen et al. [46], as the top mobile layer and the SAMs they used were quite different than those used in our study. It is worth noting that both studies were carried out independently during the same period of time. In a recent study, Ahn et al. [47] have incorporated mobile paraffin oil molecules into polymer nanolayers on Si surfaces, which have shown high wear resistance because of self-repairability due to the migration of mobile molecules into the sliding contact. Once again, we do not compare the results of the present lubrication concept used in this chapter with those of Ahn et al. [47], as the mobile molecules used and the procedure of incorporating them into the ultrathin layers are very different.

The concept of overcoating PFPE onto SAMs is similar to that of magnetic hard-disk lubrication, where a combination of both bonded and mobile PFPEs is routinely used to better protect the hard-disk surface. For example, in the studies by Katano et al. [48], Chen et al. [49], and Sinha et al. [50], a combination of both bonded and mobile PFPEs on hard-disk surfaces has shown higher wear durability than the use of either bonded or mobile PFPEs alone. In a study by Choi et al. [51], PFPE overcoating onto SAMs-modified hydrogenated amorphous carbon surface has shown higher wear durability than only a SAM-coated or PFPE-coated carbon surface.

Therefore, in the present study, we have focused on the effect of coating PFPE onto a single monolayer of SAM. Three different SAMs—OTS, aminopropyltrimethoxysilane (APTMS), and glycidoxypropyltrimethoxysilane (GPTMS)—were used. OTS is hydrophobic, with methyl terminal groups, whereas APTMS and GPTMS are hydrophilic, with amine and epoxy terminal groups, respectively. The purpose of selecting one hydrophobic and two hydrophilic SAMs was to elucidate the effect of the surface functional group on the final tribological performances after overcoating with PFPE. PFPE was chosen as the top layer because of its excellent properties, which include low surface tension, chemical and thermal stability, low vapor pressure, high adhesion to the substrate, and good lubricity [52]. The primary objective of the present study was to investigate the effects of the PFPE overcoating (mobile and partially bonded) on the wear durability of SAMs. Contact sliding tests were conducted to evaluate the wear characteristics of the PFPE overcoated onto SAM surfaces.

6.2 EXPERIMENTAL

6.2.1 MATERIALS

The substrate used for all modifications was polished single crystal silicon (100) (Engage Electronics Pte. Ltd., Singapore). Pieces measuring approximately 2 × 2 cm were cut from 4-in. Si wafers and used for surface modification. The three SAM materials (OTS, APTMS, and GPTMS) were obtained from Aldrich, whereas the PFPE (Zdol 4000, monodispersed, with a molecular weight of 4000 $g \cdot mol^{-1}$) was obtained from Solvay Solexis, Singapore. The chemical formulas of the three SAM materials and the PFPE used are shown below:

OTS:

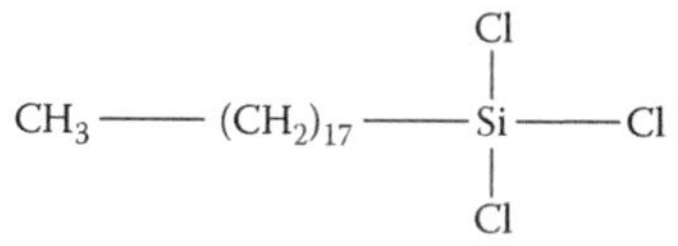

$$CH_3 - (CH_2)_{17} - \underset{\underset{Cl}{|}}{\overset{\overset{Cl}{|}}{Si}} - Cl$$

APTMS:

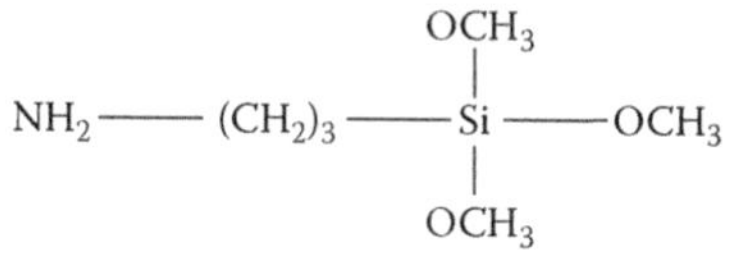

$$NH_2 - (CH_2)_3 - \underset{\underset{OCH_3}{|}}{\overset{\overset{OCH_3}{|}}{Si}} - OCH_3$$

GPTMS:

$$\overset{CH_2}{\overset{/\,\backslash}{O-CH}} - CH_2 - O - CH_2 - CH_2 - CH_2 - \underset{\underset{OCH_3}{|}}{\overset{\overset{OCH_3}{|}}{Si}} - OCH_3$$

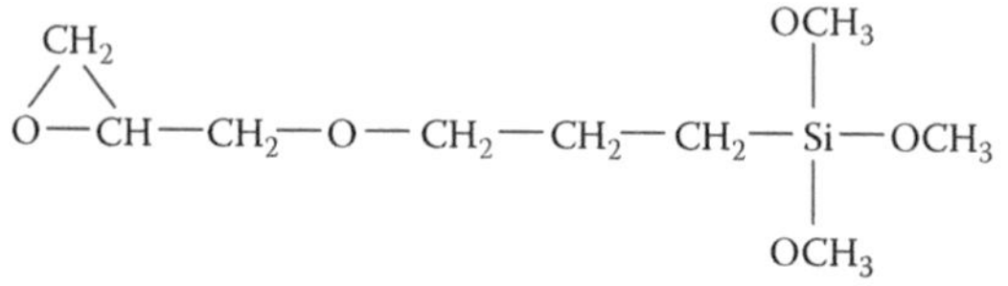

PFPE:

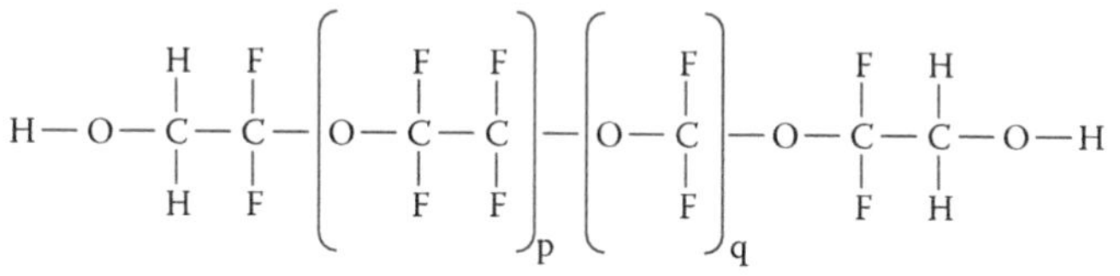

Toluene was the solvent used for SAM materials, whereas hydrofluoropolyether (H-Galden, Solvay Solexis, Singapore) was used as the solvent for PFPE. Methanol (99.8%), acetone (99.5%), and distilled water were also used for rinsing purposes in the sample preparation.

6.2.2 Preparation of SAMs

Initially, the Si substrate was subjected successively to ultrasonic cleaning with soapy water and distilled water and then was rinsed with acetone for 10 min to remove any contaminants. The cleaned samples were then immersed in a piranha solution (a mixture of 7:3 (v/v) 98% H_2SO_4 and 30% H_2O_2) at 60°C to 70°C for 50 min. Piranha treatment removes any organic contaminants and hydroxylates the Si surface. After piranha treatment, the Si was very smooth, with an rms (root mean square) roughness of 0.3 nm. The samples after piranha treatment were thoroughly rinsed successively with distilled water and acetone and dried with N_2 gas. The samples were then immersed in the respective

SAM solutions. The concentrations used for OTS, APTMS, and GPTMS were 3 mM, 5 mM, and 1 vol%, respectively, and the deposition times used for OTS and APTMS SAMs were 5–6 h and for GPTMS SAM it was 16 h. The procedure used for GPTMS SAM deposition was similar to that reported elsewhere [53], except that the deposition in our study was carried out in ambient atmosphere at 25°C and a relative humidity of ≈70%. The SAM-modified Si samples were then washed sequentially with toluene and methanol to remove any physisorbed SAM molecules and were finally dried with N_2 gas.

6.2.3 Dip-Coating of PFPE onto SAM-Modified and Unmodified Si Substrates

A custom-built dip-coating machine was used for PFPE coating onto SAM-modified and unmodified Si surfaces. A concentration of 0.2 wt% of PFPE was used. The sample was dipped in the PFPE solution and held for 1 min and withdrawn at a constant speed of 2.1 mm·s^{-1}. Tribological tests were carried out almost immediately after PFPE coating to eliminate any effect of aging at room temperature. The PFPE coated samples were heated at 150°C for 2 h in vacuum and were used for studying the effect of thermal treatment on the physical, chemical, and tribological properties.

Extensive rinsing was carried out after PFPE coating and after heat treatment, with the same solvent as used for dip-coating, to measure the atomic percentage of F (fluorine) remaining after rinsing (using x-ray photoelectron spectroscopy [XPS]), which was then used to estimate the amount/percentage of PFPE bonded to the SAM or Si surfaces. Such rinsing removes only the mobile portion of PFPE, whereas the bonded part remains firmly attached to the substrate.

6.2.4 Surface Characterization and Analysis

VCA Optima Contact Angle System (AST Products, Inc., U.S.) was used to measure static water contact angles. A water droplet of 0.5–1 µL was used for contact angle measurement, and at least five to six measurements, on three different samples, were carried out to obtain an average value. The variation in water contact angle values at various locations of a sample was within ±2°, while the measurement error was within ±1°.

The surface topography was investigated using a Multimode™ AFM (atomic force microscope) (Veeco, U.S.). Images were collected in air using a silicon tip in the tapping mode. The set-point voltage used was 1–2 V, and the scan rate was 1 Hz.

The chemical state of the sample surface before and after PFPE coating, with and without thermal treatment and after extensive rinsing of PFPE, was studied by XPS. Measurements were made on a Kratos Analytical AXIS HSi spectrometer with a monochromatized Al K_α x-ray source (1486.6 eV photons) at a constant dwell time of 100 ms and a pass energy of 40 eV. The core-level signals were obtained at a photoelectron takeoff angle of 90° (with respect to the sample surface). All binding energies (BE) were referenced to the C1s hydrocarbon peak at 284.6 eV.

6.2.5 Tribological Characterization

A commercial microtribometer (UMT-2, Universal Microtribometer, Centre for Tribology, California) was used for friction and wear tests using a ball-on-disk configuration.

A Si_3N_4 ball of 4-mm diameter was used as the counterface, whose roughness was 5 nm. The rotational speed of the spindle was 100 rpm, giving a sliding speed of 0.021 ms^{-1} at a track diameter of 4 mm. The normal load used was 5 g, which gave a contact pressure of approximately 330 MPa (calculated using the Hertzian contact model). Although the typical contact pressures in MEMS are in the range of less than one to a few MPa [54, 55], we set the contact pressure at 330 MPa to shorten the duration of the wear tests. All experiments were performed in ambient atmosphere at room temperature (23°C) and a relative humidity of approximately 70% in a class-100 clean booth. In this chapter, the initial coefficient of friction was reported after 4 s of sliding (i.e., six cycles of disk rotation) after stabilization of the sliding process. After a careful review of the literature for similar kinds of tests [16, 56], the wear durability was defined as the number of cycles after which the coefficient of friction exceeded a value of 0.3 or after a visible wear scar appeared on the substrate, whichever happened earlier. The wear durability data were obtained on at least three different samples utilizing at least two different tracks on each sample, and an average value was calculated and is reported in this chapter. The worn surfaces, after an appropriate number of sliding cycles, were observed and imaged using an optical microscope.

6.3 RESULTS

6.3.1 WATER CONTACT ANGLE RESULTS

The water contact angles on bare Si and on SAM surfaces with and without PFPE overcoat and thermally treated PFPE overcoat are listed in table 6.1. Bare Si, without any modification, shows a water contact angle value of 12°, whereas OTS, APTMS, and GPTMS SAMs surfaces show water contact angle values of 108°, 50°, and 52°, respectively. The water contact angles on Si/PFPE, Si/OTS/PFPE, Si/APTMS/PFPE, and Si/GPTMS/PFPE (as lubricated) are 66°, 114°, 114°, and 81°, respectively, and after thermal treatment they are 112°, 110°, 112°, and 107°, respectively. The water contact

TABLE 6.1
Water Contact Angles and Coefficients of Friction for Various Materials Studied

Material	Water contact angle, degrees	Coefficient of friction
Bare Si	12	0.6
Si/PFPE, as lubricated	66	0.17
Si/PFPE, thermally treated	112	0.13
Si/OTS	108	0.19
Si/OTS/PFPE, as lubricated	114	0.15
Si/OTS/PFPE, thermally treated	110	0.13
Si/APTMS	50	0.5
Si/APTMS/PFPE, as lubricated	114	0.2
Si/APTMS/PFPE, thermally treated	112	0.13
Si/GPTMS	52	0.6
Si/GPTMS/PFPE, as lubricated	81	0.1
Si/GPTMS/PFPE, thermally treated	107	0.17

angle value after PFPE overcoating (with and without thermal treatment) depends on the SAM onto which PFPE was overcoated and also depends on the chemical as well as physical interactions between the SAM molecules and the PFPE molecules. The present water contact angle values on PFPE (as lubricated) on SAMs indicate that PFPE interacts differently with different SAMs. Hydrophobicity (low surface energy) is one of the essential properties required for MEMS components, as the surfaces with high energy lead to stiction and early failure. Therefore, the PFPE-coated SAM surfaces may alleviate the problem of stiction arising from the capillary forces in the MEMS components, as they have shown low wettability/low surface energy [57, 58].

6.3.2 AFM Topography Results

Figure 6.2 shows the AFM images of bare Si and three SAM surfaces before and after PFPE overcoating. The corresponding roughness values are listed in table 6.2.

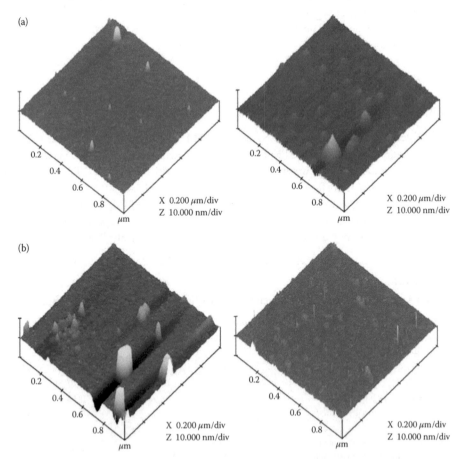

FIGURE 6.2 **(See color insert following page 80.)** AFM images of (a) bare Si, (b) Si/OTS, (c) Si/APTMS, and (d) Si/GPTMS before (left image) and after (right image) coating with PFPE. The vertical scale is 10 nm in all images.

FIGURE 6.2 (CONTINUED) **(See color insert following page 80.)** AFM images of (a) bare Si, (b) Si/OTS, (c) Si/APTMS, and (d) Si/GPTMS before (left image) and after (right image) coating with PFPE. The vertical scale is 10 nm in all images.

TABLE 6.2
Surface Roughness Values Obtained from AFM Over 1 × 1-μm Area

Material	Roughness (nm)
Bare Si	0.31
Si/PFPE	0.64
Si/OTS	1.96
Si/OTS/PFPE	0.3
Si/APTMS	1.7
Si/APTMS/PFPE	1.05
Si/GPTMS	0.74
Si/GPTMS/PFPE	0.45

After PFPE overcoating, the surface has become smooth, except in the case of bare Si. In the case of bare Si, the surface roughness has increased from 0.31 nm to 0.64 nm after PFPE overcoating, whereas in all the other cases the surface roughness has decreased after PFPE overcoating. The PFPE layer on SAM surfaces and bare Si is featureless (with the exception of when coated onto GPTMS SAM), except that a few islands are present at some random locations on the surface. PFPE overcoating onto GPTMS SAM shows the island structure, which is similar to that of GPTMS, except that the PFPE layer was smoother than the GPTMS layer. The exact reasons for the observed differences in topography after PFPE overcoating onto different SAM surfaces and Si are not clear at this moment.

6.3.3 XPS RESULTS

The purpose of XPS characterization was to identify whether or not the target film had been properly deposited, its chemical state and chemical interactions between SAM molecules and PFPE molecules, etc. For example, the amount of PFPE bonded can be obtained in the case of PFPE-overcoated SAMs with and without thermal treatment. Figure 6.3a shows the wide-scan spectra of the SAMs-modified and unmodified Si surfaces, and fig. 6.3b compares the wide-scan spectra of PFPE (as lubricated) onto SAM surfaces and Si. The wide-scan spectra of the three different SAMs qualitatively confirm the successful formation of respective SAMs on Si. For example, APTMS-modified Si shows a strong N1s peak, which must have resulted from the amine groups of APTMS molecules. The presence of the F1s peak on all PFPE-coated surfaces supports the presence of PFPE.

Table 6.3 shows the XPS atomic concentrations of F on the PFPE-coated surfaces (three SAMs and bare Si), PFPE-coated and -rinsed surfaces, and thermally treated and rinsed PFPE surfaces. From the atomic percentages of F as listed in table 6.3, the amount of PFPE bonded can be estimated [59]. The atomic percent of F is less for PFPE coated onto OTS SAM in comparison to the other two SAMs and bare Si. This suggests that the amount of PFPE present on OTS SAM was less compared with the other two SAMs and bare Si. XPS results clearly indicate that about 60% of the as-lubricated PFPE remains on the APTMS SAM surface, followed by ≈30% on the GPTMS and bare Si, whereas a negligible (≈3%) amount of PFPE remains on the OTS surfaces after rinsing. The portions of PFPE remaining on the surface after rinsing with the solvent (H-Galden) on bare Si and APTMS-modified Si are bonded through hydrogen (H) bonding, whereas in the case of Si/GPTMS, the PFPE present on the surface is bonded covalently [60, 61]. After rinsing, the entire PFPE has been removed from the OTS SAM surface, as there is no bonding between OTS SAM and PFPE molecules. Alternatively, one can say that all PFPE present on the OTS SAM is mobile in nature. Both the strongly adsorbed PFPE as well as the mobile PFPE greatly influence the tribological properties, especially the wear durability, which will be discussed in the later sections.

The amount of PFPE remaining after rinsing of thermally treated PFPE was 65% on APTMS SAM, 80% on GPTMS SAM, 15% on OTS SAM, and 67% on bare Si (table 6.3). It is clear that the thermal treatment improves the extent of bonding between PFPE and reactive SAM molecules and Si. There was a slight increase in the

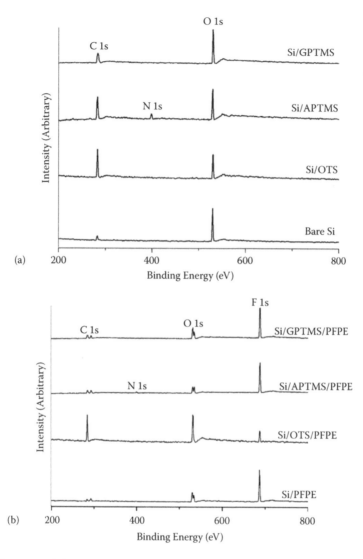

FIGURE 6.3 (a) Wide-scan spectra of SAM-modified and unmodified Si surfaces and (b) wide-scan spectra of PFPE-overcoated SAMs such as OTS, APTMS, and GPTMS and unmodified Si surface.

PFPE remaining after thermal treatment followed by rinsing in the case of OTS SAM, which must have resulted from the strongly physisorbed PFPE molecules and not the chemisorbed PFPE. Thermal treatment cannot lead to chemical bonding between methyl groups of OTS and alcohol groups of PFPE. Thermal treatment has greatly increased the amount of PFPE bonding in the case of GPTMS SAM, whereas there was not much improvement in bonding for APTMS SAM. There was an increase from 30% to 67% in the amount of PFPE bonded in the case of bare Si after thermal treatment. The exact reasons for these differences are not clear at this moment.

TABLE 6.3

Percent F Obtained from XPS Analysis of Modified and Unmodified Si Surface

Material	F conc. (At%)
Si/PFPE, as lubricated	40
Si/PFPE, as lubricated + rinsed	12
Si/PFPE, thermally treated + rinsed	26.9
Si/OTS/PFPE, as lubricated	7.81
Si/OTS/PFPE, as lubricated + rinsed	0.23
Si/OTS/PFPE, thermally treated + rinsed	1.22
Si/APTMS/PFPE, as lubricated	41.23
Si/APTMS/PFPE, as lubricated + rinsed	25
Si/APTMS/PFPE, thermally treated + rinsed	27
Si/GPTMS/PFPE, as lubricated	47.3
Si/GPTMS/PFPE, as lubricated + rinsed	12.18
Si/GPTMS/PFPE, thermally treated + rinsed	37.8

The chemical changes observed from XPS results have implications on physical properties such as water contact angle values and tribological properties, especially wear durability.

6.3.4 TRIBOLOGICAL RESULTS

The coefficient of friction values of unmodified Si and suitably modified Si obtained during sliding tests at a contact pressure of ≈330 MPa are reported in table 6.1. Without PFPE overcoating, OTS SAM shows a coefficient of friction of 0.19, whereas bare Si and the other two SAMs show a value in the range of 0.5–0.6. OTS SAM shows the lowest coefficient of friction among the three SAM surfaces studied. PFPE coating onto SAM surfaces and bare Si shows a coefficient of friction of about 0.1–0.2 and is independent of the initial coefficient of friction of the substrate before PFPE coating. We found the ranking of the coefficients of friction of different films in the following order: bare Si = Si/GPTMS > Si/APTMS > Si/APTMS/PFPE > Si/OTS > Si/PFPE > Si/OTS/PFPE > Si/GPTMS/PFPE. Thermal treatment after PFPE overcoating shows a coefficient of friction of 0.17 in the case of GPTMS, whereas it shows a value of 0.13 in the case of the other two SAMs and bare Si.

The variation of the coefficient of friction with respect to the number of sliding cycles (typical plots) of selected samples and the quantitative wear durability (along with error bars) for various samples are shown in figs. 6.4a and 6.4b, respectively. Large variations in the wear durability in some of the tests are comparable with the findings in the literature [16, 29, 62] and are due to the detrimental effect of atmospheric contaminants on thin-film properties. Bare Si and two hydrophilic SAMs (without PFPE overcoating) have failed within a few tens of cycles, whereas OTS SAM shows a wear durability of ≈1600 cycles. The tribological properties of the three SAM surfaces and bare Si are consistent with the literature data involving similar

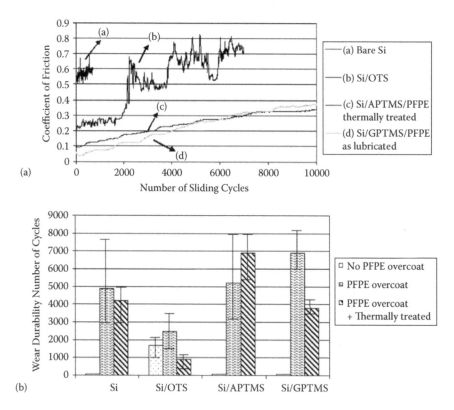

FIGURE 6.4 (a) Variation of coefficient of friction with respect to number of sliding cycles for bare Si, Si/OTS, Si/APTMS/PFPE (thermally treated), and Si/GPTMS/PFPE (as lubricated) and (b) average wear durability data of three SAM surfaces and bare Si, with and without PFPE overcoat and after thermal treatment.

kinds of sliding tests [16, 30, 63]. Overall, the PFPE overcoat has increased the wear durability of all three SAMs as well as the bare Si. PFPE coatings onto bare Si and on APTMS and GPTMS SAMs have shown appreciable improvement in wear durability, whereas PFPE coating onto OTS SAM has shown only a very minor increase. The average wear durability values of bare Si and OTS, APTMS, and GPTMS SAMs after PFPE overcoating are 5000, 2500, 5200, and 6900 cycles, respectively. The wear durability data suggest that PFPE overcoating helps in improving the tribological properties of all SAMs as well as bare Si. Moreover, the extent of improvement in wear durability is more when PFPE is coated onto hydrophilic SAMs (APTMS and GPTMS) as compared with hydrophobic SAM (OTS), even though hydrophobic SAM has shown higher wear durability than that of hydrophilic SAMs without PFPE overcoating. The ranking of the wear durability of these surfaces from the most wear durable to the least is as follows: Si/GPTMS/PFPE > Si/APTMS/PFPE > Si/PFPE > Si/OTS/PFPE > OTS SAM > APTMS SAM or GPTMS SAM or bare Si.

Thermal treatment shows an increase of ≈30% in wear durability of PFPE-coated APTMS SAM, whereas it decreases the wear durability in the case of PFPE-coated GPTMS SAM, OTS SAM, and Si surfaces. The wear durabilities of bare Si and

OTS, APTMS, and GPTMS SAMs after PFPE overcoating and followed by thermal treatment are 4200, 900, 6900, and 3800 cycles, respectively.

6.3.5 Analysis of Wear Tracks Using Optical Microscopy

Figure 6.5 shows the optical images of wear tracks for bare Si, Si/APTMS/PFPE (thermally treated), and Si/GPTMS/PFPE (as lubricated) after an appropriate number of sliding cycles. The wear track of bare Si after ≈700 sliding cycles shows the accumulation of wear debris (fragments of Si wafer) along the wear track and severe damage to the Si surface. The sharp-cornered wear debris observed along the wear track must have been produced by the fracture of the Si surface, which has typical brittle-fracture behavior. The thermally treated PFPE-coated APTMS SAM surface has shown a better performance without any signs of wear to the Si surface up to ≈14,000 sliding cycles, as evidenced by the optical micrograph shown in fig. 6.5b. It only shows signs of initiation of mild scratching inside the wear track. This supports the potential of thermally treated PFPE overcoat on APTMS SAM in enhancing wear durability. Figure 6.5c shows the wear track of Si/GPTMS/PFPE (as lubricated)

(a)

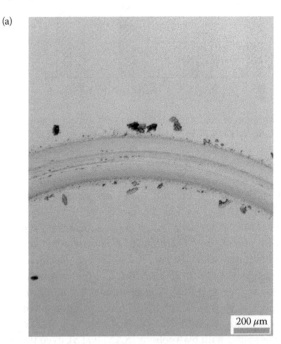

200 μm

FIGURE 6.5 Optical micrographs of worn surfaces after appropriate number of cycles: (a) bare Si, after 700 cycles of sliding, (b) Si/APTMS/PFPE (thermally treated) after ≈14000 cycles of sliding, and (c) Si/GPTMS/PFPE (as lubricated) after 5000 cycles of sliding. (Images (a) and (b) are reprinted from Satyanarayana, N., and Sinha, S. K. 2005. *J. Phys. D: Appl. Phys.* 38 (18): 3512–22. With permission from Institute of Physics Publishing.)

(b)

(c)

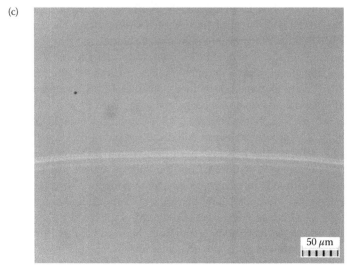

FIGURE 6.5 (CONTINUED) Optical micrographs of worn surfaces after appropriate number of cycles: (a) bare Si, after 700 cycles of sliding, (b) Si/APTMS/PFPE (thermally treated) after ≈14000 cycles of sliding, and (c) Si/GPTMS/PFPE (as lubricated) after 5000 cycles of sliding. (Images (a) and (b) are reprinted from Satyanarayana, N., and Sinha, S. K. 2005. *J. Phys. D: Appl. Phys.* 38 (18): 3512–22. With permission from Institute of Physics Publishing.)

after 5000 cycles. The wear track after 5000 cycles is very smooth in the case of Si/ GPTMS/PFPE (as lubricated), without any wear debris present along the wear track. It shows mild scratching of the surface, but wear of the Si surface has not yet started, as evidenced by the absence of wear particles in the wear track.

The above-explained wear-track images support the potential of thermally treated PFPE overcoat on APTMS SAM and Si/GPTMS/PFPE (as lubricated) in enhancing the wear durability.

6.4 DISCUSSION

6.4.1 Effect of PFPE Coating onto Bare Si

PFPE coating on Si reduces the coefficient of friction from 0.6 to 0.16, and the reasons for this behavior are the low surface energy (which eventually reduces the coefficient of friction [64]) and the little resistance offered during sliding because of the flexibility and mobility of the polymer molecules [65]. PFPE coating on Si increases the wear durability from only a few tens of cycles to ≈5000 cycles. The tribological properties of Si/PFPE observed in the present study are in good agreement with the results observed by Ruhe et al. [62].

6.4.2 Tribology of SAMs with and without PFPE Overcoat

PFPE coating onto SAM surfaces reduces the coefficient of friction (table 6.1) irrespective of the initial coefficient of friction of SAM surfaces. The reason for this behavior is that the top PFPE layer influences the initial coefficient of friction in all cases, regardless of the substrate. PFPE molecules offer a lower resistance to the shearing action and hence show a lower coefficient of friction.

It is evident from the wear-durability data that PFPE overcoating onto the APTMS and GPTMS SAMs shows considerable increase in wear resistance than that observed when it was coated onto the OTS SAM. It is speculated that the mobile PFPE molecules are initially squeezed and displaced out of the contact region during the sliding between the ball and the PFPE-coated SAM surfaces. Once the mobile PFPE is completely removed from the contact region, the hydrogen/covalently bonded PFPE (if any) lubricates the contact by boundary lubrication. After the complete depletion of both mobile and bonded PFPE, the sliding proceeds between the ball and the SAM molecules, and then the wear of the SAM follows. Therefore, because of the additional lubrication from the mobile and bonded PFPE, the PFPE coating onto the SAM surfaces results in greater improvement in wear durability.

We further speculate that when PFPE is coated onto hydrophilic SAM surfaces, a fraction of the PFPE is trapped between the SAM molecules (inside the valleys on the SAM surface), a fraction is strongly adsorbed through hydrogen/covalent bonding, and the remaining fraction is present as a mobile layer. The chemistry of the SAM and PFPE molecules, coupled with XPS data (table 6.3), suggests that a portion of PFPE is bonded through hydrogen bonding when PFPE is coated onto the APTMS SAM and bare Si, whereas some portion of PFPE is covalently bonded when PFPE is coated onto the GPTMS SAM. Moreover, in the case of APTMS SAM (and to some extent on GPTMS SAM), it is reasonable to expect that a part of the remaining portion of PFPE will be trapped in the defect regions or gaps (valleys) between various islands of SAM molecules (refer to fig. 6.2), and the remaining portion of PFPE is present as the top mobile layer. All the non-hydrogen-bonded PFPE is present as the mobile layer in the case of bare Si, as there is no trapping because of its flat and smooth surface. Similar to the case of bare Si, PFPE present on OTS SAM is also mobile in nature because, first, the surface of OTS SAM is very flat (featureless) and, second, the very high density of OTS SAM molecules gives less chance for the trapping of any PFPE molecules [23]. We strongly believe that all three portions of PFPE (the mobile, the bonded, and the trapped PFPE molecules) influence the wear durability characteristics. This is schematically illustrated in a molecular model shown in fig. 6.6, which is partly based on the model presented by Choi and Kato [66]. In the molecular model, the mobile PFPE molecules are represented by thin lines, whereas the strongly adsorbed PFPE molecules on reactive SAM surfaces are represented by thick lines. It is anticipated that the strongly adsorbed molecules act as an anchor for the mobile molecules.

The reasons for the greater increase in wear durability in the case of PFPE coating onto APTMS and GPTMS SAMs are summarized as an optimum combination of mobility, entrapment, and H-bonding/covalent bonding of PFPE. In the case of APTMS and GPTMS SAMs, the PFPE trapped in the gaps/valleys in between SAM molecules reaches the surface in a kind of "squeezing action" when a normal compressive stress is applied. The beneficial effect of the entrainment of the mobile PFPE in between the islands of APTMS/GPTMS may appear similar to the effect of boundary lubrication on etched and textured surfaces [67]. Pettersson and Jacobson [67] have observed that introducing a texture on a sliding surface improves the tri-

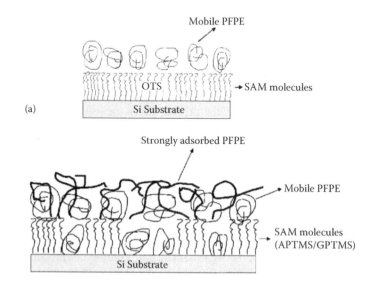

FIGURE 6.6   Molecular model of PFPE on (a) OTS SAM and (b) APTMS/GPTMS (refer to text for details). Thicker lines in (b) are used for strongly adsorbed, and the thinner lines for mobile PFPE molecules. (Reprinted from Satyanarayana, N., and Sinha, S. K. 2005. *J. Phys. D: Appl. Phys.* 38 (18): 3512–22. With permission from Institute of Physics Publishing.)

biological properties under boundary lubrication conditions. They observed that the lubricant was supplied even inside the contact by the small reservoirs (intentionally made depressions or undulations) and hence reduced the friction and increased the lifetime of the tribological contact. They further observed that the effect of texture on tribological properties greatly depends on the type of hard coating present beneath the lubricant layer and the shape and size of the depressions. However, there are few differences between the published work on texture effect under boundary lubrication conditions and the present work involving overcoating PFPE onto SAM surfaces. The amount of lubricant used in the published work on texturing was much larger than that used in the present work, and the size of the depression between the APTMS/ GPTMS SAM islands is much smaller than that on the textured surfaces used in the published work. Moreover, the chemical interactions between the PFPE and surface-reactive groups of APTMS/GPTMS also make a difference when compared with the published work, where there was no chemical interaction between the lubricant and the surface. Therefore, it is concluded that the PFPE entrainment effect in the case of APTMS/GPTMS SAMs is quite different from a simple texturing effect.

Between APTMS and GPTMS, PFPE coating onto GPTMS SAM shows higher wear durability than when PFPE is coated onto APTMS SAM. Even though the increase in wear durability is only ≈25% when PFPE is coated onto GPTMS SAM than when it is coated onto APTMS SAM, the specific chemical interaction between PFPE molecules and GPTMS molecules must have contributed to this increase in wear durability. As explained above, strong covalent bonds form between GPTMS and PFPE molecules, while hydrogen bonds form between PFPE and APTMS molecules. Moreover, it is evident from the XPS results that a lower amount/portion of PFPE is bonded with GPTMS SAM (30%), while a greater amount of PFPE is bonded with APTMS (60%). In other words, when PFPE is coated onto GPTMS and APTMS SAM surfaces, a greater amount of mobile PFPE is present in the case of PFPE-coated GPTMS SAM (70%) than that of PFPE-coated APTMS SAM (40%). Therefore, an optimum combination of the strong covalent bonding between PFPE molecules and GPTMS molecules and the higher amount of mobile PFPE might be the reasons for the higher wear durability of Si/GPTMS/PFPE than that of Si/ APTMS/PFPE. The results of the present study suggest that an optimum combination of covalent bonding between PFPE and SAM molecules and an optimum amount of mobile PFPE are essential for improved tribological properties (low coefficient of friction and high wear durability).

The reasons for the very low improvement in wear durability in the case of the PFPE-coated OTS SAM surface can be summarized as the absence of bonding between the PFPE and SAM molecules, the absence of the trapping of the PFPE molecules within the SAM, and the lower amount of PFPE present on OTS SAM. Moreover, because of the lower surface energy of OTS SAM, the PFPE molecules are easily displaced out of the wear track during sliding. PFPE coating onto bare Si also resulted in good wear durability, which is approximately equal to that of PFPE-coated APTMS SAM. A good combination of both mobile and hydrogen-bonded PFPE, in both cases, resulted in higher wear durability.

From our experimental results, it is difficult to identify whether the replenishment characteristics of PFPE contribute to the experimentally observed high wear durability for PFPE-overcoated SAMs. One needs to know the critical reflow times of PFPE on the relevant SAM surfaces to estimate their replenishment characteristics [68], but these values (theoretical or experimental) are not available. The time for one revolution during the tribological test in our study is 0.6 s, and the critical reflow time of PFPE on a particular SAM should be less than 0.6 s for the replenishment of PFPE to occur and to contribute to enhancing the wear durability. Therefore, we cannot precisely comment on the replenishment of PFPE contributing to the very high wear durability observed in the case of PFPE-overcoated SAMs.

Fluorinated PFPE molecules are very lubricious by themselves, which may be attributed to their linear and "smooth molecular profile," as evidenced by the molecular-scale friction work carried out by Tabor and coworkers on poly(tetrafluoroethy lene) (PTFE), high-density polyethylene (HDPE), and several other polymers [69]. It may be noted that PFPE and PTFE molecules are very similar in the sense that both have common CF_2 chemical groups in their backbone and both have linear molecular structure.

The absence of the wear debris on the wear track, even after many thousands of sliding cycles, is one of the advantages of the PFPE-coated surfaces, which is evident from the optical micrographs shown in fig. 6.5. Generation of wear debris hinders the smooth operation of the sliding components in MEMS or micromotors, and hence coatings that do not generate wear debris are required for smooth operation of MEMS components up to several millions of cycles at low-contact stress conditions [70].

6.4.3 EFFECT OF THERMAL TREATMENT

Thermal treatment after PFPE coating onto OTS SAM and APTMS SAM did not show any change in the water contact values, whereas thermal treatment of Si/GPTMS/PFPE and Si/PFPE increased water contact angle values from 81° to 107° and from 66° to 114°, respectively. Thermal treatment after PFPE coating influences its bonding characteristics with the underlying surface and may result in an increase in the percentage of the higher-molecular-weight polymer through evaporation of lower-molecular-weight fractions [71]. Therefore, the increase in the bonded fraction of PFPE (which eventually results in a decrease of hydroxyl groups, which usually attract water molecules) and the probable increase in the density of the polymer film must have resulted in an increase in water contact angle values on PFPE-overcoated GPTMS SAM and bare Si after thermal treatment.

The thermal treatment after PFPE coating shows coefficient of friction values (0.13–0.17) similar to that of as-lubricated PFPE.

As presented in fig. 6.4b, thermal treatment shows approximately 30% improvement in the wear durability of PFPE-coated APTMS SAM, whereas there is a reduction in wear durability in the case of OTS SAM, GPTMS SAM, and bare Si due to the thermal treatment after PFPE overcoating. From table 6.3, it is clear that there is a very minimal increase in the percentage of bonded PFPE after thermal treatment for Si/APTMS/PFPE, whereas there is an appreciable increase in bonded PFPE after thermal treatment for Si/GPTMS/PFPE and Si/PFPE. Alternatively, we can say that the thermal treatment after PFPE coating reduced the mobile fraction of PFPE on GPTMS SAM and Si. Therefore, the changes in the mobile and bonded fractions of PFPE due to thermal treatment must be the main reason for the variations in wear durability values after thermal treatment. Further studies are needed to quantify the actual effect of bonded and mobile fractions of PFPE on the final wear durability results with and without thermal treatment.

6.5 CONCLUSIONS

The following conclusions are drawn from the present study:

1. PFPE coating onto three SAM surfaces (hydrophilic and hydrophobic) increases their water contact angle values and decreases the coefficient of friction, irrespective of the initial surface characteristics of the SAM surfaces before PFPE coating.
2. PFPE overcoating onto all SAM surfaces increases the wear durability, and the extent of improvement in wear durability depends on the SAM surface properties such as surface wettability, chemical interactions between SAM molecules and PFPE molecules, and molecular packing density and order in the SAM surfaces. The extent of improvement of wear durability is very high when PFPE is coated onto reactive SAM surfaces (such as APTMS and GPTMS SAMs) and is very low when PFPE is coated onto nonpolar OTS SAM. This is mainly due to the differences in the extent of chemical interactions between PFPE and SAM molecules.
3. Thermal treatment after PFPE coating shows a similar or slightly lower coefficient of friction than the surfaces that were just coated with PFPE. Thermal treatment shows an increase in wear durability in the case of PFPE-coated APTMS SAM and is marginally reduced in the case of PFPE-coated OTS SAM, GPTMS SAM, and bare Si. Overall, Si/GPTMS/PFPE (as lubricated) and Si/APTMS/PFPE (thermally treated) show the highest wear durability among all the films studied in the present work.
4. The mechanism for improved tribological properties in the case of PFPE-coated SAM surfaces may be summarized as an optimum combination of the chemical bonding between PFPE and SAM molecules and the presence of an optimum amount of mobile PFPE.

ACKNOWLEDGMENTS

This work was supported by NUS Nanoscience and Nanotechnology Initiative (NUSNNI) research grant no. R-265-000-132-112. One of the authors (NS) wishes to acknowledge the help with a research scholarship provided by the Graduate School of Engineering, NUS.

REFERENCES

1. Madou, M. J. 1997. *Fundamentals of microfabrications*. Boca Raton, FL: CRC Press.
2. Spearing, S. M. 2000. *Acta Mater.* 48 (1): 179–96.
3. Komvopoulos, K. 1996. *Wear* 200 (1–2): 305–27.

4. Srinivasan, U., Houston, M. R., Howe, R. T., and Maboudian, R. 1998. *J. Microelectromechanical Syst.* 7 (2): 252–60.
5. Rymuza, Z. 1999. *Microsystem Technol.* 5 (4): 173–80.
6. Gabriel, K. J., Behi, F., Mahadevan, R., and Mehregany, M. 1990. *Sens. Actuators A* 21–23: 184–88.
7. Lee, A. P., Pisano, A. P., and Lim, M. G. 1992. *Mater. Res. Soc. Symp. Proc.* 276 : 67–78.
8. Bhushan, B., Israelachvili, J. N., and Landman, U. 1995. *Nature* 374 (6523): 607–16.
9. Ulman, A. 1991. *An introduction to ultrathin organic films: From Langmuir–Blodgett to self-assembly.* San Diego: Academic Press.
10. Koinkar, V. N., and Bhushan, B. 1996. *J. Vac. Sci. Technol. A* 14 (4): 2378–91.
11. Bhushan, B., Kulkarni, A. V., Koinkar, V. N., Boehm, M., Odoni, L., Martelet, C., and Belin, M. 1995. *Langmuir* 11 (8): 3189–98.
12. Clechet, P., Martelet, C., Belin, M., Zarrad, H., Jaffrezic-Renault, N., and Fayeulle, S. 1994. *Sens. Actuators A* 44 (1): 77–81.
13. Wang, W., Wang, Y., Bao, H., Xiong, B., and Bao, M. 2002. *Sens. Actuators A* 97–98: 486–91.
14. Ando, E., Goto, Y., Morimoto, K., Ariga, K., and Okahata, Y. 1989. *Thin Solid Films* 180 (1–2): 287–91.
15. DePalma, V., and Tillman, B. N. 1989. *Langmuir* 5 (3): 868–72.
16. Ruhe, J., Novotny, V. J., Kanazawa, K. K., Clarke, T., and Street, G. B. 1993. *Langmuir* 9 (9): 2383–88.
17. Colton, R. J. 1996. *Langmuir* 12 (19): 4574–82.
18. Horrison, J. A., and Perry, S. S. 1997. *MRS Bull.* 23 (6): 27.
19. Cha, K.-H., and Kim, D.-E. 2001. *Wear* 251 (1–2): 1169–76.
20. Ren, S., Yang, S., Zhao, Y., Zhou, J., Xu, T., and Liu, W. 2002. *Trib. Lett.* 13 (4): 233–39.
21. Sung, I.-H., Yang, J.-C., Kim, D.-E., and Shin, B.-S. 2003. *Wear* 255 (7–12): 808–18.
22. Satyanarayana, N., Sinha, S. K., and Srinivasan, M. P. 2004. In *Life cycle tribology, tribology and interface engineering series.* Vol. 48, 821–26, ed. D. Dowson, M. Priest, G. Dalmaz, and A. A. Lubrecht. New York: Elsevier.
23. Lio, A., Charych, D. H., and Salmeron, M. 1997. *J. Phys. Chem. B* 101 (19): 3800–5.
24. Bhushan, B., and Liu, H. 2001. *Phys. Rev. B* 63 (24): 245412-1–11.
25. Liu, H., Bhushan, B., Eck, W., and Stadler, V. 2001. *J. Vac. Sci. Technol. A* 19 (4): 1234–40.
26. Kim, H. I., Koini, T., Lee, T. R., and Perry, S. S. 1997. *Langmuir* 13 (26): 7192–96.
27. Kim, H. I., Graupe, M., Oloba, O., Koini, T., Imaduddin, S., Lee, T. R., and Perry, S. S. 1999. *Langmuir* 15 (9): 3179–85.
28. Bliznyuk, V. N., Everson, M. P., and Tsukruk, V. V. 1998. *J. Tribol.* 120 (3): 489–95.
29. Zarrad, H., Chovelon, J. M., Clechet, P., Jaffrezic-Renault, N., Martelet, C., Belin, M., Perez, H., and Chevalier, Y. 1995. *Sens. Actuators A* 47 (1–3): 598–600.
30. Ren, S., Yang, S., and Zhao, Y. 2003. *Langmuir* 19 (7): 2763–67.
31. Bhushan, B. 2004. In *Springer Handbook of Nanotechnology*, ed. B. Bhushan, 831–60. Heidelberg: Springer.
32. Tsukruk, V. V. 2001. *Adv. Mater.* 13 (2): 95–108.
33. Zhang, S., and Lan, H. 2002. *Tribol. Int.* 35 (5): 321–27.
34. Mino, N., Ogawa, K., Minoda, T., Takatsuka, M., Sha, S., and Moriizumi, T. 1993. *Thin Solid Films* 230 (2): 209–16.
35. Tsukruk, V. V., Everson, M. P., Lander, L. M., and Brittain, W. J. 1996. *Langmuir* 12 (16): 3905–11.
36. Nakano, M., Ishida, T., Numata, T., Ando, Y., and Sasaki, S. 2003. *Jpn. J. Appl. Phys.* 42 (7B): 4734–38.

37. Xiao, X., Hu, J., Charych, D. H., and Salmeron, M. 1996. *Langmuir* 12 (2): 235–37.
38. Deng, K., Collins, R. J., Mehregeng, M., and Sukenik, C. N. 1993. *J. Electrochem. Soc.* 142 (4): 1278–85.
39. Dugger, M. T., Senft, D. C., and Nelson, G. C. 2000. *ACS Symp. Ser.* 741: 455–73.
40. Alley, R. L., Howe, R. T., and Komvopoulos, K. 1995. US Patent 5,403,665.
41. Lander, L. M., Brittain, W. J., DePalma, V. A., and Girolmo, S. R. 1995. *Chem. Mater.* 7 (8): 1437–39.
42. DeBoer, M. P., and Mayer, T. M. 2001. *MRS Bull.* 26 (4): 302–4.
43. Satyanarayana, N., and Sinha, S. K. 2005. *J. Phys. D: Appl. Phys.* 38 (18): 3512–22.
44. Satyanarayana, N., Gosvami, N. N., Sinha, S. K., and Srinivasan, M. P. 2007. *Phil. Mag.* 87 (22): 3209–3227.
45. Julthongpiput, D. 2003. Design of nanocomposite polymer coatings for MEMS applications. Ph.D. thesis, Iowa State Univ.
46. Eapen, K. C., Patton, S. T., Smallwood, S. A., Phillips, B. S., and Zabinski, J. S. 2005. *J. Microelectromechanical Syst.* 14 (5): 954–60.
47. Ahn, H.-S., Julthongpiput, D., Kim, D.-I., and Tsukruk, V. V. 2003. *Wear* 255 (7–12): 801–7.
48. Katano, T., Oka, M., Nakazawa, S., Aramaki, T., and Kusakawa, K. 2003. *IEEE Trans. Magn.* 39 (5): 2489–91.
49. Chen, C.-Y., Bogy, D., and Bhatia, C. S. 2001. *Tribol. Lett.* 10 (4): 195–202.
50. Sinha, S. K., Kawaguchi, M., Kato, T., and Kennedy, F. E. 2003. *Tribol. Int.* 36 (4–6): 217–25.
51. Choi, J., Morishita, H., and Kato, T. 2004. *Appl. Surf. Sci.* 228 (1–4): 191–200.
52. Liu, H., and Bhushan, B. 2003. *Ultramicroscopy* 97 (1–4): 321–40.
53. Luzinov, I., Julthongpiput, D., Liebmann-Vinson, A., Cregger, T., Foster, M. D., and Tsukruk, V.V. 2000. *Langmuir* 16 (2): 504–16.
54. Lumbantobing, A., Kogut, L., and Komvopoulos, K. 2004. *J. Microelectromechanical Syst.* 13 (6): 977–87.
55. Williams, J. A., and Le, H. R. 2006. *J. Phys. D: Appl. Phys.* 39 (12): R201–14.
56. Eapen, K. C., Patton, S. T., and Zabinski, J. S. 2002. *Tribol. Lett.* 12 (1): 35–41.
57. Mastrangelo, C. H. 1997. *Tribol. Lett.* 3 (3): 223–38.
58. Maboudian, R., and Howe, R. T. 1997. *Tribol. Lett.* 3 (3): 215–21.
59. Zhu, L., Zhang, J., Liew, T., and Ye, K. D. 2003. *J. Vac. Sci. Technol. A* 21 (4): 1087–91.
60. Ellis, B. 1993. In *Chemistry and technology of epoxy resins*, ed. Bryan Ellis. New York: Marcel Dekker.
61. Elender, G., Kuhner, M., and Sackmann, E. 1996. *Biosens. Bioelectron.* 11 (6–7): 565–77 (1996).
62. Ruhe, J., Novotny, V., Clarke, T., and Street, G. B. 1996. *J. Tribol.* 118 (3): 663–68.
63. Sidorenko, A., Ahn, H. S., Kim, D. I., Yang, H., and Tsukruk, V. V. 2002. *Wear* 252 (11–12): 946–55.
64. Makkonen, L. 2004. In *27th Annual Meeting of the Adhesion Society*, 399–401. Wilmington, NC.
65. Mate, C. M. 1992. *Phys. Rev. Lett.* 68 (22): 3323–26.
66. Choi, J., and Kato, T. 2003. *IEEE Trans. Magn.* 39 (5): 2444–46.
67. Pettersson, U., and Jacobson, S. 2003. *Tribol. Int.* 36 (11): 857–64.
68. Tani, H., and Matsumoto, M. 2001. *J. Tribol.* 123 (3): 533–40.
69. Pooley, C. M., and Tabor, D. 1972. *Proc. R. Soc. (Lond.) A* 329 (1578): 251–74.
70. Patton, S. T., Cowan, W. D., Eapen, K. C., and Zabinski, J. S. 2000. *Tribol. Lett.* 9 (3–4): 199–209 (2000).
71. Ruhe, J., Blackman, G., Novotny, V. J., Clarke, T., Street, G. B., and Kuan, S. 1994. *J. Appl. Polym. Sci.* 53 (6): 825–36.

7 Surface Forces, Surface Energy, and Adhesion of SAMs and Superhydrophobic Films
Relevance to Tribological Phenomena

Ya-Pu Zhao, Li-Ya Guo, Jun Yin, and Wei Dai

ABSTRACT

Because of the very large surface-area-to-volume ratios of micro-devices, adhesion/stiction has been considered the most important failure mode and the major obstacle for the commercialization of micro-electromechanical systems (MEMS). In this chapter, most important surface forces are introduced. The physical origin and mathematical models of these surface forces are presented. Then, adhesion effects such as wetting and surface energy, which are related to these surface forces, are extensively discussed. Self-assembled monolayers (SAMs) have recently received considerable attention as molecular-level lubricants in MEMS. The structure and the surface characteristics of SAMs are introduced. Experiments, molecular dynamics (MD) simulations, and theoretical models on the adhesion force between the atomic force microscope (AFM) tip and sample are discussed in detail. Finally, the adhesion problems related to super-hydrophobic films are discussed.

7.1 INTRODUCTION

Microelectromechanical systems (MEMS) are based on the integration of mechanical elements, sensors, actuators, and electronics on a common silicon substrate through microfabrication technology. Because MEMS devices are manufactured using batch fabrication techniques similar to those used for integrated circuits, unprecedented levels of functionality, reliability, and sophistication can be achieved on a small silicon chip at a relatively low cost. Therefore, MEMS are rapidly being developed for their wide potential use. MEMS-based sensors are a crucial component in automotive electronics, medical equipment, and smart portable electronics such as cell phones, personal digital assistants (PDAs), hard disk drives, computer peripherals, and wireless devices. However, the large surface-area-to-volume ratio raises serious adhesive and frictional problems for their operations [1, 2]. High adhesion forces between micromachine surfaces occur immediately after the release process and during operation. The relatively low kinetic energies of MEMS (typically in the range of 10^{-18}–10^{-16} J) may not be sufficient to overcome release-related and in-use stiction. Thus the nature and magnitudes of adhesion forces arising at micromachine interfaces are of great importance to the design of antistiction microdevices [3].

In conventional lubrication, a low-friction substance (solid or liquid) is applied to the system by external means, an approach that cannot be adopted in a MEMS device due to its very small dimensions. This is because bulk liquid lubricants may introduce capillary and viscous shear mechanisms resulting in energy dissipation and, eventually, seizure due to excessive viscous drag forces. To resolve the adhesion and friction problems of MEMS, modification of the surfaces of the microdevices is strongly suggested. Therefore, the study of surface properties is quite important. Polymer thin films with potential as protective coatings, adhesion-resistant materials, and boundary lubricants have attracted much attention. Polymers containing polar groups are capable of forming stable thin films via the multiple functional groups strongly attached to the substrate [4–11]. However, due to the strong polar groups present in the polymer, such a polymer surface would be hydrophilic. In other words, these polymer films would possess high surface energy, which could increase both the adhesion and friction, especially on a microscale. Fortunately, polymer films with low surface energy can be achieved by converting the exposed groups from polar to apolar groups. The adsorption of low-surface-energy self-assembled monolayers (SAMs) is shown to be an effective surface modification technique for controlling adhesion and friction at MEMS interfaces [3].

SAMs can be prepared using different types of surfactants and substrates, and SAMs of alkylsilane and fluorosilane compounds with chain lengths ranging from C_{10} to C_{18} have been useful in the reduction of friction and adhesion in MEMS. With well-ordered structures, SAMs have been shown to be particularly useful for studies of many surface interaction phenomena, such as wetting, adhesion, nucleation, and growth, and are shown to function as a protective lubricant layer. SAMs can be spontaneously formed by immersion of an appropriate substrate into a solution of an active surfactant in an organic solvent. This simple process makes SAMs inherently manufacturable and thus technologically interesting for building efficient lubricants. Because a large number of properties can be designed into SAMs, they offer almost unlimited possibilities for studies to correlate many surface and film parameters with friction and adhesion behavior. SAMs have been demonstrated to solve the "stiction" problem by both reducing interfacial adhesion and providing very low coefficients of friction, and thus they have received considerable attention.

For the development of directed self-assembly and other soft nanotechnological tools, it is required to measure, quantify, and understand the surface forces that control the assembly of nanosized objects. Many of these nanotribological studies have been carried out using an atomic force microscope (AFM). Therefore, a fundamental understanding of the nature and origin of surface forces, especially van der Waals, electrostatic, and polymer-induced forces, will allow a greater control of these technological processes. These forces are important in future applications of nanotechnology, since these interactions between the surfaces control phenomena such as dispersion, agglomeration, adhesion, coating, and polishing, which play a crucial role in materials processing.

7.2 SURFACE FORCES

When we consider the long-range interactions between macroscopic bodies (such as colloidal particles) in liquids, we find that the two most important forces are the van der Waals forces and electrostatic forces, although in the shorter distance, solvation forces often dominate over both. In this section, some important types of surface forces are discussed, and it should be helpful for calculation of surface forces.

7.2.1 VAN DER WAALS FORCES

Among the various surface forces, van der Waals forces are ubiquitous. Van der Waals forces act between all atoms and molecules, even between totally neutral ones such as helium, carbon dioxide, and hydrocarbons. Therefore, van der Waals forces play a dominant role in adhesion, adsorption, wetting and spreading of liquids, etc. Van der Waals forces are always attractive for like materials. Attractive van der Waals forces have been used to explain why neutral chemically saturated atoms congregate to form liquids and solids. Attractive van der Waals forces are difficult to measure quantitatively with any surface-force technique, since the forces increase by orders of magnitude close to the surface. This causes a "jump in" of the measuring probe, and no data can be collected. Hamaker also predicted repulsive van der Waals forces in his papers from 1937. Repulsive van der Waals forces are expected to be observable in certain combinations of dissimilar materials. Using an AFM, repulsive van der Waals forces can be measured with greater precision than attractive van der Waals forces [12].

The van der Waals forces between atoms or molecules are the sum of three different forces, all proportional to $1/r^6$, where r is the distance between the atoms or molecules. The corresponding potentials are the orientation or Keesom potential, the induction or Debye potential, and the dispersion or London potential, respectively. For two dissimilar polar molecules, the van der Waals potential is

$$w_{vdW}(r) = -C_{vdW}/r^6 = -\frac{C_{ind} + C_{orient} + C_{disp}}{r^6} \tag{7.1}$$

$$= \frac{1}{(4\pi\varepsilon)^2 r^6}\left[\left(u_1^2\alpha_{02} + u_2^2\alpha_{01}\right) + \frac{u_1^2 u_2^2}{3k_BT} + \frac{3\alpha_{01}\alpha_{02}h\nu_1\nu_2}{2(\nu_1 + \nu_2)}\right]$$

where C_{vdW} is the van der Waals coefficient, u_1 and u_2 are the dipole moments of the molecules, α_{01} and α_{02} are the electronic polarizabilities of the molecules, k_B is the Boltzmann's constant, T is the temperature, ν_1 and ν_2 are the orbiting frequencies of the electrons, h is the Planck's constant, and ε is the dielectric constant of the medium.

Here we assume that the interaction between different bodies is nonretarded and additive, and thus the nonretarded van der Waals interaction energies between bodies of different geometries can be calculated. For example, the interaction energy of a molecule at a distance D away from the planar surface is

$$w(D) = -2\pi C\rho \int_{z=D}^{z=\infty} dz \int_{x=0}^{x=\infty} \frac{x\,dx}{(z^2 + x^2)^3} = -\frac{\pi C\rho}{6D^3} \tag{7.2}$$

where ρ is the number density of molecules in the solid.

Hamaker introduced a simple procedure for calculating the van der Waals energy between two macroscopic bodies by summing the energies of all the atoms in one body with all the atoms in the other. For example, the van der Waals energy for the interaction between a sphere and a flat surface separated by a distance D is given by $w = -A_H R/6D$, where R is the radius of the sphere and the Hamaker constant A_H is defined by $A_H = \pi^2 C \rho_1 \rho_2$, where ρ_1 and ρ_2 are the numbers of atoms per unit volume in the two bodies and C is the coefficient in the atom–atom pair potential. Typical values for the Hamaker constants of condensed phases are about 10^{-19} J for interactions across vacuum. However, since Hamaker's method ignores the influence of nearby atoms on the pair of interacting atoms, it is difficult to calculate the Hamaker constant accurately when a medium separates the two interacting bodies.

Lifshitz considered the multibody interaction by treating the interacting bodies and intervening medium as continuous phases; therefore, the Lifshitz theory is particularly suitable for analyzing the interactions of different phases across a medium. Hamaker constants calculated on the basis of the Lifshitz theory can be given as [13]

$$A_H = A_{H,\nu=0} + A_{H,\nu>0} \tag{7.3}$$

$$\cong \frac{3}{4}k_B T \frac{\varepsilon_1 - \varepsilon_3}{\varepsilon_1 + \varepsilon_3} \frac{\varepsilon_2 - \varepsilon_3}{\varepsilon_2 + \varepsilon_3} + \frac{3h\nu_e}{8\sqrt{2}} \frac{\left(n_1^2 - n_3^2\right)\left(n_2^2 - n_3^2\right)}{\sqrt{n_1^2 - n_3^2}\sqrt{n_2^2 - n_3^2}\left[\sqrt{n_1^2 - n_3^2} + \sqrt{n_2^2 - n_3^2}\right]}$$

where ν_e is the mean absorption frequency, and ε_i and n_i are the static dielectric constants and refractive indices of the materials, respectively.

So far we have assumed that the molecules stay so close to each other that the propagation of the electric field is instantaneous. In fact, the molecular charge distribution of molecule A is constantly varying due to its internal electronic motions, and this variation generates an electric field. The field expands with the speed of light, c, and polarizes the second molecule B, which, in turn, causes an electric field that reaches molecule A with the speed of light. Covering the distance d between molecules A and B takes a time $\Delta t = d/c$. The interaction takes place only if Δt is smaller than the time that the dipole moment changes, which is on the order of $1/\nu$. If the first dipole changes faster than Δt, the interaction becomes weaker. For two molecules in free space, retardation effects begin at separations above 5 nm. At distances beyond about 5 nm, the dispersion contribution to the total van der Waals force begins to decay more rapidly due to retardation effects. Particularly in a medium where the speed of light is slower, retardation effects become operative at smaller distances, and they become quite important when macroscopic bodies or surfaces interact in a liquid medium. So there is no simple equation for calculating the van der Waals forces at all separations.

Strictly, the Hamaker constant is never truly constant. Görner and Pich's exact solution of the retarded van der Waals energy between two spherical particles is compared with the fully nonretarded Hamaker approximation, with the retarded approximation based on Derjaguin's method for spheres as described by Gregory and on Vincent's short-range retarded formula [14]. For small particle sizes, the short-range retarded formula from Vincent is superior over the other approximations, whereas Gregory's retarded expression is the most suitable mainly for large spheres over the whole range of separation distances [13–16].

7.2.2 Electrostatic Force

As far as we know, charged objects exert a force on one another. If the charges are at rest, then this force between them is known as the electrostatic force. An interesting characteristic of the electrostatic forces is that they can be either attractive or repulsive, unlike the gravitational force, which is always attractive. Electrostatic forces are present between ions, between ions and permanent dipoles, and between ions and nonpolar molecules in which a charge induces a dipole moment. Most surfaces in contact with a highly polar liquid acquire a surface charge, either by dissociation of ions from the surface into the solution or by preferential adsorption of certain ions from the solution. The electrostatic double-layer force just arises because of the surface charges at the interface. The surface charge is balanced by a layer of oppositely charged ions in the solution at some distance from the surface, which is called the Debye length. Together, the ions and charged surface are known as the electrical double layer. For a monovalent salt it is

$$\lambda_D = \sqrt{\frac{\varepsilon\varepsilon_0 k_B T}{e^2 N_A \sum_i z_i^2 M_i}} \tag{7.4}$$

where ε_0 is the vacuum permittivity, e is the electronic charge, M_i is the molar concentration, and z_i is valency. The Debye length falls with increasing ionic strength of the ions in solution. For example, for 1:1 electrolytes at 25°C, $\lambda_D = 0.304 \text{nm} / \sqrt{M_{1:1}}$ [15, 16].

The electrostatic double-layer force can be calculated using the continuum theory, which is based on the theory of Gouy, Chapman, Debye, and Hückel for an electrical double layer. The Debye length relates the surface charge density σ_c of a surface to the electrostatic surface potential ψ_0 via the Grahame equation, which for 1:1 electrolytes can be expressed as

$$\sigma_c = \sqrt{8\varepsilon\varepsilon_0 k_B T} \sinh\left(e\psi_0 / 2k_B T\right) \times \sqrt{M_{1:1}} \tag{7.5}$$

The Debye length also determines the range of the electrostatic double-layer interaction, which is an entropic effect that arises upon decreasing the thickness of the liquid film containing the dissolved ions between two charged surfaces. The long-range electrostatic interaction energy between two similarly charged molecules or surfaces is typically repulsive and is roughly an exponentially decaying function of distance D:

$$E(D) \approx +C_{ES} e^{-\frac{D}{\lambda_D}} \tag{7.6}$$

where C_{ES} is a constant that depends on the geometries of the interacting surfaces, on their surface charge density, and on the solution conditions, and it can be determined by solving the Poisson–Boltzmann equation (7.7), which is a second-order differential equation and depends only on the properties of the surfaces

$$\frac{d^2\psi}{dx^2} = -\frac{ze\rho_0}{\varepsilon\varepsilon_0} e^{-ze\psi/k_BT} \tag{7.7}$$

where ψ is the electrostatic potential and ρ_0 is the number density of ions of valency z at point $x = 0$ between two surfaces (fig. 7.1).

To solve this equation, certain boundary conditions have to be assumed. Two boundary conditions, which determine the two integration constants, are often used: either it is assumed that upon approach the surface charges remain constant or that the surface potentials remain constant. These boundary conditions have a strong influence on the electrostatic force at distances below roughly $2\lambda_D$. At smaller separations, one must use the numerical solution of the Poisson–Boltzmann equation to obtain the exact interaction potential, for which there are no simple expressions.

At small separations, as $D \rightarrow 0$, the double-layer pressure at constant surface charge becomes infinitely repulsive and independent of the salt concentration. Thus the repulsive pressure approaches infinity according to

$$P = -\frac{2\sigma_c k_B T}{zeD} \tag{7.8}$$

Taking into account an AFM tip with a parabolic end of radius of curvature R and a flat surface, the force between them is given by [13]

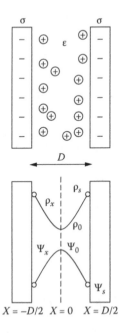

FIGURE 7.1 Two negatively charged surfaces of surface charge density σ separated by a distance D in water (top). The counterion density profile and potential are shown schematically in the bottom. (From Israelachvili, J. N. 1985. *Intermolecular and surface forces.* 2nd ed. London: Academic Press. With permission.)

$$F_{el}^{cp} = \frac{2\pi R \varepsilon \varepsilon_0}{\lambda_D} \left[2\psi_S \psi_T e^{-D/\lambda_D} - \left(\psi_S^2 + \psi_T^2\right) e^{-2D/\lambda_D} \right] \tag{7.9}$$

Assuming constant potentials of the sample ψ_S and the tip ψ_T, for constant charge conditions the electrostatic double-layer force is

$$F_{el}^{cc} = \frac{2\pi R \lambda_D}{\varepsilon \varepsilon_0} \left[2\sigma_S \sigma_T e^{-D/\lambda_D} + \left(\sigma_S^2 + \sigma_T^2\right) e^{-2D/\lambda_D} \right] \tag{7.10}$$

where σ_S and σ_T are the surface charge densities of the sample and the tip, respectively.

There are three limitations for the equations above: it is assumed that the surface potentials are below about 50 mV; the radius of curvature has to be large ($R \gg \lambda_D$); and the separation D should satisfy $D \geq \lambda_D$.

The total interaction between any two surfaces must also include the van der Waals attraction, which is largely insensitive to variations in electrolyte concentration and pH, and so may be considered as fixed for any particular solute–solvent system. Further, the van der Waals attraction wins out over the double-layer repulsion at small distances, since it is a power-law interaction, whereas the double-layer interaction energy remains finite or rises much more slowly as $D \to 0$. This is the theoretical prediction that forms the basis of the so-called Derjaguin–Landau–Verwey–Overbeek (DLVO) theory (illustrated in fig. 7.2) [15]. In the DLVO theory, the interaction between two particles is assumed to consist of two contributions: the van der Waals attraction and the electrostatic double-layer repulsion. At low salt concentration, the

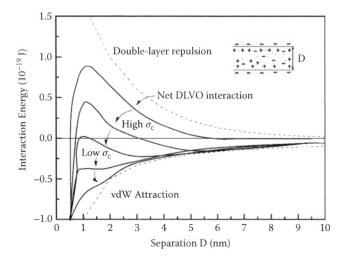

FIGURE 7.2 Schematic plots of the DLVO interaction potential energy between two flat, charged surfaces as a function of the surface separation D. The van der Waals attraction together with the repulsive electrostatic double-layer force at different surface charge σ_c determine the net interaction potential in aqueous electrolyte solution. (From Bhushan, B., ed. 2004. *Springer handbook of nanotechnology.* Heidelberg: Springer. With permission.)

double-layer repulsion is strong enough to keep the colloidal particles apart. With increasing salt concentration, the electrostatic repulsion is more screened. At a certain concentration the van der Waals attraction overcomes the repulsive electrostatic barrier and coagulation sets in.

7.2.3 Solvation and Hydration Forces

The theories of van der Waals and double-layer forces are both continuum theories wherein the intervening solvent is characterized solely by its bulk properties such as refractive index, dielectric constant, and density. When a liquid is confined within a restricted space, it ceases to behave as a structureless continuum. At small surface separations, the van der Waals force between two surfaces is no longer a smoothly varying attraction; instead, there arises an additional "solvation" force that generally oscillates between attraction and repulsion with distance, with a periodicity equal to some mean dimension of the liquid molecules.

Solvation forces are often well described by an exponentially decaying oscillating function of the form [17]

$$f = f_0 \cos\left(\frac{2\pi x}{r}\right)\exp\left(-\frac{x}{\lambda_s}\right) \tag{7.11}$$

where r is the molecular diameter, λ_s is the decay length, and x is the distance between two flat surfaces, and it may be either active or passive. Consider hydrogen bonding of water with a hydrophilic surface; in this case a force needs to be applied to break the hydrogen bonds and push the solvent out of the way, which is an example of active solvation force. Now consider the orientation of octacyclotetramethylsiloxane (OCTMS) against a mica substrate, where essentially a force needs to be exerted to push out layers of molecules that are trapped close to a surface, which is an example of passive solvation force [17].

The two major forces between two surfaces or macromolecules in liquids are the attractive van der Waals force, which is always present, and the repulsive electrical double-layer force, which depends on the existence of charged surface groups [18]. When two hydrated surfaces are brought into contact, repulsive forces of about 1–3-nm range can dominate over both. These forces are termed "hydration forces." Hydration forces were originally proposed by Langmuir and Derjaguin when they found that even uncharged molecules and particles were often miscible [19, 20]. This hydration force arises whenever water molecules bind to surfaces containing hydrophilic groups (fig. 7.3) (i.e., certain ionic, zwitterionic, or H-bonding groups), and then the strength depends on the energy needed to disrupt the ordered water structure and to ultimately dehydrate two surfaces as they approach each other [21].

Hydration forces are relatively short-ranged, so that at salt concentrations below 0.1 M they can easily be distinguished from the longer range electrostatic and van der Waals forces [13]. Early measurements of these forces between lecithin and other

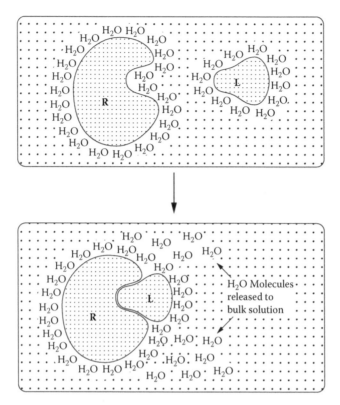

FIGURE 7.3 Schematic illustration of a receptor molecule (R) binding to a ligand group (L) in an aqueous medium, showing that the hydration force arises from the strongly bound and oriented first layer of water molecules on surfaces. (From Israelachvili, J. N., and Wennerstrom, H. 1996. *Nature* 379: 219. With permission.)

uncharged bilayers in aqueous solutions showed that, at distance D below about 1–3 nm, they decayed exponentially with distance, and the repulsive force per unit area, or pressure P, between two surfaces is given as

$$P = Ae^{-D/\lambda_H} \qquad (7.12)$$

and the hydration energy density can be written as

$$w_{Hyd} = w_0 e^{-D/\lambda_H} \qquad (7.13)$$

The maximum range of interbilayer hydration forces is 1–3.5 nm, below which they rise steeply and exponentially with a decay length λ_H of about 0.2–0.3 nm, which is about the size of a water molecule. However, from eqs. (7.12) and (7.13), we can notice that both the hydration force and hydration energy will be asymptotic to a finite value as $D \rightarrow 0$, and experimental data show that the hydration energy keeps on increasing as D decreases. So the expression proposed for hydration energy is as follows [21]:

$$w_{Hyd} = w_0\left(1+\frac{\lambda_H}{D}\right)e^{-D/\lambda_H} \tag{7.14}$$

In contrast to the electrostatic double-layer force, hydration forces tend to become stronger and longer ranged with increasing salt concentration, especially for divalent cations [13].

The origin of hydration force is not clear, and there are mainly two different points of view. At first, it was assumed that the hydration force originated from the energy needed to dehydrate interacting surfaces that contained ionic or polar species [22]. But in 1990 a completely different explanation for the origin of hydration force was proposed: "It is concluded that the short-range repulsion between amphiphilic surfaces is mainly of entropic origin" [23].

A question is whether or not the hydration force should be considered as a solvation force. It was reported [24] that the forces between bilayers in a variety of non-aqueous solvents were very similar to those measured in water, while other data do not appear to support the modified solvent-structure origin of hydration forces [23].

7.2.4 HYDROPHOBIC FORCE

For almost 70 years, researchers have attempted to understand the hydrophobic effect (the low solubility of hydrophobic solutes in water) and the hydrophobic interaction or force (the unusually strong attraction of hydrophobic surfaces and groups in water) [25]. The immiscibility of inert substances with water is known as the "hydrophobic effect." So a hydrophobic surface is inert to water; it cannot bind to water molecules via ionic or hydrogen bonds. The strongly attractive hydrophobic force has many important manifestations and consequences, such as the low solubility or miscibility of water and oil molecules, micellization, protein folding, strong adhesion and rapid coagulation of hydrophobic surfaces, nonwetting of water on hydrophobic surfaces, and hydrophobic particle attachment to rising air bubbles [15]. Israelachvili and Pashley [26] measured the hydrophobic force between two macroscopic curved hydrophobic surfaces in water and found that, in the range 0–10 nm, the force decayed exponentially, with a decay length of about 1 nm. A short-range attraction at separations <20 nm is thought to contain more information about the true hydrophobic interaction [25].

The hydrophobic effect is related to the water structure adopted around nonpolar molecules. The strong inclination of water molecules to form H bonds with each other influences the interactions with nonpolar molecules that are incapable of forming H bonds. The nonpolar molecules affect the water structure around them by reorientation of the water molecules. Because of these phenomena, it is proposed that the hydrophobic effect has an entropic nature, but like the repulsive hydration force, the origin of the hydrophobic force is still unknown. The hydration and hydrophobic forces are not of a simple nature. These interactions are probably the most important, yet the least understood, of all the forces in aqueous solutions.

7.2.5 STERIC FORCES

When a polymer is adsorbed onto a surface, it generally does not simply lie over the surface, but some parts of the polymer adsorb onto the surface, while other parts extend away from the surface into the medium. Then the polymer assumes the shape of a random coil in the absence of segment–segment interactions in the solvent. This implies that the distance between the two ends fluctuates and will practically never be equal to the length of the stretched polymer. To characterize the size of such a randomly coiled chain, we take the mean square of the end-to-end distance. The square root of this value,

$$\sqrt{\langle r_p^2 \rangle}$$

is often called the size of a polymer [16]. Its root-mean-square radius is known as the radius of gyration,

$$R_g = \sqrt{\langle r_p^2 \rangle / 6}$$

When two polymer-covered surfaces approach each other, they experience a force once the outer segments begin to overlap (i.e., once the separation is below a few R_g). This interaction usually leads to a repulsive force because of the unfavorable free energy associated with compressing the chains against the surfaces. Thus, for a polymer to provide a strong repulsion between two approaching surfaces, the polymer must satisfy two criteria, i.e., it must adsorb strongly onto the surface as well as extend away from the surface. For a homopolymer, these criteria are mutually exclusive, but for a copolymer, these criteria are well met.

When two layers of a graft or a block copolymer overlap, the steric interaction is given as

$$F(D) = \frac{\beta k_B T}{s^3} \left[\left(\frac{2\delta}{D} \right)^{9/4} - \left(\frac{D}{2\delta} \right)^{3/4} \right] \tag{7.15}$$

where D is the separation between the surfaces, δ is the thickness of the polymer layer, s is the separation between the terminally attached polymer chains, and β is a fitting parameter.

7.3 SURFACE ENERGY AND ADHESION

For most physical properties of matter it is a good approximation to assume the bodies to be homogeneous, which means that every element of volume has the same properties as any other. This assumption of homogeneity is bound to break down near the surface of a body. The direction of the normal to the surface provides a polar direction from the interior to the exterior, so that the properties of the body just beneath the surface may become different from those in the interior [27].

A distinction can be made between low-energy and high-energy solid surfaces. The surface energies of organic compounds, such as polymers, are usually less than 100 mJ/m^2. Metals, metal oxides, and ceramics are typically greater than 500 mJ/m^2

[28]. The surface energy has an immediate effect on the adhesion between surfaces, which nowadays is an important issue in MEMS.

7.3.1 SURFACE ENERGY

The nature of the chemical bonding of atoms at the surface is different from the bonding of atoms in the interior. The surface atom has fewer neighboring atoms and experiences fewer interaction forces from its surroundings than the atom in the interior. Because of this, the potential energy of the surface atom is higher than that of the bulk atom [28]. If we want to bring an atom from the interior of the bulk to the surface, this energy difference has to be overcome, and this energy difference is measured by the surface energy γ.

Now we will consider the internal energy of a one-component solid and consider a reversible transformation in which heat, stress, and particle number can be varied. The internal energy change reads

$$dU = \left[\frac{\partial U}{\partial S}\right]_{V,N} dS + \left[\frac{\partial U}{\partial V}\right]_{N,S} dV + \left[\frac{\partial U}{\partial N}\right]_{V,S} dN \tag{7.16}$$

$$dU = TdS - PdV + \mu dV \tag{7.17}$$

which defines the temperature T, the pressure P, and the chemical potential μ of the bulk. The extensive property $U(\lambda S, \lambda V, \lambda N) = \lambda U(S, V, N)$ leads to the Euler equation $U = TS - PV + \mu N$ [29]. To create a new surface, for example, by cleaving, requires energy that is proportional to the amount of the surface area A, i.e., the additional area created:

$$U = TS - PV + \mu N + \gamma A \tag{7.18}$$

The surface energy can be defined as the reversible work of forming a unit area of surface at constants T, V, μ, and number of components, and is given as

$$\gamma = \frac{dU}{dA} \tag{7.19}$$

Shuttleworth [30] gives a relation between surface stress and surface energy as

$$\sigma^E = \gamma^E + \frac{\partial \gamma^E}{\partial \varepsilon} \tag{7.20}$$

in Eulerian coordinates and

$$\sigma^L = \frac{\partial \gamma^L}{\partial \varepsilon} \tag{7.21}$$

in Lagrangian coordinates, which is a more general form. Equation (7.20) can be rewritten as

$$\sigma^E - \gamma^E = \frac{\partial \gamma^E}{\partial \varepsilon} \tag{7.22}$$

which is a thermodynamic driving force to move atoms from the bulk into the surface layer, and shows that the difference between the surface stress and surface energy is equal to the variation of surface energy with respect to the elastic strain of the surface. For most solids, $\partial \gamma^L / \partial \varepsilon \neq 0$. In fact, $\partial \gamma^L / \partial \varepsilon$ is usually of the same order of magnitude as γ^E and can be positive or negative. For a clean surface, γ^E is always positive. Then, σ^E is also generally of the same order of magnitude as γ^E and can be positive or negative [31]. Thus the stretching of a wire under reversible conditions would imply that interior atoms would move into the surface as needed so that the increased surface area is not accompanied by any changes in specific surface properties.

A linearized constitutive relation of the interface has been given by Gurtin and Murdoch [32, 33]. For an isotropic interface relative to the reference configuration, the linearized constitutive relation for the interfacial Piola–Kirchhoff stress of the first kind can be written as

$$\mathbf{S}_s = [\gamma_0 \mathbf{1} + (\lambda_L + \gamma_0)(\mathrm{tr}\varepsilon_s) - d_0 \Delta T]\mathbf{1} + 2(\mu_L - \gamma_0)\varepsilon_s + \gamma_0 \mathbf{u} \nabla_{0s} \tag{7.23}$$

Here $\gamma_0 \mathbf{1}$ is the residual interface stress at reference configuration, λ_L and μ_L are the Lamé constants of the interface, ΔT is the difference in temperature from the reference value, d_0 is the coefficient related to the thermal expansion, $\mathbf{u}$ denotes the displacement from the reference configuration, and ε_s is the interfacial strain.

Berger et al. [34] measured the surface stress changes and kinetics by micromechanical sensors during the self-assembly process of $HS\text{-}(CH_2)_{n-1}\text{-}CH_3$ for $n = 4, 6, 8, 12,$ and 14 on gold, where n is the number of carbon atoms in the alkyl chain. The sensor deflection was shown to be a function of time for experiments with alkanethiols of various chain lengths (fig. 7.4a) and has been fitted to a Langmuir adsorption isotherm model as

$$\Theta \propto 1 - \exp(-\kappa t) \tag{7.24}$$

where Θ is the coverage, t is the time, and κ is the reaction rate. During the adsorption, a compressive surface stress was observed that is saturated at a permanent value after some time. Stoney's formula is used to relate the sensor curvature radius R to the surface stress σ as

$$\sigma = \frac{E t_s^2}{6R(1 - \nu)} \tag{7.25}$$

where E is Young's modulus, ν is Poisson's ratio of the sensor material, t_s is sensor thickness, $R^{-1} = 3\Delta z / 2L^2$, L is length, and Δz is deflection. As fig. 7.4b shows, the saturated surface stress increased linearly with chain length. The authors attributed the linear dependence on the chain length to electrostatic dipole repulsion forces. However, all the chemisorbed alkanethiols caused compressive surface stress during self-assembly. Thermal effects are negligible for the transient bending caused by the

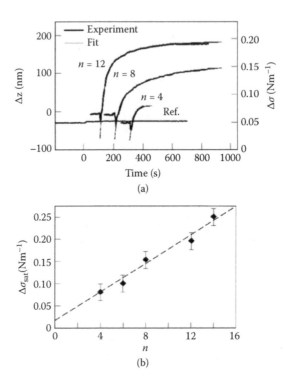

FIGURE 7.4 (a) Deflection, Δz, and change in surface stress, $\Delta\sigma$, of a gold-coated AFM microcantilever are plotted as a function of time after exposure to vapors of alkanethiols with different chain lengths. (b) Adsorption-induced surface stress at saturation coverage ($\Delta\sigma_{sat}$) is plotted as a function of alkyl chain length for $n = 4, 6, 8, 12$, and 14. (From Berger, R., et al. 1997. *Science* 276: 2021. Reprinted with permission from AAAS.)

reaction heat, on the order of 0.5 pm. The gravimetric deflection resulting from the molecular loading is calculated to be 5 pm, so it is surprising that all the chemisorbed alkanethiols caused compressive surface stress during self-assembly. The authors did not provide the answer, and here we wish to explain their results on the basis of the physical origin of the surface stress. The physical origin of the surface stress can be understood from the fact that the chemical bonding of atoms at the surface is different from the bonding of atoms in the interior, and the surface atoms would have an equilibrium interatomic distance different from that of the interior atoms. When the alkanethiols are adsorbed onto the SiN_x cantilever with a 20-nm gold receptor layer, the head group Au^+-S^- of the SAM breaks the equilibrium by adding the local electron density around the atoms near the surface. Since the surface atoms now sit in a higher average charge-density environment than the optimal value, the response of the surface atoms would be to increase the interatomic distance in order to achieve a new balance, which at the same time exerts a compressive stress on the surface.

Surface energy is most commonly quantified using contact angle goniometry. Wetting is quantitatively defined with reference to a liquid droplet resting on a solid surface, as shown in fig. 7.5. The tensions at the three-phase contact point are indicated such that S/V represents the solid/vapor interface, L/V is the liquid/vapor

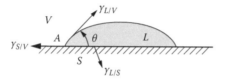

FIGURE 7.5 A liquid droplet on a solid surface with a contact angle θ.

interface, and S/L is the liquid/solid interface. Thomas Young described surface energy as a result of the interplay between the forces of cohesion and the forces of adhesion, which, in turn, dictates whether wetting would occur [35, 36]. If wetting occurs, the droplet will spread out. In most cases, the droplet beads to some extent, and the surface energy of the system can be determined by calculating the contact angle formed where the droplet makes contact with the solid. The Young's equation, relating the various tensions to the equilibrium contact angle θ, can be written as

$$\gamma_{sv} = \gamma_{sl} + \gamma_{lv} \cos\theta \tag{7.26}$$

The term γ_{sv} represents the surface energy of the solid substrate, and it is important for determining how a liquid droplet wets the solid surface. If the surface is hydrophobic, the contact angle of a droplet is larger. Hydrophilicity is indicated by smaller contact angles and higher solid surface energy [37, 38].

7.3.2 CONTACT ANGLES

When a droplet is placed on a solid surface, either it does not move and has a definite contact angle between the liquid and the solid phase, or it spreads across the surface to form a wetting film [39]. This criterion is expressed by defining a parameter termed the "equilibrium spreading coefficient," $S_{L/S}$, where

$$S_{L/S} = \gamma_{sv} - \gamma_{sl} - \gamma_{lv} \tag{7.27}$$

Hence, a liquid will spread spontaneously and completely to wet a solid surface when $S \geq 0$. It is also possible to make a liquid spread across a solid surface even when $\theta > 0°$, but this requires the application of a pressure or a force to the liquid to spread it forcibly [28].

Generally, a surface is called hydrophilic when the contact angle of water is smaller than 90°, while a surface is hydrophobic when the contact angle of water is larger than 90°. If the contact angle is larger than 150°, the surface is termed superhydrophobic. Since superhydrophobicity is usually achieved from the effect of the solid–liquid–air composite surface, the topography of the surface must be considered as an important criterion for superhydrophobicity. The contact line may be pinned by the sharp edges of the asperities, so the optimized asperities, such as hemispherically topped cylindrical and pyramidal asperities, are recommended [40]. Furthermore, the contact line density and the asperity height are also important criteria for superhydrophobic surfaces. Interaction of a liquid with asperities must direct surface forces upward at the contact line, and the surface forces can suspend the liquid against the downward pull of gravity or other body forces. To prevent the liquid protruding between the asperities from contacting the underlying solid, the asperities must be tall enough [41].

The hydrophilic surfaces of glass and silicon are high energy and can be wetted by all liquids. It is possible to lower the surface energy by coating these surfaces with a hydrophobic molecular layer of the type $-(CH_2)-$ or $-(CF_2)-$ [42]. Surfaces of extremely low energies can be created by this approach.

The Young's equation can only be applied to a smooth and homogeneous surface. On rough solid surfaces, Wenzel has shown that the surface roughness may alter the apparent advancing contact angle, θ_w. This change in the contact angle is expressed by

$$\cos\theta_w = r_w \cos\theta \qquad (7.28)$$

where r_w is the roughness factor, which is defined as the ratio of the actual area of a rough surface to the projection area of the solid. Since r_w is always larger than 1, if on a smooth surface θ is less than 90°, then roughening the surface will result in θ_w being even smaller. This will obviously increase the apparent surface energy of the solid surface and, consequently, also increase the extent of wetting. On the other hand, if for a smooth surface θ is greater than 90°, roughening the surface will increase the contact angle θ_w still further and, therefore, decrease the degree of wetting [28].

Wenzel's equation applies to what is called "homogeneous wetting," and it can only be applied to homogeneous, rough surfaces. Surfaces having heterogeneous character can be modeled with the Cassie–Baxter model [43]:

$$\cos\theta_c = f_1 \cos\theta_1 + f_2 \cos\theta_2 \qquad (7.29)$$

where f_1 and f_2 are the fractions of the surface occupied by surface types having contact angles θ_1 and θ_2 ($f_1 + f_2 = 1$).

It is recognized that water spreading does not occur on lotus plant leaves, but instead forms a spherical droplet. This is the classic heterogeneous surface with trapped air in the surface grooves, where now f_2 is the fraction of open area. Then the relationship becomes

$$\cos\theta_c = f_1 \cos\theta_1 - f_2 \qquad (7.30)$$

If the surface is sufficiently rough, and we let f be the wetted part, eq. (7.30) can be written as

$$\cos\theta_c = r_w f \cos\theta + f - 1 \qquad (7.31)$$

When f decreases, the water contact angle of a hydrophobic surface increases. Therefore, the roughness ratio also affects the wetting properties of a heterogeneous rough surface.

With the development of mathematical geometry, the concept of fractals has been brought in to clarify the relationship between the roughness factor and the surface topography. The box-counting method is used to calculate the fractal dimension D^f for the rough solid surface ($2 \leq D^f < 3$). Then the roughness factor r_w can be expressed in terms of a linear dimension λ, which is in dimensionless form, as a ratio to the sample length:

$$r_w = \lambda^{2-D^f} \tag{7.32}$$

Let L and l be the upper and the lower limiting lengths of the fractal behavior, respectively. Then the Wenzel equation can be expressed by [44]

$$\cos \theta_f = \left(\frac{L}{l}\right)^{D^f - 2} \cos \theta \tag{7.33}$$

for the apparent contact angle θ_f on fractal surfaces. Fractals are also used for composite rough surfaces, and the Cassie–Baxter equation can be expressed as [45]

$$\cos \theta_f = f_1 \left(\frac{L}{l}\right)^{D^f - 2} \cos \theta - f_2 \tag{7.34}$$

As the scale of the wetting is smaller, the line tension τ introduced by Gibbs [46] affects the contact angle of nanodroplets. The modified Young's equation [47] is derived as

$$\cos \theta = \frac{\gamma_{sv} - \gamma_{sl}}{\gamma_{lv}} - \frac{\tau}{\gamma_{lv} R} \tag{7.35}$$

where R is the radius of the three-phase contact circle. Both positive and negative values of τ have been reported by experimental measurements, and the theoretical estimation is on the order of 10^{-12} N. Molecular dynamics (MD) simulations have been used to validate this theoretical model [48], and a good result was obtained when sessile argon droplets were on an ideal solid surface that is relatively more hydrophobic. The contact angle increases with decreasing drop size, depending on the interaction between the liquid molecules and the solid substrate, and vice versa. However, the experimental results [49] show that, although the relation between the apparent contact angle and the radius of the droplet obeys eq. (7.35), τ is in the range of 10^{-10} N, which is larger than theoretically expected. Also, the mesoscale substrate defects may have a strong effect on wetting whenever the droplet size becomes sufficiently small. So the effect of the line tension calculated as the free-energy correction due to the modification of the droplet profile close to the contact line by the van der Waals forces is still controversial.

Electrowetting is the phenomenon of enhancing the wettability of a solid surface by applying an electric field, because the external electric field can alter the surface tension at the solid–liquid interface. This is expressed by the Lippmann equation [50]:

$$\sigma_c = -\frac{\partial \gamma_{sl}}{\partial V} \tag{7.36}$$

where σ_c is the surface charge density of the solid–liquid interface and V is the voltage applied across the interface. On a smooth and homogeneous solid surface, the Young–Lippmann equation describes the contact angle θ_e for electrowetting as

$$\cos\theta_e = \cos\theta_0 + \eta \tag{7.37}$$

where $\eta = \varepsilon\varepsilon_0 V^2/(2d\gamma_{L/V})$ is the dimensionless electrowetting number, measured as the ratio between the electrostatic energy per unit area and the liquid–vapor surface tension [51]. If the surface is rough enough to trap air in the grooves, the electrowetting contact angle on this heterogeneous surface can be predicted by the extended Young–Lippmann equation on the Cassie–Baxter model as

$$\cos\theta_c = r_w f_1\left(\cos\theta + f_1\eta\right) - f_2 \tag{7.38}$$

which applies before the liquid penetrates into the grooves.

7.3.3 ADHESION

Adhesion is the molecular force of attraction between unlike materials. The strength of attraction is determined by the surface energy of the materials. The higher the surface energy, the greater is the molecular attraction; while the lower the surface energy, the weaker is the attractive force. The adhesion energy, or the work of adhesion, W_{ad}, is defined as the work per unit area that needs to be provided to separate reversibly a solid–liquid interface so as to create solid–vapor and liquid–vapor interfaces [52]. Thus

$$W_{ad} = \gamma_{S/V} + \gamma_{L/V} - \gamma_{S/L} \tag{7.39}$$

This work can be calculated if the potential-energy functions for the different phases are known. Equation (7.39) may be combined with various semiempirical equations. If Antonow's rule applies, one obtains $W_{ad} = 2\gamma_{S/V}$, which agrees well with the benzene–water example [39]. In addition, the surface energies can be related as

$$\gamma_{S/L} = \gamma_{S/V} + \gamma_{L/V} - 2\varphi\sqrt{\gamma_{S/V}\gamma_{L/V}} \tag{7.40}$$

where the interaction parameter φ is virtually unity for intermolecular forces of dispersion type but is less than unity for immiscible phases in general. From eq. (7.40), one can obtain the Good–Girifalco equation [39] as

$$W_{ad} = 2\sqrt{\frac{\gamma_{S/V}}{\gamma_{L/V}}} \tag{7.41}$$

where φ has been approximated as unity. Zisman and coworkers [39] observed that $\cos\theta$ was usually a monotonic function of $\gamma_{L/V}$ for a homologous series of liquids. They proposed the following relationship:

$$\cos\theta = 1 - \beta(\gamma_{LIV} - \gamma_c) \tag{7.42}$$

where γ_c is the critical surface energy for wetting and is a quantity characteristic of a given solid. For Teflon, the representative γ_c is taken to be about 18 mJ/m^2 and is regarded as characteristic of a surface consisting of -CF$_2$- groups. The β term usually has a value of about 0.03–0.04.

Taking into account the Young's equation, eq. (7.39) can be rearranged to form the Young–Dupre equation:

$$W_{ad} = \gamma_{LIV}(1 + \cos\theta) \tag{7.43}$$

Combining it with the Zisman's eq. (7.42), one obtains

$$W_{ad} = \gamma_{LIV}(2 + \beta\gamma_c) - \beta\gamma_{LIV}^2 \tag{7.44}$$

For a given solid, the work of adhesion goes through a maximum

$$\frac{(2 + \beta\gamma_c)^2}{4\beta} \quad \text{at} \quad \gamma_{LIV} = \frac{2 + \beta\gamma_c}{2\beta}.$$

For example, the critical surface energy γ_c of -CH$_3$- group is 22 mJ/m^2 at 20°C [39]; for the low-energy surfaces, β is about 0.04, and thus the maximum work of adhesion is approximately 51.84 mJ/m^2.

Modern theories of adhesion mechanics of two contacting solid surfaces are based on the Johnson, Kendall, and Roberts (JKR theory), or on the Derjaguin–Muller–Toporov (DMT) theory. The JKR theory is applicable to easily deformable, large bodies with high surface energy, whereas the DMT theory better describes very small and hard bodies with low surface energy. The JKR theory gives an important result about the adhesion or "pull-off" force. That is, the adhesion force F_{ad} is related to the work of adhesion, W_{ad}, and the reduced radius, R, of the AFM tip-surface contact as

$$F_{ad} = -\tfrac{3}{2}\pi R W_{ad} \tag{7.45}$$

The most direct way to measure the adhesion of two solid surfaces is to suspend one of them from a spring and measure the adhesion force needed to separate the two bodies from deflection of the spring [53, 54]. Usually, an AFM can measure the adhesion force conveniently. The AFM utilizes a sharp probe, a tip on the end of a cantilever that bends in response to the force between the tip and the sample, moving over the surface of the sample in a raster scan. A typical force–displacement curve obtained by using an AFM is depicted in fig. 7.6. The force between the AFM tip and the sample surface is quantified along the vertical axis, while the horizontal axis shows the surface displacement coordinates [55]. When the AFM tip approaches the surface, there is no detectable interaction; however, when the tip gradually gets close to the surface, the tip "jumps in" to the surface due to the attractive force; as the tip

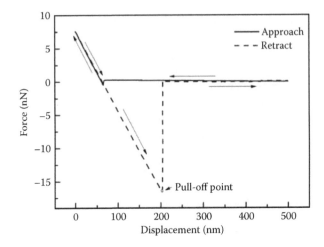

FIGURE 7.6 Force–displacement curves for silicon nitride tip and mica substrate.

is further pushed into the surface, the interaction force becomes repulsive, and the repulsive force is replaced by an attractive force again, which represents the adhesion force when the tip is withdrawn. This is called a "pull-off" behavior and, therefore, the adhesion force is defined as the "pull-off" force.

7.4 SELF-ASSEMBLED MONOLAYERS

Molecular self-assembly, the spontaneous formation of molecules into covalently bonded, well-defined, stable structures, is a very important concept in biological systems and has increasingly become the focus of synthetic sophisticated research [56]. In 1946, Zisman published the preparation of a monomolecular layer by adsorption of a surfactant onto a clean metal surface [57]. However, the potential of self-assembly was not recognized at that time, and the techniques for surface analysis were limited, so the development of self-assembly was slow. In 1980, Sagiv reported SAMs with the adsorption of octadecyltrichlorosilane (OTS-$C_{18}H_{37}SiCl_3$) on a silicon substrate [58]. In 1983, Nuzzo and Allara [59] successfully prepared SAMs of alkanethiolates on gold. As a potential molecular-level lubricant, SAMs have attracted much attention [60–62] and have been demonstrated to be capable of effectively reducing the friction and adhesion in MEMS [3, 61, 63, 64].

7.4.1 FORMATION AND STRUCTURE OF SAMS

SAMs are ordered molecular assemblies formed by spontaneous adsorption of an active surfactant onto the surfaces of appropriate substrates. The molecules consist of three building blocks: an active head group that binds strongly to the substrate, a terminal (end) group that constitutes the outer surface of the film, and an alkyl backbone chain that connects the head and terminal groups [15, 56, 65, 66]. As shown in fig. 7.7 [61], each part of a SAM has a significant effect on its tribological and mechanical properties. The substrate should have a high surface energy so that

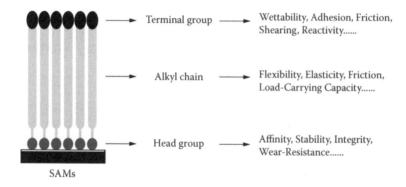

FIGURE 7.7 Effects of the compositions and structures of SAMs on the tribological properties.

there will be a strong tendency for molecules to adsorb on its surface. Because of the exothermic head-group–substrate interactions, molecules try to occupy every available binding site on the surface, and during this process they generally push together molecules that have already adsorbed. The process results in the formation of crystalline molecular assemblies. The exothermic interactions between molecular chains are van der Waals or electrostatic-type, with energies on the order of a few (<10) kcal/mol. The molecular chains in SAMs are not perpendicular to the surface; the tilt angle depends on the anchor group, as well as on the substrate and the spacer group. For example, the tilt angle for alkanethiolate on Au is typically about 30°–35° with respect to the substrate normal [15].

SAMs can be prepared using different types of molecules and different substrates. Widespread examples are adsorption of a surfactant onto a clean metal surface. SAMs are usually produced by immersing a substrate in a solution containing a precursor that is reactive to the substrate surface, or by exposing the substrate to vapors of the reactive chemical species. The backbone chain of SAMs is typically an alkyl chain or is made of a derivatized alkyl group. The functionality of the SAM surface can be controlled by choosing different terminal groups or by chemical derivations of terminal groups. Multiple layers can also be constructed by functionalizing the terminal group and then depositing another monolayer on top of the previous one [67]. Relationships between the chemical structures and tribological properties of SAMs have been extensively studied to obtain boundary lubricants with good lubrication, high adhesion resistance, and strong antiwear ability [68–74]. For example, a novel self-assembled dual-layer film as a potential lubricant for MEMS was successfully designed and prepared by introducing hydrogen bonds and two-layer structure in the film, which not only greatly reduces the friction and adhesion, but also possesses a relatively high antiwear property [73].

The Langmuir growth curve is the first approximation for the adsorption, assuming an irreversible statistical adsorption process [75] expressed as

$$\frac{d\Theta}{dt} \propto 1 - \Theta \tag{7.46}$$

which yields a time-dependent coverage as

$$\Theta = 1 - \exp(-ck_a t) \tag{7.47}$$

If the growth nucleates only after a certain time, t_c, the simple growth law can be given by taking into account phenomenologically a time offset:

$$\Theta = 1 - \exp\left[-ck_a\left(t - t_c\right)\right] \tag{7.48}$$

Also, a diffusion-limited first-order Langmuir model (LD) was suggested as

$$\Theta = 1 - \exp(-ck_a^{LD} t^{0.5}) \tag{7.49}$$

which alters the time t dependence of eq. (7.48) to a $t^{0.5}$ dependence. Furthermore, a second-order nondiffusion-limited model (SO) was proposed as

$$\Theta = 1 - \frac{1}{1 + ck_a^{SO} t} \tag{7.50}$$

Here, c is the concentration of the adsorbing species, k_a is the adsorption rate constant, and k_a^{LD} and k_a^{SO} are the characteristic growth rate parameters. Dannenberger et al. [75] studied thiolate formation of n-alkanethiols $(CH_3(CH_2)_{m-1}SH)$ adsorbing on polycrystalline gold. Only the Langmuir kinetics could explain the experimental data. However, small but significant differences between the experimental data and the Langmuir model assuming statistical adsorption were identified. The Langmuir growth law is strictly valid only if the adsorbate molecules do not interact with each other [76]. A modified Kisliuk model accounting for the different adsorption sites was introduced. Different sticking coefficients for areas already covered and for those still free of thiols can be taken into account in the modified Kisliuk model. Assuming that the desorption rate from a precursor state is much higher than the chemisorption rate, one obtains

$$\frac{d\Theta}{dt} = ck_a(1 - \Theta)(1 + k_E \Theta) \tag{7.51}$$

where k_E is the sticking probability. The value for $k_E \Theta$ describes the deviation from the Langmuir kinetics due to changes in the overall sticking coefficient by the adsorbed thiolate. Integration of eq. (7.51) yields

$$\Theta = \frac{e^{ck_a t(1+k_E)} - 1}{e^{ck_a t(1+k_E)} + k_E} \tag{7.52}$$

As indicated by the solid line in fig. 7.8, eq. (7.52) represents a significantly better description [75].

Analogous to eq. (7.47), eq. (7.52) assumes that the rate of desorption is negligible. A study of adsorption kinetics in case of alkanoic acid monolayers on alumina was carried out by Chen and Frank [39, 77]. They found that the Langmuir kinetic

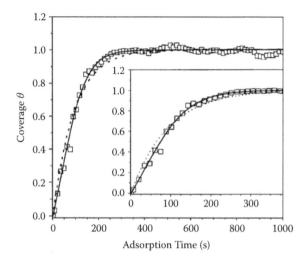

FIGURE 7.8 Time-dependent coverage determined from second harmonic generation (SFG) measurements (75) and fits based on Langmuir kinetics (dotted line, eq. [7.48]) and Kisliuk model (solid line, eq. [7.52]). (From Xiao, X. D., et al. 1996. *Langmuir* 12: 235. With permission.)

equation fits well the adsorption data for linear fatty acids having 12 or more carbons, i.e.,

$$\frac{\partial \Theta}{\partial t} = k_a c(1 - \Theta) - k_d \tag{7.53}$$

where k_d is the desorption rate constant. The ratio of rates depends on the free energy of adsorption ΔG as

$$b = \frac{k_a}{k_d} \propto e^{(-\Delta G)} \tag{7.54}$$

The adsorption energy is found to increase linearly with the alkyl chain length as

$$-\Delta G^0 = -\Delta G^{0,h} + N_c W \tag{7.55}$$

where N_c is the number of carbons in the chain and W is the energy per methylene group. Therefore, chemisorption controls the SAMs created from alkylthiols or silanes, but it is often preceded by a physical adsorption step.

7.4.2 TRIBOLOGICAL STUDIES OF SAMs

Tribological properties of several SAMs have been studied. Octadecyltrichlorosilane (OTS) was prepared on a single-crystal silicon wafer (112). Structure and morphology characterizations were done using contact-angle goniometry, ellipsometry, Fourier transform infrared spectroscopy, and AFM. It was observed that the SAM of OTS on the silicon wafer was quite smooth and homogeneous. Due to the wear of the OTS monolayer and the formation of the transfer film on the counterpart ball, the friction coefficient gradually increases from 0.06 to 0.13 with increasing sliding cycles and then stays constant at a normal load of 0.5 N. At a relatively high load of 1 N, the friction coefficient rises sharply to 0.62 even after sliding for only a few cycles, which is the same as in the case of a bare silicon substrate sliding against the ceramic counterface. These results show that the SAM of OTS has a poor load-carrying capacity as well as poor antiwear ability. Thus, the OTS monolayers can be a potential lubricant only at very low normal loads [78].

A novel self-assembled monolayer was successfully prepared on a silicon substrate coated with amino-group-containing polyethyleneimine (PEI) by the chemical adsorption of stearic acid (STA) molecules [72]. The formation and structure of the STA–PEI film were characterized by the same means as above. As fig. 7.9 shows, the hydrophilic silicon surface possesses the highest adhesion force of 70.4 nN due to the high capillary forces and chemical forces between the Si_3N_4 tip and the SiO_2 surface, whereas the PEI coating records a considerably lower adhesion force of 17.2 nN as compared with that of the silicon surface. The STA–PEI thin film has a considerably reduced adhesion force of only about 7.1 nN due to its strong hydrophobicity, which helps to eliminate the capillary force, as well as due to the negligible chemical force between the STA molecules and the Si_3N_4 tip [72]. The PEI coating shows relatively high adhesion and friction and poor antiwear ability. However, the STA–PEI film possesses good adhesion resistance and high load-carrying capacity as well as good antiwear ability. A C_{60} thin film chemisorbed onto PEI was also prepared [71]. Due to the hydrophobicity and low surface energy, the C_{60} film possessed an adhesion force of about only 7.1 nN. It also has good friction reduction, load-carrying capacity, and antiwear ability. The self-assembled dual-layer film of STA–PEI and C_{60}–PEI should find promising application in the lubrication and protection of MEMS devices.

7.4.2.1 Molecular Dynamics (MD) Simulation

The MD method deals with the simultaneous motion and interaction of atoms or molecules [79]. A wide range of potentials have been employed in the studies of

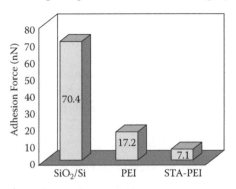

FIGURE 7.9 Adhesion forces between an AFM tip and the surfaces of the hydroxylated silicon (SiO_2/Si), PEI, and STA-PEI films [72]. (From Ren, S. L., Yang, S. R., and Zhao, Y. P. 2004. *Langmuir* 20: 3601. With permission.)

tribology and adhesion [15]. The Lennard–Jones (LJ) potential is a two-body potential that is commonly used for interactions between atoms or molecules with closed electron shells. It is applied not only to the interaction between noble gases, but also to the interaction between different segments on a polymer. The 12-6 LJ potential has the form

$$U(r_{ij}) = 4\varepsilon^{LJ}\left[\left(\frac{\sigma^{LJ}}{r_{ij}}\right)^{12} - \left(\frac{\sigma^{LJ}}{r_{ij}}\right)^6\right] \qquad (7.56)$$

where r_{ij} is the distance between the particles i and j, ε^{LJ} is the LJ interaction energy, and σ^{LJ} is the LJ diameter. The exponents 12 and 6 used above are very common, but other values may be chosen, depending on the system of interest.

The Morse potential is a convenient model for the potential energy of a diatomic molecule. It is a better approximation for the vibrational fine structure of the molecule because it implicitly includes the effect of bond breaking. The Morse potential energy function is of the form

$$V(r) = D_e(1 - e^{-a(r_{ij}-r_e)})^2 \qquad (7.57)$$

where r_{ij} is the distance between the atoms, r_e is the equilibrium bond distance, D_e is the equilibrium dissociation energy of the molecule (measured from the potential minimum), and a is a parameter controlling the width of the potential well.

For high computational efficiency, a pair potential such as the LJ or the Morse potential is used. With the increasing demand on accuracy and available computational power, many-body potentials such as the Finnis–Sinclair potential and the EAM (embedded atom method) have been commonly used [79].

For example, the EAM includes a contribution in the potential energy associated with the cost of "embedding" an atom in the local electron density ρ_i produced by surrounding atoms. The total potential energy U is approximated by

$$U = \sum_i \tilde{F}_i(\rho_i) + \sum_i \sum_{j<i} \phi_{ij}(r_{ij}) \qquad (7.58)$$

where F_i is the embedding energy, whose functional form depends on the particular metal. The pair potential $\Phi_{ij}(r_{ij})$ is a doubly screened short-range potential reflecting the core–core repulsion. The EAM has been particularly successful in reproducing experimental vacancy formation energies and surface energies, even though the potential parameters were only adjusted to reproduce bulk properties [15].

A typical mechanical model of the simulation system is shown in fig. 7.10 [80]. In simulations of adhesion, the tip is dragged by the support z_M through the spring k_z at a velocity v while the support x_M, y_M is fixed. Two interfaces in a vacuum created by CH_3/CH_3 and OH/OH have been studied. Figure 7.11 shows that at the pull-off stage, the molecular adhesion force reaches -0.9 nN. When CH_3- terminal groups are replaced by OH groups, the adhesion force increases by about four times due to the formation of hydrogen bonds among the OH groups.

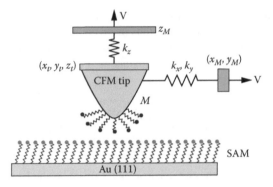

FIGURE 7.10 **(See color insert following page 80.)** Mechanical model for the simulation system [80]. The tip was dragged by the support (z_M) through the spring (k_z) at a velocity v while the support (x_M, y_M) was fixed. (From Ghoniem, N. M., et al. 2003. *Philos. Mag.* 83: 3475. With permission.)

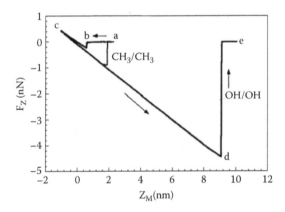

FIGURE 7.11 Force–distance curve vs. support position for the OH/OH contact pair and CH_3/CH_3 pair [80]. (From Ghoniem, N. M., et al. 2003. *Philos. Mag.* 83: 3475. With permission.)

A novel hybrid molecular simulation technique was developed to simulate AFM over experimental timescales. This method combines a dynamic element model for the tip–cantilever system in AFM and an MD relaxation approach for the sample. The hybrid simulation technique was applied to investigate the atomic scale friction and adhesion properties of SAMs as a function of chain length [81]. The Ryckaert–Bellmans potential, harmonic potential, and Lennard-Jones potential were used. The Ryckaert–Bellmans potential, which is for torsion, has the form

$$U_{\text{torsion}}(\phi) = a_0 + a_1 \cos(\phi) + a_2 \cos^2(\phi) + a_3 \cos^3(\phi) + a_4 \cos^4(\phi) + a_5 \cos^5(\phi) \quad (7.59)$$

where Φ is the torsion angle and a_0 to a_5 are constants. The harmonic potential for bond-angle bending has the form

$$U_{\text{bend}}(\theta_{ijk}) = 0.5 k_\theta (\theta_{ijk} - \theta_0)^2 \quad (7.60)$$

where θ_{ijk} is the bond angle of C-C-C or S-C-C, while k_θ and θ_0 are the corresponding force constant and equilibrium angle, respectively.

Pseudo atoms on different chains and those on the same chain interact with each other via a cut-and-shift Lennard–Jones potential. Lorenz et al. [82] modeled adhesion contact and friction for perfluorinated alkylsilane SAMs. The OPLS (optimized potential for liquid simulations) all-atom force-field parameters were used. The total potential energy for the system, $U^{\text{tot}}(r)$ is represented as a sum of nonbond interactions $U^{\text{NB}}(r_{ij})$ as well as energy contributions due to the distortion of bonds $U^{\text{bond}}(r_{ij})$, bond angles $U^{\text{bend}}(\theta_{ijk})$, and torsion angles $U^{\text{tors}}(\Phi_{ijkl})$ and is given by

$$U^{\text{tot}}(r) = U^{\text{NB}}(r_{ij}) + U^{\text{bond}}(r_{ij}) + U^{\text{bend}}(\theta_{ijkl}) + U^{\text{tors}}(\phi_{ijkl}) \tag{7.61}$$

where

$$U^{\text{NB}}(r_{ij}) = U^{\text{vdw}}(r_{ij}) + U^{qq}(r_{ij}) \tag{7.62}$$

$$U^{\text{bond}}(r_{ij}) = k^{\text{bond}}(r_{ij} - r_0)^2 \tag{7.63}$$

$$U^{\text{bend}}(\theta_{ijk}) = k^{\text{bend}}(\theta_{ijk} - \theta_0)^2 \tag{7.64}$$

$$U^{\text{tors}}(\phi_{ijkl}) = \sum k_n \cos(\phi_{ijkl})^n \tag{7.65}$$

where r_{ij} is the distance of the bond connecting atoms i and j; θ_{ijk} is the angle formed by atoms i, j, and k; and Φ_{ijkl} is the torsion angle formed by atoms i, j, k, and l. The nonbond energy $U^{\text{NB}}(r_{ij})$ is the sum of two-body repulsion and dispersion energy term $U^{\text{vdW}}(r_{ij})$ plus electrostatic interactions $U^{qq}(r_{ij})$. Here, the standard LJ 12-6 potential was used.

Hu et al. [83] simulated the frictional behavior of SAMs, with emphasis on a comparison of the performances of commensurate and incommensurate in relative sliding. As the further insight to the fundamental issue revealed, the friction depended on the strength of interfacial interaction and the structural commensurability of sliding bodies [83]. Results show that the shear stress on commensurate SAMs exhibits a clean pattern, manifesting the atomic stick–slip friction that may result from the periodic storing and releasing of mechanical energy during the sliding of highly ordered commensurate surfaces. Random fluctuations and a much lower average value of the shear stress are observed for incommensurate sliding. Chandross et al. [84, 85] also simulated the stick–slip behavior for well-ordered SAMs on crystalline and amorphous SiO_2 substrates. Stick–slip was observed in the former one.

The frictional behavior of SAMs as a function of normal load and sliding speed was also simulated. The friction behaviors at a fixed normal load of 39 nN and 300 K, but at various sliding speeds, have been simulated for both commensurate and incommensurate cases [83]. As fig. 7.12 shows, friction in commensurate sliding

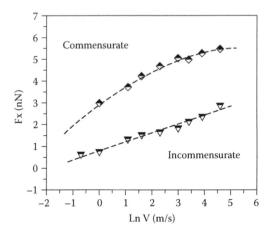

FIGURE 7.12 Friction forces obtained at fixed load of 39 nN and 300 K, displayed as a logarithmic function of sliding velocity at the commensurate and incommensurate sliding [83]. (From Lorenz, C. D., et al. 2005. *Tribol. Lett.* 19: 93. With permission.)

is much higher than that in an incommensurate case under the same conditions. The logarithmic dependence of friction on velocity was also found in experiments on various systems. Results show that simulation techniques are very powerful for interpreting experimental results and predicting properties.

Some computational studies predicted that the friction of SAMs had a characteristic dependence on temperature [86]. The effect of temperature on friction between a Si_3N_4 tip and a dodecanethiol SAM on Au (111) was studied using friction force microscopy (FFM) combined with a temperature regulator module [87]. Frictional properties were investigated as a function of effective normal load and temperature. The observed friction was found to be an approximately linear function of load. The results of measurements for friction at various temperatures plotted as the coefficient of friction versus temperature show that the friction strongly depends on the temperature (fig. 7.13). The friction increased in the low-temperature region (130–230 K) and decreased in the high-temperature region (>250 K). That is, the friction has a maximum around 240 K. This result is quite consistent with the prediction by computational studies [87].

7.4.2.2 Adhesion of SAMs

It is well known that the adhesion of SAMs is mainly controlled by the chemical characteristics of the terminal group and the environment, and has been studied extensively [67–70, 88–90]. A representative study on the effect of the terminal group on adhesion was done by Frisbie and coworkers [68, 89]. By using an AFM, they investigated the adhesion force between the probe tip and sample surfaces, both molecularly modified by SAMs with different terminal groups, and found it to be in the order: COOH/COOH > CH_3/CH_3 > COOH/CH_3. The influence of relative humidity on adhesion was studied in an environmentally controlled chamber at 22°C [91]. The results showed that for Si(111), Au(111), DHBp(4,4′-dihydroxybiphenyl), and MHA(16-mercaptohexadecanoic

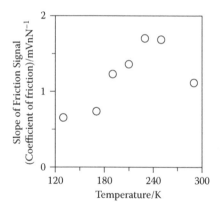

FIGURE 7.13 Temperature dependence of friction of $CH_3(CH_2)_{11}SH$ is shown as a plot of the slope of the friction signal as a function of the coefficient of friction vs. temperature [87]. (From Ohzono, T., and Fujihira, M. 2000. *Phys. Rev. B* 62: 17055. With permission.)

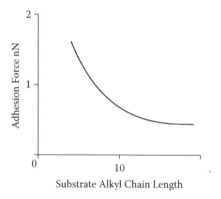

FIGURE 7.14 Drop in adhesion force with increasing chain length [92]. (From Noy, A., et al. 1995. *J. Am. Chem. Soc.* 117: 7943. With permission.)

acid thiol), the adhesion force increased with relative humidity; for BPT (1,1′-biphenyl-4-thiol) and BPTC (cross-linked BPT), the adhesion force increased only slightly with relative humidity when the relative humidity was higher than 40%. Actually, in addition to the terminal group and the environment, the molecular chain length also plays a significant role in the adhesion of SAMs.

Some experimental investigations [92] have been carried out. A simple dialkyl sulfide monolayer formed on gold surfaces gives a uniform film leading to low adhesion. The result shows that as the chain length is increased, the adhesion force falls systematically, roughly with the square root of chain length, as shown in fig. 7.14. However, if the dialkyl molecule has one chain much longer than the other, then the adhesion force can increase, because the longer chain can then interpenetrate the gaps on the opposite film, and the adhesion force is then found to increase with chain length. Van der Vegte et al. [93] reported such a result. In their experiment, by varying the length of one alkyl chain from 10 to 18 carbon atoms, while keeping

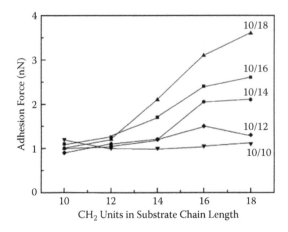

FIGURE 7.15 Average adhesion force for every probe–substrate combination: 10/10 dialkyl sulfide-coated tip (▼), 10/12 tip (◆), 10/14 tip (●), 10/16 tip (■), and 10/18 tip (▲) [93]. (From Ren, S. L., Yang, S. R., and Zhao, Y. P. 2004. *Appl. Surf. Sci.* 227: 293. With permission.)

the other alkyl chain length constant at 10 carbon atoms, the adhesion force between the dialkyl sulfide-coated probes and substrates was found to be dependent on the degree of asymmetry of the two chain lengths of the dialkylsulfide compounds. An increasing chain length of the molecules on the tip from 10/12 to 10/18 results in an increasingly stronger dependence of the adhesion force on the chain length of the molecules on the substrate (fig. 7.15).

A few molecular simulations [81, 94–97] have been carried out with respect to the adhesion and friction of SAMs. Lorenz et al. [82] utilized an OPLS all-atom force-field to study the force between two separated SAM films on amorphous SiO_2. It is reported that as the separation is decreased, the normal pressure first becomes slightly negative due to attractive van der Waals interactions and then becomes positive as repulsive interactions dominate at short separations. A hybrid molecular simulation approach has been applied to investigate dynamic adhesion and friction between an AFM tip and a SAM substrate. For the interactions of SAM chains, the united-atom (UA) model [98] was used. It is found that the elastic modulus is chain-length independent; however, under shear, the effective shear modulus is found to be chain-length dependent. Surface energies calculated for SAM surfaces also compare well with those determined by experiments. By using MD simulations, Chandross et al. [84] reported that the adhesion force between two monolayers at the same separation distance increases monotonically with decreasing chain length.

A quasi-continuum model [99] was proposed to calculate the adhesion force between the AFM tip and SAMs on a substrate by integrating the LJ potential for the molecular interactions. According to this theoretical model, the adhesion force decreases with increasing chain length of SAMs, which agrees well with the experimental results, as fig. 7.16 shows. When the radius of curvature of the tip is varied, the adhesion force will have a small increment, but there is no effect on the general tendency of the curve, which indicates that the size of the tip does not influence the adhesion force. The adhesion force calculated is on the order of nanonewtons, which

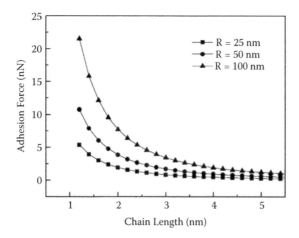

FIGURE 7.16 Adhesion force vs. chain length for silicon nitride tip and mica substrate for different tip radii R [99]. (From Hautman, J., and Klein, M. L. 1989. *J. Chem. Phys.* 91: 4994. With permission.)

is similar to the results of Morales-Cruz et al. [100], and so it is in agreement with the actual value.

Barbero and Evangelista [101] proposed a continuum model to describe the orientational states of a self-assembled system in which the membrane is formed by rodlike molecules in contact with an isotropic solid substrate. The intermolecular and the molecule–substrate interactions are modeled by an LJ-like potential. The total free energy was evaluated as follows:

$$F = F_0 + K_1 \frac{\partial \theta}{\partial x} + \frac{1}{2} \left[K_{2x} \left(\frac{\partial \theta}{\partial x} \right)^2 + K_{2y} \left(\frac{\partial \theta}{\partial y} \right)^2 \right] \tag{7.66}$$

where F_0 is the uniform part of the intermolecular interaction energy; θ is the angle by which a molecule of the membrane may bend under the action of the interactions; and K_1, K_{2x}, and K_{2y} are elastic coefficients. The total free energy contains the usual quadratic terms in the deformation tensor and the linear terms. The linear terms, which are known in the magnetic theory as the Lifshitz invariants, are responsible for ground states that can be periodically deformed. According to the authors' analysis, the elastic coefficients are positive increasing functions in the entire range of values of the dimensionless surface molecular density. As the surface molecular density increases, the attractive part of the LJ potential favors a homogeneous alignment, indicating that spatial deformations are very expensive from the point of view of energy.

Stiffness and damping constants of SAMs were measured using an apparatus designed by Devaprakasam and Biswas [102]. It shows that both the stiffness and damping constants of polytetrafluoroethylene (PTFE) increase with penetration of this relatively thick PTFE film. The nondimensional damping constant-to-stiffness ratios of perfluorooctyltrichlorosilane (FOTS) and perfluorooctadecylacid (PODA) as a

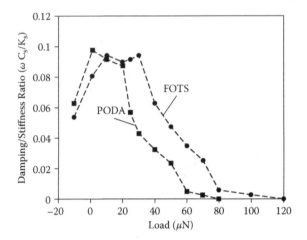

FIGURE 7.17 Damping/stiffness ratio of perfluorooctyltrichlorosilane (FOTS) (●) and per-fluorooctadecylacid (PODA) (■) as a function of load [102]. (From Barbero, G., and Evange-lista, L. R. 2004. *Phys. Rev. E* 70: 041407. With permission.)

function of load show a sharp transition from a viscous to a purely elastic behavior at a compression load that is specific and different for these different molecules (fig. 7.17).

7.4.3 SUPERHYDROPHOBIC FILMS

The wettability of solid surfaces is an important issue from both theoretical and applied points of view. Generally, the wettability of solid surfaces is controlled by both the topography and the chemical composition; both the roughness and hetero-geneity of the contacting surfaces greatly affect the contact angle and its hysteresis [103–105]. Recently, much attention has been paid to surfaces with superhydropho-bic properties. Superhydrophobic surfaces with water contact angles higher than 150° are arousing much interest because they will bring great convenience in daily life as well as in many industrial processes [45].

7.4.3.1 Mechanism of Superhydrophobic Films

In the simplest case, the wettability of a solid surface is commonly evaluated by the contact angle given by Young's equation. Then Wenzel proposed a model describ-ing the contact angle on a rough surface, and Cassie proposed an equation for the surface composed of solid and air. The concepts of these equations are outlined in fig. 7.18 [106].

The water contact angles on smooth hydrophobic surfaces are generally not higher than 120°. For example, the contact angles of water on self-assembled mono-layers of long-chain hydrocarbon and fluorocarbon are about 112° and 115°, respec-tively [107]. To reach the extreme values of the contact angle, i.e., more than 150°, surface roughness is often adjusted to amplify the surface hydrophobicity. An appro-priate surface roughness or structure is an important factor in fabricating the biomi-metic superhydrophobic surfaces [108].

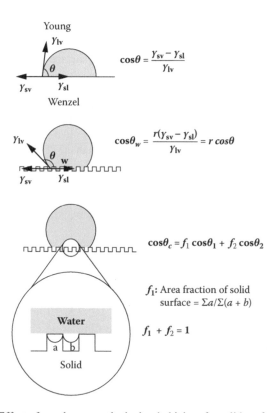

FIGURE 7.18 Effect of roughness on the hydrophobicity of a solid surface: Young's equation evaluates the contact angle θ; Wenzel proposed a model describing the contact angle θ_w on a rough surface; Cassie proposed an equation for the heterogeneous surface [106]. (From Yang, S. Y., et al. 1999. *J. Pet. Sci. Eng.* 24: 63. With permission.)

A fully superhydrophobic film should exhibit both high-contact-angle and small-contact-angle hysteresis. The hysteresis is defined as the difference between the advancing and receding angles. The superhydrophobic films in nature have attracted researchers' attention for many years. The study of the lotus leaf, which is one of the ordinary superhydrophobic films, shows that the novel topography of the surface, which consists of micro- and nanoscale hierarchical structures, makes the water droplets roll off the surface easily. Jiang et al. [45, 109] have fabricated many kinds of superhydrophobic films by creating a similarly rough structure on different solid surfaces. Another conventional way to attain superhydrophobicity is to decrease the surface energy of the solids. The water contact angle on the solids of lower surface energy is larger. As the molecular configuration can be modified by heat, pH, light, etc., and if this change results in alteration of the surface energy, switched wettability from superhydrophobic to superhydrophilic can be achieved. For example, every azobenzene unit of the azobenzene polyelectrolytes has two configurations. The *trans* has a small dipole moment and a low surface free energy, while the *cis* form presents a bigger dipole moment and a high surface energy. The azobenzene film will transform the *trans*-rich surface to a polar *cis*-rich surface after ultraviolet irradiation, and can be reversed by irradiation with visible light. Thus a large change of the contact angle from superhydrophobic to wettable on the azobenzene film is exhibited by combining the microstructures on the surface at the same time [110].

7.4.3.2 Recent Studies on Superhydrophobic Films

Generally, there are two ways to produce superhydrophobic surfaces. One is to create a rough structure on a hydrophobic surface, and the other is to modify a rough surface with materials of low surface energy [45]. Tadanaga et al. [111] prepared transparent, super-water-repellent coatings of alumina on glass plates by a combination of physical and chemical approaches. The contact angle for water on the films was 165°, and the transmittance for visible light was higher than 92%. Onda et al. [44] prepared a fractal surface by dipping a glass plate into a molten alkylketene dimer (AKD) and achieved a contact angle of about 174°.

Ren et al. [112] reported that a novel superhydrophobic ultrathin film was prepared by chemically adsorbing stearic acid (STA) onto a polyethyleneimine (PEI)-coated aluminum wafer. By immersing the freshly polished Al substrates in boiling water for 0, 10, 60, and 300 s, STA monolayers on PEI-coated aluminum substrates with root-mean-square surface roughnesses of 1.2, 3.5, 7.2, and 21.3 nm were prepared. The formation and the structure of the films were characterized by water contact angle measurements, ellipsometry, Fourier transform infrared spectroscopy, and x-ray photoelectron spectroscopy. The water contact angles on these samples increased with the surface roughness and were 105°, 116°, 146°, and 166°, respectively. On the other hand, the contact angle of water on the polished Al surface was about 62°; it decreased to less than 5° after the Al was boiled in water or coated with PEI, indicating that both the rough and PEI-coated Al surfaces were strongly hydrophilic [113]. Once the STA monolayer was formed on the PEI coating, the static contact angle for water on the surface of this ultrathin organic film was measured to be as high as 166°. The results on the relation between the superhydrophobicity and surface nanostructures also show that both the rough needlelike geometric nanostructures and hydrophobic materials are essential to the formation of a superhydrophobic surface. The effect of relative humidity on the adhesion and friction was also investigated. The STA superhydrophobic film possessed good lubricity compared with the other surfaces, and the adhesion force was greatly decreased by increasing the surface roughness of the Al wafer to reduce the contact area between the AFM tip and the sample surface. Thus it might be feasible and rational to prepare a surface with good adhesion resistance and good lubricity by properly controlling the surface morphology and the composition. These findings on the relation between nanostructures and superhydrophobicity are instructive in devising ways to reduce the stiction problem in MEMS [93].

7.5 CONCLUSION

An extensive review has been presented on surface forces, surface energy and stress, wetting and adhesion of SAMs, and superhydrophobic film. Relevant theoretical models, experimental results, and MD simulation results are discussed in detail.

Emphases of this review are placed on the mechanical problems of SAMs and related surface properties. It is expected that this review will assist in understanding the mechanics of adhesion related to SAMs and superhydrophobic films.

ACKNOWLEDGMENT

This work was supported by the National Basic Research Program of China (973 Program, Grant No. 2007CB310500) and the National Natural Science of China (NSFC, Grant No. 10772180).

REFERENCES

1. Maboudian, R., and Carraro, C. 2004. *Annu. Rev. Phys. Chem.* 55: 35.
2. Zhao, Y. P., Wang, L. S., and Yu, T. X. 2003. *J. Adhesion Sci. Technol.* 17 : 519.
3. Komvopoulos, K. 2003. *J. Adhesion Sci. Technol.* 17: 477.
4. Chen, J. Y., Huang, L., Ying, L. M. G., Luo, B., Zhao, X. S., and Cao, W. X. 1999. *Langmuir* 15: 7208.
5. Decher, G., Hong, J. D., and Schmitt, J. 1992. *Thin Solid Films* 210: 831.
6. Ferreira, M., and Rubner, M. F. 1995. *Macromolecules* 28: 7107.
7. Lvov, Y., Decher, G., and Mohwald, H. 1993. *Langmuir* 9: 481.
8. Mueller, A., Kowalewski, T., and Wooley, K. L. 1998. *Macromolecules* 31: 776.
9. Tsukruk, V. V., Rinderspacher, F., and Bliznyuk, V. N. 1997. *Langmuir* 13: 2171.
10. Zhang, X. Y., Klein, J., Sheiko, S. S., and Muzafarov, A. M. 2000. *Langmuir* 16: 3893.
11. Zhang, X. Y., Wilhelm, M., Klein, J., Pfaadt, M., and Meijer, E. W. 2000. *Langmuir* 16: 3884.
12. Lee, S., and Sigmund, W. M. 2001. *J. Colloid Interface Sci.* 243: 365.
13. Butt, H. J., Cappella, B., and Kappl, M. 2005. *Surf. Sci. Rep.* 59: 1.
14. Bowen, W. R., and Jenner, F. 1995. *Adv. Colloid Interface Sci.* 56: 201.
15. Bhushan, B., ed. 2004. *Springer handbook of nanotechnology.* Heidelberg: Springer.
16. Israelachvili, J. N. 1985. *Intermolecular and surface forces.* 2nd ed. London: Academic Press.
17. Luckham, P. F. 2004. *Adv. Colloid Interface Sci.* 111: 29.
18. Israelachvili, J. N., and Wennerstrom, H. 1996. *Nature* 379: 219.
19. Israelachvili, J. N., and Wennerstrom, H. 1992. *J. Phys. Chem.* 96: 520.
20. Leikin, S., Parsegian, V. A., Rau, D. C., and Rand, R. P. 1993. *Annu. Rev. Phys. Chem.* 44: 369.
21. Yang, C. Y., and Zhao, Y. P. 2004. *J. Chem. Phys.* 120: 5366.
22. Israelachvili, J. N., and Pashley, R. M. 1983. *Nature* 306: 249.
23. Israelachvili, J. N., and Wennerstroem, H. 1990. *Langmuir* 6: 873.
24. McIntosh, T. J., Magid, A. D., and Simon, S. A. 1989. *Biochemistry* 28: 7904.
25. Meyer, E. E., Rosenberg, K. J., and Israelachvili, J. 2006. *Proc. Natl. Acad. Sci. U.S.A.* 103: 15739–15746.
26. Israelachvili, J. N., and Pashley, R. M. 1982. *Nature* 300: 341.
27. Gomer, R., and Smith, C. S. 1953. *Structure and properties of solid surfaces.* Chicago: The University of Chicago Press.
28. Kinloch, A. J. 1987. *Adhesion and adhesives: Science and technology.* London: Chapman & Hall.
29. Venables, J. A. 2000. *Introduction to surface and thin film processes.* Cambridge: Cambridge University Press.
30. Muller, P., and Saul, A. 2004. *Surf. Sci. Rep.* 54: 157.
31. Cammarata, R. C. 1994. *Prog. Surf. Sci.* 46: 1.

32. Gurtin, M. E., and Murdoch, A. I. 1975. *Arch. Rat. Mech. Anal.* 57: 291.
33. Murdoch, A. I. 2005. *J. Elasticity* 80: 33.
34. Berger, R., Delamarche, E., Lang, H. P., Gerber, C., Gimzewski, J. K., Meyer, E., and Guntherodt, H. J. 1997. *Science* 276: 2021.
35. Gupta, P., Ulman, A., Fanfan, S., Korniakov, A., and Loos, K. 2005. *J. Am. Chem. Soc.* 127: 1, 4.
36. Lawrence, J. 2004. *Proc. R. Soc. Lond.* 460: 1723.
37. Mittal, K. L., ed. 2003. *Contact angle, wettability and adhesion.* Vol. 3. Leiden: VSP.
38. Mittal, K. L., ed. 2006. *Contact angle, wettability and adhesion.* Vol. 4. Leiden: VSP.
39. Adamson, A. W., and Gast, A. P. 1997. *Physical chemistry of surfaces.* 6th ed. New York: Wiley-Interscience.
40. Nosonovsky, M., and Bhushan, B. 2005. *Microsystem Technologies* 11: 535.
41. Extrand, C. W. 2004. *Langmuir* 20: 5013.
42. de-Gennes, P. G., Brochard-Wyart, F., and Quéré, D. 2002. *Capillarity and wetting phenomena.* New York: Springer.
43. Cassie, A. B. D., and Baxter, S. 1944. *Trans. Faraday Soc.* 40: 546.
44. Onda, T., Shibuichi, S., Satoh, N., and Tsujii, K. 1996. *Langmuir* 12: 2125.
45. Feng, L., Li, S. H., Li, Y. S., Li, H. J., Zhang, L. J., Zhai, J., Song, Y. L., Liu, B. Q., Jiang, L., and Zhu, D. B. 2002. *Adv. Mater.* 14: 1857.
46. Gibbs, J. W. 1961. *The scientific papers.* New York: Dover Publications.
47. Boruvka, L., and Neumann, A. W. 1977. *J. Chem. Phys.* 66: 5464.
48. Guo, H. K., and Fang, H. P. 2005. *Chin. Phys. Lett.* 22: 787.
49. Checco, A., Schollmeyer, H., Daillant, J., Guenoun, P., and Boukherroub, R. 2006. *Langmuir* 22: 116.
50. Lippmann, M. G. 1875. *Ann. Chim. Phys.* 5: 494.
51. Mugele, F., Klingner, A., Buehrle, J., Steinhauser, D., and Herminghaus, S. 2005. *J. Phys.: Condensed Matter* 17: S559.
52. Lee, L. H., ed. 1971. *Recent advances in adhesion.* New York: Gordon and Breach Science Publisher.
53. Mukhopadhyay, A., Zhao, J., Bae, S. C., and Granick, S. 2003. *Rev. Sci. Instrum.* 74: 3067.
54. Veeramasuneni, S., Yalamanchili, M. R., and Miller, J. D. 1996. *J. Colloid Interface Sci.* 184: 594.
55. Wei, Z., and Zhao, Y. P. 2004. *Chin. Phys.* 13: 1320.
56. Ulman, A. 1996. *Chem. Rev.* 96: 1533.
57. Bigelow, W. C., Pickett, D. L., and Zisman, W. A. 1946. *J. Colloid Interface Sci.* 1: 513.
58. Sagiv, J. 1980. *J. Am. Chem. Soc.* 102: 92.
59. Nuzzo, R. G., and Allara, D. L. 1983. *J. Am. Chem. Soc.* 105: 4481.
60. Schreiber, F. 2004. *J. Phys.: Condensed Matter* 16: R881.
61. Tsukruk, V. V. 2001. *Adv. Mater.* 13: 95.
62. Ulman, A. 1995. *MRS Bull.* 20: 46.
63. Maboudian, R., Ashurst, W. R., and Carraro, C. 2000. *Sensors Actuators A* 82: 219.
64. Rymuza, Z. 1999. *Microsystem Technologies* 5: 173.
65. Bittner, A. M. 2006. *Surf. Sci. Rep.* 61: 383.
66. Ulman, A. 1991. *An introduction to organic ultrathin films: From Langmuir to self-assembly.* Boston: Academic Press.
67. Wang, M. J., Liechti, K. M., Srinivasan, V., White, J. M., Rossky, P. J., and Stone, M. T. 2006. *J. Appl. Mech. Trans. ASME* 73: 769.
68. Frisbie, C. D., Rozsnyai, L. F., Noy, A., Wrighton, M. S., and Lieber, C. M. 1994. *Science* 265: 2071.
69. Kim, H. I., and Houston, J. E. 2000. *J. Am. Chem. Soc.* 122: 12045.
70. Lee, S., Puck, A., Graupe, M., Colorado, R., Shon, Y. S., Lee, T. R., and Perry, S. S. 2001. *Langmuir* 17: 7364.

71. Ren, S. L., Yang, S. R., and Zhao, Y. P. 2004. *Langmuir* 20: 3601.
72. Ren, S. L., Yang, S. R., Wang, J. Q., Liu, W. M., and Zhao, Y. P. 2004. *Chem. Mater.* 16: 428.
73. Ren, S. L., Yang, S. R., and Zhao, Y. P. 2003. *Langmuir* 19: 2763.
74. Xiao, X. D., Hu, J., Charych, D. H., and Salmeron, M. 1996. *Langmuir* 12: 235.
75. Dannenberger, O., Buck, M., and Grunze, M. 1999. *J. Phys. Chem. B* 103: 2202.
76. Schreiber, F. 2000. *Prog. Surf. Sci.* 65: 151.
77. Chen, S. H., and Frank, C. W. 1989. *Langmuir* 5: 978.
78. Ren, S. L., Yang, S. R., Zhao, Y. P., Zhou, J. F., Xu, T., and Liu, W. M. 2002. *Tribol. Lett.* 13: 233.
79. Ghoniem, N. M., Busso, E. P., Kioussis, N., and Huang, H. C. 2003. *Philos. Mag.* 83: 3475.
80. Leng, Y. S., and Jiang, S. Y. 2002. *J. Am. Chem. Soc.* 124: 11764–11770.
81. Jiang, S. Y. 2002. *Mol. Phys.* 100: 2261.
82. Lorenz, C. D., Webb, E. B., Stevens, M. J., Chandross, M., and Grest, G. S. 2005. *Tribol. Lett.* 19: 93.
83. Hu, Y. Z., Zhang, T., Ma, T. B., and Wang, H. 2006. *Comput. Mater. Sci.* 38: 98.
84. Chandross, M., Grest, G. S., and Stevens, M. J. 2002. *Langmuir* 18: 8392.
85. Chandross, M., Webb, E. B., Stevens, M. J., Grest, G. S., and Garofalini, S. H. 2004. *Phys. Rev. Lett.* 93: 166103-1–166103-2.
86. Ohzono, T., and Fujihira, M. 2000. *Phys. Rev. B* 62: 17055–17071.
87. Fujita, M., and Fujihira, M. 2002. *Ultramicroscopy* 91: 227.
88. Nakano, M., Ishida, T., Numata, T., Ando, Y., and Sasaki, S. 2003. *Jpn. J. Appl. Phys. Part 1* 42: 4734.
89. Noy, A., Frisbie, C. D., Rozsnyai, L. F., Wrighton, M. S., and Lieber, C. M. 1995. *J. Am. Chem. Soc.* 117: 7943.
90. Ren, S. L., Yang, S. R., and Zhao, Y. P. 2004. *Appl. Surf. Sci.* 227: 293.
91. Bhushan, B., and Liu, H. W. 2001. *Phys. Rev. B* 63: 245412-1–245412-11.
92. Kendall, K. 2001. *Molecular adhesion and its applications.* New York: Kluwer Academic/Plenum.
93. van der Vegte, E. W., Subbotin, A., Hadziioannou, G., Ashton, P. R., and Preece, J. A. 2000. *Langmuir* 16: 3249.
94. Hung, S. W., Hwang, J. K., Tseng, F., Chang, J. M., Chen, C. C., and Chieng, C. C. 2006. *Nanotechnology* 17: S8.
95. Jang, S. S., Jang, Y. H., Kim, Y. H., Goddard, W. A., Flood, A. H., Laursen, B. W., Tseng, H. R., Stoddart, J. F., Jeppesen, J. O., Choi, J. W., Steuerman, D. W., Delonno, E., and Heath, J. R. 2005. *J. Am. Chem. Soc.* 127: 1563.
96. Jang, Y. H., Jang, S. S., and Goddard, W. A. 2005. *J. Am. Chem. Soc.* 127: 4959.
97. Zheng, J., Li, L. Y., Tsao, H. K., Sheng, Y. J., Chen, S. F., and Jiang, S. Y. 2005. *Biophys. J.* 89: 158.
98. Hautman, J., and Klein, M. L. 1989. *J. Chem. Phys.* 91: 4994.
99. Guo, L. Y., and Zhao, Y.-P. 2006. *J. Adhesion Sci. Technol.* 20: 1281.
100. Morales-Cruz, A. L., Tremont, R., Martinez, R., Romanach, R., and Cabrera, C. R. 2005. *Appl. Surf. Sci.* 241: 371.
101. Barbero, G., and Evangelista, L. R. 2004. *Phys. Rev. E* 70: 041407-1–041407-7.
102. Devaprakasam, D., and Biswas, S. K. 2005. *Rev. Sci. Instrum.* 76: 035102-1–035102-7.
103. Decker, E. L., Frank, B., Suo, Y., and Garoff, S. 1999. *Colloids Surf. A* 156: 177.
104. Kwok, D. Y., and Neumann, A. W. 1999. *Adv. Colloid Interface Sci.* 81: 167.
105. Yang, S. Y., Hirasaki, G. J., Basu, S., and Vaidya, R. 1999. *J. Pet. Sci. Eng.* 24: 63.
106. Nakajima, A., Hashimoto, K., and Watanabe, T. 2001. *Monatsh. Chem.* 132: 31.
107. Srinivasan, U., Houston, M. R., Howe, R. T., and Maboudian, R. 1998. *J. Microelectromechanical Syst.* 7: 252.

108. Guo, Z. G., Zhou, F., Hao, J. C., and Liu, W. M. 2006. *J. Colloid Interface Sci.* 303: 298.
109. Feng, L., Li, S. H., Li, H. J., Zhai, J., Song, Y. L., Jiang, L., and Zhu, D. B. 2002. *Angew. Chem. Int. Ed.* 41: 1221.
110. Jiang, W. H., Wang, G. J., He, Y. N., Wang, X. G., An, Y. L., Song, Y. L., and Jiang, L. 2005. *Chem. Commun.* 3550.
111. Tadanaga, K., Katata, N., and Minami, T. 1997. *J. Am. Ceramic Soc.* 80: 1040.
112. Ren, S. L., Yang, S. R., Zhao, Y. P., Yu, T. X., and Xiao, X. D. 2003. *Surf. Sci.* 546: 64.
113. Ren, S. L., Yang, S. R., and Zhao, Y. P. 2004. *Acta Mechanica Sinica* 20: 159.

NOMENCLATURE

A area
A_H Hamaker constant
c concentration of the adsorbing species
C coefficient in the atom–atom pair potential
$C_{vdW}, C_{ind}, C_{orient}, C_{disp}$ coefficients for van der Waals force, induction, orientation, and dispersion, respectively
d_0 coefficient related to the thermal expansion
D distance
D_e equilibrium dissociation energy of the molecule
D^f fractal dimension
e electronic charge
E Young's modulus
f fractions of wetted part
f_1, f_2 fractions of the surface occupied by different surface types
F_{ad} adhesion force
F_{el} double-layer force
F_i embedding energy
k_a adsorption rate constant
k_d desorption rate constant
k_E sticking probability
k_B Boltzmann's constant
k_a^{LD}, k_a^{SO} characteristic growth rate parameters of Langmuir model (LD) and second-order nondiffusion-limited model (SO)
L length
M_i molar concentration
n_1, n_2, n_3 refractive indices
P repulsive pressure
r molecular radius
r_e equilibrium bond distance
r_{ij} distance between particles
r_w roughness factor
R curvature radius
R_g radius of gyration of polymer
s separation between the terminally attached polymer chains
$S_{L/S}$ spreading coefficient
t time

t_s sensor thickness

T temperature

u displacement from the reference configuration

u_1, u_2 dipole moments of the molecules

U internal energy

V voltage

w interaction energy

W_{ad} work of adhesion

z_i valency

α_{01}, α_{02} electronic polarizabilities of the molecules

β coefficient

γ, $\gamma_{S/V}$, $\gamma_{L/S}$, $\gamma_{L/V}$ surface and interface energies

γ_c critical surface energy

δ thickness of the polymer layer

ε_0 vacuum permittivity

ε dielectric constant of the medium

ε_1, ε_2, ε_3 static dielectric constants

ε_{LJ} LJ interaction energy

ε_s interfacial strain

η dimensionless electrowetting number

Θ coverage

θ, θ_c, θ_f, θ_w, θ_1, θ_2 equilibrium, Cassie–Baxter, fractal, and Wenzel contact angles and contact angles of different surface types

θ_0 equilibrium angle

θ_{ijk} bond angle formed by atoms i, j, k

κ reaction rate

λ linear dimension

λ_D Debye length

λ_H decay length of hydration force

λ_L Lamé constant

λ_s decay length of solvation force

μ chemical potential

μ_L Lamé constant

ν Poisson's ratio

ν_1, ν_2 orbiting frequencies of electron

ν_e mean absorption frequency

ψ, ψ_0, ψ_s, ψ_T electrostatic potential

ρ, ρ_1, ρ_2 density

ρ_0 number density of ions

ρ_l local electron density

σ surface stress

σ_c, σ_S, σ_T surface charge densities of the surface, sample, and tip

σ_{LJ} LJ diameter

φ interaction parameter

Φ torsion angle

8 Molecular Understanding of Nanofriction of Self-Assembled Monolayers on Silicon Using AFM

Sharadha Sambasivan, Shuchen Hsieh,
Daniel A. Fischer, and Stephen M. Hsu

ABSTRACT

Friction at the nanoscale has become a significant challenge for microsystems, including MEMS, NEMS and other devices. At nanoscale, lateral loading often causes component breakage and loss of functions in devices; therefore, accurate measurement and understanding of nanofriction are critical in device reliability and durability. Since silicon-based devices offer overwhelming cost advantages in manufacturing, the issue of controlling friction on silicon using nanometer thin films has attracted significant interest recently, in addition to lubrication by a thin layer of water in rare cases. Due to the small scale of measurements, the atomic force microscope (AFM) has been the instrument of choice to measure nanofriction. At the same time, a detailed surface characterization of the monolayer film structure is needed. This chapter explores the use of near-edge x-ray absorption fine structure (NEXAFS) spectroscopy and Fourier transform infrared (FTIR) spectroscopy to ascertain the order of the film. A series of n-alkyltrichlorosilanes self-assembled monolayer films with various chain lengths (C5 to C30) were prepared on silicon (100) surfaces. Nanofriction measurements were conducted using an atomic force microscope (AFM). Results showed that the lowest friction was obtained with a C12 film with higher friction values observed for C5 and C30 films. The film structure and order of organization were probed with an x-ray absorption technique (NEXAFS) in conjunction with Fourier transform infrared (FTIR) spectroscopy. It was observed that C12, C16, and C18 films were more ordered (molecular orientation of the carbon backbone nearly perpendicular to the surface) than those of C5 and C30 films. The combination of these two techniques provides complementary evidence that both film order and organization strongly influence nanofriction.

8.1 INTRODUCTION

8.1.1 MONOLAYER THIN FILMS AND THEIR EFFECTS ON FRICTION

Monomolecular layers have been created ever since the invention of the Langmuir–Blodgett trough in the 1930s to study surfactants in water. Bowden and Tabor [1] conducted the first fundamental study of monolayers and their effect on friction in the 1940s. They showed that monolayers of pure fatty acids on glass surfaces (flat-on-flat contacts) reduced friction, and increasing the alkyl chain length reduced friction up to C_{18}, and thereafter, the friction was constant and steady. These films functioned successfully under low load, but at higher loads, they failed rapidly. Subsequent to their study, the experimental results have been repeatedly confirmed by many others studying the fundamentals of lubrication science. Since one of the basic premises of lubrication is sacrificial wear, i.e., forming a lubricating film that is easily sheared off during contacts to protect the substrate surface, monolayer film studies remain, for the large part, a study of basic molecular structural influence on both friction and wear. Monolayer film lubrication, as perceived by lubrication scientists, does not carry practical significance. It is only in the past decade, with the development of magnetic hard disk technology, microelectromechanical systems (MEMS), and nanoelectromechanical systems (NEMS), that the notion that monolayer lubrication might be practical in lubricating micro- and nanoscale devices has come into existence.

As demonstrated by the magnetic hard disk technology, a monolayer or a sub-monolayer of perfluoroalkylether (PFPE) is capable of protecting a hard disk for the duration of an estimated 7 y of usage [2]. In order to accomplish this seemingly impossible feat, the film has to have strong chemical bonding with the surface to withstand repeated shear; it must be able to repair itself after a contact by diffusing back into the contact zone and reorganizing itself; it must have chemical stability to resist oxidation from tribochemistry; it must produce no waste products (evaporate into air as decomposition gases); and it must provide a lubricant reservoir via surface texturing to replenish lost lubricant molecules, etc. [3].

The issue of having to reorganize itself to re-form a functional film has led to the intense interest in examining self-assembled monolayers (SAMs), which are films of alkyl-based surfactants. They are formed by adsorbing linear alkyl molecules with a functional end group onto an energetically uniform surface to form a solid structure [4]. The degree of close-packing depends on the uniformity of the surface defect site distribution, purity of the molecules (monodispersity), cross-linking of the chains, impurity level in the solution, and the time it takes to form the film. On ideal surfaces, SAMs are self-limiting to a monolayer.

Silicon is the basic material used in electronic applications, and the well-developed CMOS (complementary metal oxide semiconductor) process gives it an important cost advantage over any other materials for device fabrication. The chemistry of silica is well known [5], and silicon surfaces are known for their surface heterogeneity in terms of defects, crystallographic planes, and steps. Therefore, for most silicon surfaces, it is difficult to have a well-packed perfect monolayer. Most of the studies in the literature have focused on a small domain that appears to be well organized. In controlling friction, uniform homogeneous coverage actually is not required, because as long as there are enough patches of film in the contact region, the friction will be reduced. However, the nanomechanical properties of the film will be influenced by the order and disorder of the film structure. One of the key parameters, for example, is the real area of contact between the probe and the film as determined by the elastic modulus of the film, which is a function of defect population and orderliness of the organization. At the nanoscale, adhesion is a function of contact area, and friction is influenced strongly by the adhesion forces it has to overcome to commence motion.

In general, SAMs have been explored for use in hydrophobic films, barrier films, and as organic spacers. With proper modifications, SAMs can also be used as molecular recognition biosensors, as lubricating films in digital mirror devices, and for dip-pen surface patterning of organic molecules [6]. Another emerging application for SAMs is barrier films for devices to control adhesion and hydrophobicity of the components. In MEMS applications, SAMs have been used as lubricating films to reduce friction between moving parts in devices [7, 8].

8.1.2 Chain Length Effects on Friction

In understanding how SAMs can influence friction at the nanoscale, the issue of chain length is, perhaps, the most interesting. From Bowden and Tabor's work [1], we expect that friction will decrease as the chain length increases until a characteristic chain length is reached, and then the friction level will be constant. However, there are several exceptions, and the relationship between frictional properties and chain length is very system specific [9]. For this reason, we decided to investigate systematically the effect of chain length on the friction of a series of SAMs.

We have conducted friction measurements via the AFM (atomic force microscopy) technique on a series of trichlorosilanes with different chain lengths on silicon wafer surfaces and found the friction to be the lowest for the C_{12} silane film. For SAM films with chain lengths greater than C_{12}, (C_{16}, C_{18}, and C_{30}), friction increased with chain length [10]. This result is contrary to the long-established observation of longer chain length producing lower friction in microscale experiments [11]. However, AFM experiments on alkylsilanes on mica by Lio and coworkers [12] showed that C_{12} and C_{18} chains had lower friction than C_8 or C_6. In order to resolve these contradictory results, this chapter examines the structure and order of these films in an attempt to explain the friction results. Detailed experimental results of the friction measurements will be presented elsewhere [10]. Because of the molecular scale of these films, detailed characterizations of the order and structure of SAMs are difficult to achieve. At the same time, SAMs have been widely used in many applications and are emerging as a key tool in microelectronics, NEMS fabrication, and device patterning [13] due to their self-repairing and organizing abilities.

8.1.3 Order and Structure of Self-Assembled Monolayer Films

The order of the film structure is controlled by the packing density of the molecules and the spacing between the molecules on the surface. We cannot measure directly the packing density of the molecules, but if the surface is energetically uniform, then the defect population can be linked to the degree of molecular orientation or alignment in a particular direction on the surface. The molecular orientation of alkanethiol

SAMs on gold surfaces has been studied [2, 14], but the results have not been linked to the nanomechanical properties (such as frictional characteristics) of the films.

Synchrotron-based NEXAFS (near-edge x-ray absorption fine structures) is a versatile tool to measure quantitatively the degree of molecular orientation of SAMs [15–17]. NEXAFS is a nondestructive spectroscopic method in which soft x-rays are absorbed, causing resonant excitations of core K- or L-shell electrons to unoccupied (antibonding) molecular orbitals. The initial-state K-shell excitation gives NEXAFS its elemental specificity, e.g., carbon, oxygen, and substrate copper [18, 19], while the final-state unoccupied molecular orbitals give NEXAFS its bonding or chemical selectivity. Linearly polarized soft x-rays from a synchrotron light source can be exploited to measure molecular orientation in a film. The NEXAFS resonance intensities vary as a function of the direction of the incident polarized x-rays relative to the axes of the σ^* and π^* orbitals [20]. This variation in resonance intensity allows a measurement of the average molecular orientation of the alkyl chain in n-alkyltrichlorosilanes (R-SiCl$_3$, where R contains 5 to 30 carbon atoms) SAMs with respect to the surface. Due to high surface sensitivity, molecular orientation sensitivity, and rapid data acquisition and analysis, the NEXAFS technique affords a clearer molecular ordering information than infrared reflection or absorption spectroscopy. Recently, NEXAFS has been employed to determine the orientation of organized assemblies of organic molecules, such as self-assembled monolayers of oriented fatty acids [14] and the surfaces of thin polymer films [15, 21], and it has evolved into a routine synchrotron soft x-ray spectroscopic tool. In this chapter, we report on molecular orientation and structure utilizing NEXAFS of n-alkyltrichlorosilane SAMs of various carbon lengths on silicon [22]. The results from NEXAFS were also compared with other qualitative techniques such as Fourier transform infrared (FTIR) spectroscopy and AFM. The present work is directed toward understanding the correlation of SAM molecular order and friction measured via the AFM technique.

8.2 EXPERIMENTAL

8.2.1 MATERIALS

Polished test-grade silicon (100) wafers were purchased from Virginia Semiconductor (Fredericksburg, VA). Self-assembled monolayers on Si (100) substrates were prepared using n-pentyltrichlorosilane (C$_5$ or C5), n-decyltrichlorosilane (C$_{10}$ or C10), n-dodecyltrichlorosilane (C$_{12}$ or C12), n-hexadecyltrichlorosilane (C$_{16}$ or C16), n-octadecyltrichlorosilane (C$_{18}$ or C18), and n-triacontyltrichlorosilane (C$_{30}$ or C30) purchased from Gelest (Morrisville, PA). Solvents such as hexadecane (C$_{16}$H$_{34}$), chloroform (CHCl$_3$) (99% anhydrous), isopropyl alcohol (dehydrated), and ethanol (200 proof) were purchased from Aldrich and used without further purification. Water was deionized with a Milli-Q Academic water purification system manufactured by Millipore (Billerica, MA).

8.2.2 SUBSTRATE PREPARATION

The silicon (100) wafers were first cleaned using a surfactant, Micro 90, purchased from Cole-Palmer (Vernon Hills, IL), followed by ethanol rinse two to three times.

Then the Si substrates were dried under a stream of nitrogen gas. These wafers were then treated using a plasma cleaner purchased from Harrick Plasma (Ithaca, NY) for 2 min to increase the OH$^-$ concentration at the top surface.

8.2.3 SAMs Preparation and Characterization

The SAMs were formed on the clean Si (100) substrates by first immersing them into solutions of 1 mM n-alkyltrichlorosilanes in hexadecane. The samples were left in the solutions for 24 h, after which they were removed and sonicated sequentially in chloroform, isopropyl alcohol, and deionized water for 15 min each. Finally, the samples were annealed in an oven for 10 min at 115°C. The SAMs were initially characterized utilizing the advancing water contact angle described in detail elsewhere [4]. The advancing water contact angle measured (with an error of ±1.5°) for the C_5 SAM film was 93.2°, for C_{10} it was 96.5°, for C_{12} it was 95.1°, for C_{16} it was 100.2°, for C_{18} it was 95.4°, and for the C_{30} SAM film it was 101.4°, which indicates near-monolayer film coverage in all these cases. The SAMs thus prepared were characterized for frictional properties with an applied force in the nano-Newton range using a novel multiscale friction AFM apparatus, the details of which can be found elsewhere [10, 23]. New sets of prepared SAMs were stored for less than a week in sealed glass bottles or Petri dishes until they were ready to be characterized using NEXAFS.

8.2.4 NEXAFS Experiment

Molecular order characterization at the SAMs surface (top 3–5 nm) was performed using an x-ray absorption technique of NEXAFS, the details of which are reported elsewhere [24]. NEXAFS at the carbon K-edge was carried out at U7A NIST beamline of the National Synchrotron Light Source at Brookhaven National Laboratory [25]. The incident soft x-rays are linearly polarized and monochromatic, with an energy resolution of 0.2 eV at 300 eV. Multiple samples approximately 1 × 1 cm on the Si wafer were freshly prepared with SAMs of different chain lengths and were loaded on a stainless-steel bar and introduced into the vacuum chamber through a sample load-lock system. Samples were precisely aligned with the aid of a fully automated, computerized sample manipulator stage. Energy resolution and photon energy calibration of the monochromator were performed by comparing the C=O 1s to the π^* peak of the gas-phase carbon monoxide with electron energy loss reference data [26]. The partial-electron-yield (PEY) NEXAFS measurements at carbon K-edge from 270 eV to 330 eV were obtained with a channeltron at a negative 150-V bias on the entrance grid to enhance surface sensitivity (probe depth of approximately 3 to 5 nm) [24]. Under these bias conditions, the partial-electron-yield measurement is dominated by the Auger yield from the sample, and low-energy electrons and other background photoelectrons are rejected (repelled away) due to the high negative bias electric field. The sample was charge neutralized using a low-energy electron flood gun, which added no noticeable background to the NEXAFS spectrum. Electron-yield data are sensitive to incident x-ray beam intensity fluctuations and monochromator absorption features. Hence, these spectra were normalized to the incident beam intensity, I_0, by collecting the total electron yield intensity from a clean, gold-coated, 90% transmitting grid placed in the incoming x-ray beam path.

8.3 RESULTS AND DISCUSSION

8.3.1 Nanofriction Measurements

Nanofrictional measurements were performed using an atomic force microscope (AFM) with a sharp tip. Table 8.1 summarizes the results of the coefficient of friction with a nominal applied load of 40 nN for various SAM systems examined in the present study. It is observed that bare silicon has a much higher coefficient of friction compared with any of the trichlorosilane monolayer SAM systems on Si. It is also observed that C_{12} film exhibits the lowest coefficient of friction, followed by C_{10}, C_{16}, C_{18}, C_{30}, and C_5. The observed trend for the coefficient of friction value versus the chain length is quite contrary to the established observations of longer chain lengths leading to lower friction values [11, 27]. In order to understand the differences and examine the correlation of coefficient of friction with molecular structure, these films were characterized by FTIR and NEXAFS. This chapter focuses on the use of NEXAFS to determine the molecular orientation and the degree of order in the film organization to explain the friction measurements.

8.3.2 NEXAFS Measurements

For each sample, the NEXAFS spectra at the carbon K-edge were obtained at various angles θ, ranging from 20° (glancing incidence) to 90° (normal incidence), where θ is measured between the sample normal and polarization vector E of the incident x-ray beam. Figure 8.1 shows NEXAFS PEY spectra for a C_5 trichlorosilane SAM, a C_{12} trichlorosilane SAM, and a C_{30} trichlorosilane SAM at normal incidence when $\theta = 90°$ between E and the sample normal (shown as solid curves) and at glancing incidence when $\theta = 20°$ between E and surface normal (shown as dashed curves). The NEXAFS PEY spectra of the trichlorosilane SAMs are dominated by the C–H resonance Rydberg peak, which was observed at 287.3 eV. This peak represents the relaxation of the C 1s core-electron, which is excited to the C–H σ^* molecular orbital followed by emission of Auger electrons, which are detected by the partial-electron-yield detector. The other prominent peak at 292.7 eV has been assigned to the C–C σ^* resonance peak [28]. A weak peak at 285 eV has been

TABLE 8.1
Coefficient of Friction at a Load of 40 nN as Determined via AFM for Trichlorosilane SAMs

SAMs system	Applied load (nN)	Coefficient of friction
Bare Si	40.0	0.85
C5	27.0	0.58
C10	8.8	0.19
C12	7.5	0.16
C16	9.6	0.21
C18	14.0	0.31
C30	16.0	0.35

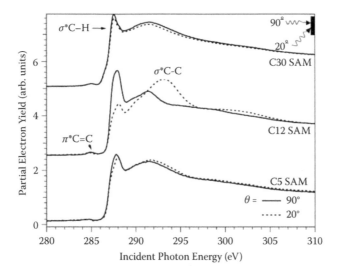

FIGURE 8.1 NEXAFS carbon K-edge partial electron yield spectra for a C_5 trichlorosilane SAM, C_{12} trichlorosilane SAM, and C_{30} trichlorosilane SAM at normal incidence when $\theta = 90°$ (solid curve) and at glancing incidence when $\theta = 20°$ (dashed curve) between E and surface normal.

assigned to the transition of C 1s electrons excited to C-C π^*, which originates due to the presence of small amounts of unsaturation or impurity in the SAM.

All of the PEY NEXAFS spectra presented in this chapter are first normalized by incident photon intensity, and then the background subtraction is performed at the preabsorption edge region (below 275 eV), followed by the postabsorption edge (above 325 eV) normalization, which makes the absorption above 325 eV equal to unity. Under these normalization methods, it is valid to compare the absorption features among SAMs directly, and the changes in the NEXAFS spectra directly relate to the differences in the SAM chain lengths. It is observed that in the case of C_5 and C_{30} trichlorosilane SAMs, the normal incidence ($\theta = 90°$) and glancing incidence ($\theta = 20°$) spectra look very similar. However, the C_{12} SAM shows different C-H and C-C intensities at normal and glancing incidence angles. In the C_{12} system, the C-H resonance peak (287.3 eV) is pronounced when θ is 90° (normal incidence), and the C-C resonance peak (292.7 eV) is pronounced when θ is 20° (glancing incidence). Hence, the resonance intensities exhibit a strong angular dependence on the linearly polarized x-ray angle of incidence θ. The molecular orientation of these SAMs on the surface can be determined from the variation of the C-H and the C-C resonance intensities that correspond to dipoles defined by the spatial orientation of the final state orbital.

Previous NEXAFS studies on *n*-octadecyltrichlorosilane (OTS-C18) on silicon [29] and mica [30] indicate that monolayer films of OTS are highly oriented. The angular dependence of the resonances is indicative of a preferential molecular orientation of the polymeric C-C chains perpendicular to the surface (standing up), and the C-H bond plane is parallel to the surface. It should be noted that the orientation in the form of tilt angle or order calculated from NEXAFS are average values.

With NEXAFS, it is not possible to discriminate between the case of all chains homogeneously tilted by the same angle and the case of a partially disordered system with a broad distribution of tilt angles. Nevertheless, a high degree of molecular orientation indicates a high degree of organizational order in the SAMs, and various chain lengths of SAMs can be compared with one another to draw conclusions about their relative organizational order with confidence.

8.3.3 SURFACE MOLECULAR ORDER DETERMINATION FROM NEXAFS DATA

Resonances are quantified by first subtracting the background corresponding to the excitation edge of carbon, then fitting a series of Gaussian shapes to the spectra, and finally integrating the peak areas using the method proposed by Outka and others [31]. The resulting PEY NEXAFS intensities or peak areas have a $\sin^2(\theta)$ dependence. Figure 8.2 shows the normalized PEY NEXAFS peak intensities vs. $\sin^2(\theta)$ for (A) C_5, (B) C_{12}, and (C) C_{30} SAMs. The open circles represent C-C σ^* intensity, and the filled squares represent the C-H σ^* intensity. Smaller slopes, as seen in the C_5 and C_{30} cases in the NEXAFS intensity vs. $\sin^2(\theta)$, indicate a more random distribution of various alkyl chains, whereas in the C_{12} SAM film, the alkyl chains are highly oriented.

Molecular orientation can be quantified using the dichroic ratio, $R = (I_{90°} - I_{0°})/(I_{90°} + I_{0°})$, where $I_{90°}$ and $I_{0°}$ are, respectively, the NEXAFS peak intensities at 90° and 0° incidence, linearly extrapolated to zero, which is a $\sin^2(\theta)$ value of zero, as shown in fig. 8.2 [32, 33]. The dichroic ratio R can vary between +0.75 and −1.00, where a more positive R value for the C-H σ^* plane indicates increased tilt away from the substrate, while a more negative R for the C-C σ^* axis indicates greater surface normality. The dichroic ratio is a direct measure of molecular orientation and order. Table 8.2 summarizes the dichroic ratios calculated from C-H σ^* resonance for various carbon chain lengths of SAMs investigated in the present study. From the summary of dichroic ratio values, it was observed that C_5 and C_{30} were the most randomly oriented (little order), followed by C_{10}, C_{18}, and C_{16}. Interestingly, C_{12} was the most oriented among all the SAMs. Dichroic ratios measured from C-C σ^* resonances are smaller than those measured from C-H σ^* resonances, which is similar to the result of Hähner et al. [30] and is explained by the probability that C-C resonances are more likely to be affected by Gauche defects. Hähner and coworkers [30] also observed that C_{18} was the most ordered system in the NEXAFS studies of dialkylammonium SAMs on mica. We believe that differences between our work and Hähner's work may be due to different grafting or packing densities of the SAMs. It has been shown by Genzer et al. [34] that grafting densities vary significantly in semifluorinated mono-, di-, and trichlorosilane SAMs on Si. They also observed that the difference in grafting densities resulted in significantly different molecular orders as estimated via NEXAFS. It is well known that the packing within a SAM is caused by a complex interaction of the chemical nature of the SAM and its bonding to the substrate, which influences its molecular organization [35]. The differences in substrate and dialkylammonium binding sites may explain some of the differences observed between Hähner's work and the present study.

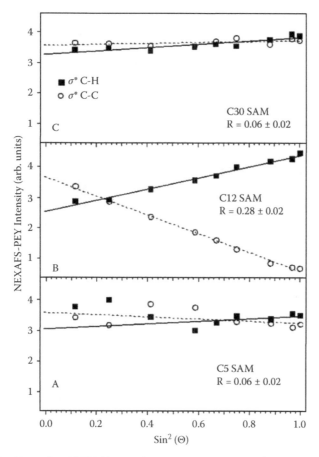

FIGURE 8.2 Normalized PEY NEXAFS peak intensities vs. $\sin^2(\theta)$ for (A) C_5, (B) C_{12}, and (C) C_{30} trichlorosilane SAMs. The open circles represent C-C σ^* intensity, and the filled squares represent the C-H σ^* intensity. Dichroic ratios indicated as R are calculated from the NEXAFS intensity at 90° and the interpolated NEXAFS intensity at 0°.

TABLE 8.2
Dichroic Ratio from C-H σ^* Resonances as Measured via NEXAFS for Trichlorosilane SAMs on Silicon (100)

SAMs system	Dichroic ratio (R)
Bare Si	0.002 ± 0.001
C5	0.07 ± 0.03
C10	0.21 ± 0.05
C12	0.28 ± 0.02
C16	0.24 ± 0.03
C18	0.22 ± 0.03
C30	0.06 ± 0.03

8.3.4 FTIR MEASUREMENTS ON SAMs

Conformation-sensitive regions in the infrared (IR) spectra have been routinely utilized to estimate qualitatively the molecular order in the micrometer range in self-assembled monolayers (SAMs) [36]. The frequency shifts in the IR asymmetric CH_2 and symmetric CH_2 peaks illustrate the changes in conformational order in SAMs when compared with the reference. We applied complementary FTIR spectroscopy to the n-trichlorosilane SAM systems to check the trends in the order qualitatively and to compare with the NEXAFS results.

Figure 8.3 shows the FTIR spectra in the reflection mode for the trichlorosilane SAMs on silicon. These SAMs were prepared in the same way as for the NEXAFS studies. The asymmetric CH_2 band maximum is observed in the range of 2927–2912 cm^{-1}. Symmetric CH_2 stretches are observed at 2853–2846 cm^{-1}. The fingerprinting position of the CH_2 symmetric and antisymmetric stretching peak maxima provide qualitative information about the conformational order of the alkyl chains. For completely disordered and random coil-like structures, the frequency of the CH_2 antisymmetric stretching band is close to that of a liquid alkane ($\upsilon_a = 2924$ cm^{-1}). For well-ordered systems, the frequency is shifted to lower wavenumbers ($\upsilon_a = 2915$–2918 cm^{-1}), which are similar to the absorption frequency of crystalline alkanes. It can be inferred from our FTIR measurements that C_5 film was completely disordered, and C_{10} and C_{12} films were partially ordered. However, from NEXAFS we found that C_{12} was the most ordered system, which also supports past studies on trichlorosilane SAMs on Si [4, 12, 34]. In these studies, it was concluded that C_{12}–C_{18} SAMs were all highly ordered and possessed a nearly similar degree of molecular orientation, within the limits of experimental error. FTIR measurements revealed that C_{16}, C_{18}, and C_{30}

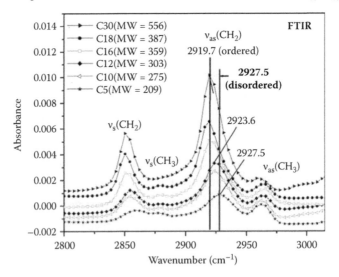

FIGURE 8.3 **(See color insert following page 80.)** FTIR spectra in reflection mode of trichlorosilane SAMs with carbon chain lengths C_5–C_{30} on silicon. The asymmetric CH_2 band maxima are observed in the range of 2927–2912 cm^{-1}. Symmetric CH_2 stretches, observed at 2853–2846 cm^{-1}, are also indicated for various SAMs.

were ordered, but could not distinguish the degree of order between C_{30} and C_{18}, similar to the observation by Srinivasan and coworkers on a titania substrate [36].

FTIR measurements can be seen to complement the NEXAFS technique, which has unequivocally proved to be extremely sensitive and effective in estimating molecular order.

8.3.5 Correlation of Molecular Order and Nanofriction Results

The results from the AFM nanofriction studies and for the dichroic ratio (representing molecular order in SAMs) are plotted as a function of SAM carbon chain length in fig. 8.4. This figure shows the coefficient of friction as a function of carbon chain length (plotted on the right y-axis) and the dichroic ratio (plotted on the left y-axis). The dichroic ratio represents the molecular orientation as calculated from the C-H σ* resonance of SAMs, as measured by NEXAFS. From this graph, it is observed that the optimum (lowest) value of the coefficient of friction for the C_{12} system is primarily caused by the film organization order and the molecular chain alignment, and this optimum also has the highest dichroic ratio. A similar trend in friction measurements was observed by Lio et al. [12] for alkanethiols on gold and alkylsilanes on silica.

Tables 8.1 and 8.2 summarize, respectively, the coefficient of friction measurements from AFM for different trichlorosilane SAMs and the corresponding dichroic ratios obtained from NEXAFS measurements. In both tables, the given values are compared with those for uncoated (bare) silicon. A remarkable correlation is seen in the trends for the order in the dichroic ratio and the coefficient of friction. SAMs with higher dichroic ratios have lower coefficients of friction. Combining the measurements from NEXAFS and FTIR with the AFM measurements of friction coefficient, the C_{12} film emerges as the most ordered system with the lowest friction. This conclusion is also supported by molecular dynamics simulations [37]. Mikulski and

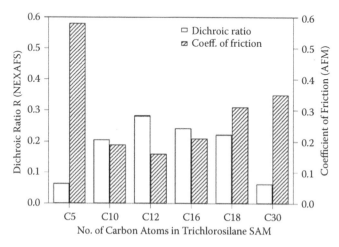

FIGURE 8.4   Coefficient of friction (patterned bars) measured via AFM (right y-axis) and dichroic ratio (blank bars, left y-axis) measured via NEXAFS spectroscopy as a function of carbon chain length of SAMs

coworkers [38] examined the tribological behavior of *n*-alkane monolayer systems using molecular dynamics simulations. Their studies revealed that pure and highly ordered monolayers exhibit lower normal forces relative to mixed alkane monolayers by a carbon tip. Mixed alkane monolayers are relatively more disordered and show additional pulling force, moving the tip toward the surface in comparison with those of ordered monolayers, which show a lower net coefficient of friction. These results, taken together, illustrate that molecular organization is an important factor for controlling friction properties and may indeed be the deciding factor in choosing the most suitable material for coating silicon devices for nanoscale lubrication.

8.3.6 THERMAL STABILITY OF SAM FILMS

Film stability is important in ensuring the durability of a lubricating film during the device manufacturing steps. The packaging temperatures for most of the devices are very high, and survival of the film during the packaging process is a critical design consideration. We performed a series of in situ thermal NEXAFS studies on C_{16} SAM starting from room temperature (25°C) and recorded NEXAFS at $\theta = 55°$, which represents a polarization-independent angle. Upon increasing the temperature to 50°C, 100°C, and 150°C, and then holding these temperatures for 5 min, we observed no loss of SAM film based on NEXAFS peak intensities in the C-C and C-H regions. Furthermore, NEXAFS angular scans were obtained, and the dichroic ratio was calculated at 200°C, 250°C, 300°C, and 350°C. A plot of the dichroic ratio from C-H σ^* resonance with respect to temperature is shown in fig. 8.5. From this study, we observed that the dichroic ratio, and hence molecular order, remained nearly constant up to 200°C and was similar in value to that observed at room temperature. Upon increasing the temperature further to 300°C, the disorder increases, as indicated by the lowering of the dichroic ratio. Upon further heating to 350°C, decomposition of the film occurs, as seen in the NEXAFS peak signature. At this temperature, the dichroic ratio drops to 0.065 ± 0.03, which is similar to the highly

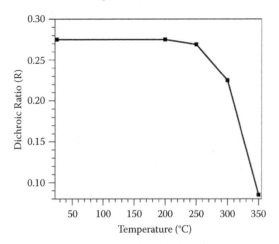

FIGURE 8.5 Temperature dependence of the dichroic ratio calculated from the C-H σ^* NEXAFS intensity for the C_{16} SAMs system.

disordered C_5 or C_{30} films. AFM studies were performed on this degraded sample to estimate the coefficient of friction, which was found to be 0.45, similar to the values seen for C_5 and C_{30} (as seen from Table 8.1) and approaching the value of the coefficient of friction for uncoated (bare) Si. It can be concluded from the preliminary thermal study that, upon heating the SAM films above 200°C, the order of the films begins to degrade, and heating to 350°C results in loss of the film.

8.4 CONCLUSIONS

N-alkyltrichlorosilane self-assembled monolayer films with different chain lengths (C_n films, where $n = 5$–30) were characterized by NEXAFS, FTIR, and AFM to understand the relationship between molecular organization and friction characteristics. NEXAFS results showed that n-alkylsilane chains with 12 to 18 carbon atoms formed extremely ordered systems. The C_{12} SAM film had the highest molecular order and also had the lowest friction coefficient, as measured via AFM. The self-assembled monolayers of C_5 and C_{30} were disordered. The C_{10} SAM film was partially ordered. FTIR spectroscopic measurements revealed similar qualitative order trends that were affirmed by the more quantitative NEXAFS technique. Thermal studies on C_{16} films indicated an increased disorder upon heating, and the films began to degrade at temperatures $\geq$350°C. This study has suggested that the molecular organization of the surfactant film (which is critically dependent on chain length) is an important factor for controlling and predicting frictional properties. The information afforded by these complementary techniques has provided a unique opportunity to interpret the frictional properties of the monolayer films in relation to the molecular assembly and chain length.

ACKNOWLEDGMENTS

Identification of specific commercial equipment, instruments, or materials in this chapter is not intended to imply recommendation or endorsement by the National Institutes of Standards and Technology, nor is it intended to imply that these materials or equipment are necessarily the best available for the stated purposes.

NEXAFS experiments were carried out at the National Synchrotron Light Source, Brookhaven National Laboratory, and were supported by the U.S. Department of Energy, Division of Materials Sciences and Division of Chemical Sciences, under contract no. DE-AC02-98CH10886.

The authors thank Dr. Jan Genzer at NCSU for his suggestions on angular distribution plots. The author SS deeply acknowledges the help of Dr. Dean Delongchamp with the dichroic ratio calculations. All of the authors acknowledge the help of Dr. Zugen Fu with the NEXAFS instrumentation.

REFERENCES

1. Bowden, E. P., and Tabor, D. 1950. *The friction and lubrication of solids.* Oxford: Oxford University Press.

2. Cheng, T., Zhao, B., Chao, J., Meeks, S. W., and Velidandea, V. 2001. *Tribol. Lett.* 9: 181.
3. Hsu, S. M. 2004. *Tribol. Int.* 37: 553.
4. Nuzzo, R. G., and Allara, D. L. 1983. *J. Am. Chem. Soc.* 105: 105.
5. Iler, R. K. 1979. *The chemistry of silica.* New York: John Wiley & Sons.
6. Aviram, A., and Ratner, M. A. 1974. *Chem. Phys. Lett.* 29: 277.
7. Ying, Z. C., and Hsu, S. M. 2004. *Annual Reports, NISTIR* 7124: 12.
8. Hsu, S. M. 2004. *Tribol. Int.* 37: 537.
9. Brewer, N. J., Beake, B. D., and Leggett, G. J. 2001. *Langmuir* 17 (6): 1970.
10. Hsieh, S., Hsu, S. M., and Sengers, J. Submitted to Langmuir.
11. Major, R.C., Kim, H. I., Houston, J. E., and Zhu, X. Y. 2003. *Tribol. Lett.* 14: 237.
12. Lio, A., Charych, D. H., and Salmeron, M. 1997. *J. Phys. Chem. B* 101: 3800.
13. Ulman, A. 1991. *An introduction to ultrathin organic films: From Langmuir–Blodgett to self-assembly.* Boston: Academic Press.
14. Dubois, L. H., and Nuzzo, R. G. 1992. *Annu. Rev. Phys. Chem.* 43: 437.
15. Hähner, G., Kinzel, M., Wöll, C., Grunze, M., Scheller, M. K., and Cederbaum, L. S. 1991. *Phys. Rev. Lett.* 67: 851.
16. Genzer, J., and Efimenko, K. 2000. *Science* 290: 2130.
17. Kraptechov, D. A., Ma, H., Jen, A. K. Y., Fischer, D. A., and Loo, Y.-L. 2005. *Langmuir* 21: 5887.
18. Stöhr, J. 1992. *NEXAFS spectroscopy.* Ed. G. Ertl, R. Gomer, and D. L. Mills. New York: Springer.
19. Fischer, D. A., Hu, Z. S., and Hsu, S. M. 1997. *Tribol. Lett.* 3: 41.
20. Stöhr, J., Samant, M. G., Luning, J., Callegari, A. C., Chaudhari, P., Doyle, J. P., Lacey, J. A., Lien, S. A., Purushothaman, S., and Speidell, J. L. 2001. *Science* 292: 2299.
21. Genzer, J., Sivaniah, E., Kramer, E. J., Wang, J., Körner, H., Char, K., Ober, C. K., DeKoven, B. M., Bubeck, R. A., Fischer, D. A., and Sambasivan, S. 2000. *Langmuir* 16: 1993.
22. Sambasivan, S., Hsieh, S., Fischer, D. A., and Hsu, S. M. 2006. *J. Vac. Sci. Technol. A* 24: 1484.
23. Hsu, S. M. 2003. Ceramics Division Annual Reports. NISTIR 7014, Gaithersburg, MD: NIST.
24. Fischer, D. A., Sambasivan, S., Fu, Z., Yoder, D., Genzer, J., Effimenko, K., Moodenbaugh, A. R., Mitchell, G. M., and DeKoven, B. M. 2003. In *Proc. 8th Int. Conf. Synchrotron Radiation Instrumentation*, San Francisco.
25. Fischer, D. A., and DeKoven, B. M. 1996. *National Synchrotron Light Source Newsletter.* http://www.nsls.bnl.gov/newsroom/publications/newsletters/1996/96-nov.pdf.
26. Hitchcock, A. P., and Brion, C. E. 1980. *J. Electron. Spectrosc. Rel. Phenom.* 18: 1.
27. Xiao, X.-D., Hu, J., Charych, D. H., and Salmeron, M. 1996. *Langmuir* 12: 235.
28. Dhez, O., Ade, H., and Urquhart, S. 2003. *J. Electron. Spectrosc. Relat. Phenom.* 128: 85.
29. Bierbaum, K., Kinzler, M., Wöll, C., Grunze, M., Hähner, G., Heid, S., and Effenberger, F. 1995. *Langmuir* 11: 512.
30. Hähner, G., Zwahlen, M., and Caseri, W. 2005. *Langmuir* 21: 1424.
31. Outka, D., Stöhr, J., Rabe, J. P., and Swalen, J. D. 1994. *J. Chem. Phys.* 88: 4076.
32. Wu, W. L., Sambasivan, S., Wang, C. Y., Wallace, W. E., Genzer, J., and Fischer, D. A. 2003. *Eur. Phys. J. E* 12: 127.
33. Delongchamp, D., Sambasivan, S., Fischer, D. A., Lin, E. K., Chang, P., Murphy, A. R., Fréchet, J. M. J., and Subramanian, V. 2005. *Adv. Mater.* 17: 2334.
34. Genzer, J., Efimenko, K., and Fischer, D. A. 2002. *Langmuir* 18: 9307.
35. Grunze, M. 1993. *Phys. Scr. T* 49B: 711.

36. Srinivasan, G., Pursch, M., Sander, L. C., and Muller, K. H. 2004. *Langmuir* 20: 1746.
37. Chandross, M., Webb III, E. B., Stevens, M. J., and Grest, G. S. 2004. *Phys. Rev. Lett.* 93: 166103.
38. Mikulski, P. T., Gao, G., Chateauneuf, G. M., and Harrison, J. A. 2005. *J. Chem. Phys.* 122: 024701.

9 Mechanism of Friction Reduction by MoDTC/ ZDDP Tribofilms and Associated Nanometer- Scale Controlling Factors

Jiping Ye

ABSTRACT

This chapter demonstrates that macroscopic effects can be clarified from the perspective of nanoproperties and microscopic phenomena by evaluating quantitatively and qualitatively the mechanical/frictional/structural properties on a nanometer scale. A prime example is given by showing how to explain the difference in the macro-scale frictional behavior between the MoDTC/ZDDP and ZDDP tribofilms originating from the MoDTC and ZDDP surfactants, which are well-used engine oil additives in automotive field. The primary emphasis is on how to combine nanoprobing techniques such as AFM phase image method, nanoindentation and nanoscratch methods with beam-based analytical methods such as AES and XPS to determine the nanometer-scale controlling factors involved in reducing friction.

9.1 INTRODUCTION

Tribology is the science and technology concerning the interactive behavior between two contacting surfaces in repetitive relative motion. Surfactants, structural coatings, surface chemical modifications, and/or changes in the lubrication environment make it possible to form new nanometer-scale materials on sliding surfaces, which exhibit novel tribological behavior or ultrahigh tribological performance. The nanostructure, nanoproperties, and nanofunctions of surface materials have attracted considerable attention, inasmuch as they act as extremely important factors in controlling macroscale tribological performance. For the purpose of elucidating these controlling factors and their correlations, it is indispensable to obtain direct surface information, including various structural and physical properties or characteristics as well as the chemical state on a nanometer scale [1–3]. Conventional analytical techniques for acquiring such information, however, generally use various kinds of beams. These include optical light beams (optical microscopy, Raman spectroscopy, infrared spectroscopy, etc.), x-ray beams (x-ray photon electron spectroscopy [XPS], x-ray diffraction, etc.), electron beams (electron microscopy, Auger electron spectroscopy [AES], electron probe microanalysis spectroscopy [EPMA], etc.), and ion beams (secondary ion mass spectrometry [SIMS], Rutherford backscattering spectrometry [RBS], and particle-induced x-ray emission spectrometry [PIXES]) [1–3]. These beam-based analytical methods can provide information on the morphology, composition, texture, microstructure, crystal structure, and chemical structure of surface materials. Because it is difficult for these methods to provide direct information about physical properties in a small analysis area, indirect analytical information based on prior knowledge or past experience generally has to be used for explaining material phenomena.

In recent years, scanning probe microscopy (SPM) has prompted an intense interest not only in directly obtaining the three-dimensional morphology, but also in obtaining various physical properties on a nanometer scale. With SPM, physical properties correspond to the interaction force between the sample surface and a sharp probing tip. The interaction force increases greatly as the tip approaches the surface and comes in contact with it, or decreases as the tip moves away from the sample surface. As shown in fig. 9.1, atomic force microscopy (AFM) and nanoindentation make use of the same principle, i.e., the probe–surface interaction changes depending on the distance from the surface; the measurement operation can be controlled by detecting the distance when varying the interaction or by detecting the interaction when varying the distance [1]. Thus, AFM and nanoindentation can be regarded as sibling tools in the SPM family. This nanoprobing principle enables SPM to measure various mechanical, electrical, magnetic, thermal, and optical properties of all kinds of solid-state materials, including metals, semiconductors, ceramics, polymers, as well as structural or functional materials. These nanometer-scale properties, referred to here as analytical properties, constitute the direct analytical information inherent to microstructural differences. Analytical properties can act as nanometer-scale controlling factors and can serve as the basis for verifying phenomena presumed with conventional beam-based analytical methods. In tribological studies in particular, analytical properties represent local mechanical/structural values on a nanometer scale without any influence from the surface roughness. Thus, probing techniques are powerful tools for ascertaining the real nature of sliding surface materials.

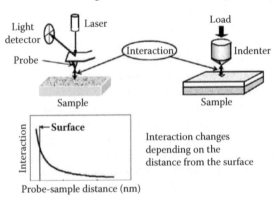

FIGURE 9.1   Schematic of a probe microscope.

In the tribological field, for example, demands for improving the fuel economy of vehicles have been increasing steadily in recent years because of environmental concerns and the need for energy conservation. Improvement of engine oil performance is particularly important because of its critical role in reducing friction, which translates directly into better fuel economy. One effective method of modifying engine oil containing the widely used zinc dithiophosphate (ZDDP) additive is to add the molybdenum dithiocarbamate (MoDTC) friction modifier [4–10], which is a well-used surfactant in the automotive field. The MoDTC additive was found to be a pressure-resistant and antiwear additive for engine oil in the 1950s, and its low frictional characteristic was discovered in the 1970s. As a result, the MoDTC additive was recognized in the 1980s as the most important friction modifier of the 20th century and went into commercial production in the 1990s.

It is well known that engine oil additives form tribochemically reacted films (often called tribofilms) on sliding steel surfaces. Many reports have revealed that a MoDTC/ZDDP tribofilm originating from both the MoDTC and ZDDP additives possesses a much lower friction coefficient than a ZDDP tribofilm formed only from the ZDDP additive. Since both tribofilms form on sliding steel surfaces under boundary lubrication but exhibit obviously different friction behaviors [9, 10], the source of this difference has attracted considerable attention not only from the standpoint of industrial value, but also with regard to scientific understanding. In the past two decades, various extensive studies have been made to elucidate why the MoDTC/ZDDP tribofilm shows a lower level of friction than the ZDDP tribofilm alone [11–15]. However, these numerous studies have all been based on indirect information obtained with beam-based analytical methods. So far, it was thought to be experimentally difficult to estimate directly the nanostructure and nanometer-scale mechanical properties of tribofilms having a thickness of only a few tens of nanometers. That is why the mechanisms of the friction reduction presumed based on the beam-based analytical methods are still unclear and lack a unified consensus.

In this chapter, we are concerned with the difference in the macroscale frictional behavior between the MoDTC/ZDDP and ZDDP tribofilms formed on sliding steel surfaces under boundary lubrication. We describe several novel and practical SPM techniques, such as the AFM phase image method and the nanoindentation and nanoscratch methods. The primary emphasis is on how to apply these methods to determine nanometer-scale structural and mechanical properties. By combining a nanoprobing technique with a beam-based analytical method, we explain how to determine the nanometer-scale controlling factors involved in reducing friction. The relationships between nanometer-scale properties, micrometer-scale phenomena, and macroscale frictional effect are explained by elucidating the friction reduction due to the MoDTC additive.

9.2 MICROSTRUCTURE AND PRESUMPTIONS FROM BEAM-BASED ANALYSIS

9.2.1 MACROSCALE FRICTIONAL BEHAVIOR

As mentioned above, tribofilms form from oil additives on sliding steel surfaces during sliding tests conducted in engine oil. In the research laboratory, MoDTC/ZDDP and

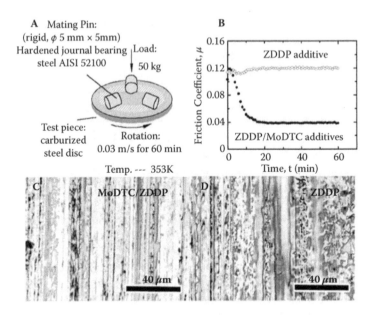

FIGURE 9.2 (A) Schematic of pin-on-disc test rig and test condition, (B) friction coefficients of the MoDTC/ZDDP and ZDDP tribofilms, and (C) and (D), respectively, morphologies of the ZDDP and MoDTC/ZDDP tribofilms on the discs.

ZDDP tribofilms are usually formed with a pin-on-disc test rig like that shown in fig. 9.2A [10]. The tribofilms discussed in this chapter were formed at 353 K under pressure of 0.7 GPa and at a sliding speed of 0.03 m/s during a 60-min test in engine oil. The pins were made of hardened journal bearing steel AISI 52100, and the disks were made of carburized steel. The disc and pin surfaces were superfinished to roughness levels of R_a 10 nm and R_a 40 nm, respectively. Two different kinds of engine oils were used. One was SAE 5W-30 grade oil with the ZDDP antiwear additive, and the other was 5W-20 grade oil with the ZDDP additive and the MoDTC friction modifier. The friction coefficients of the MoDTC/ZDDP and ZDDP tribofilms are shown in fig. 9.2B. As shown in figs. 9.2C and 9.2D, MoDTC/ZDDP and ZDDP tribofilms formed on the steel discs. The friction coefficient of the MoDTC/ZDDP tribofilm decreased markedly from 0.12 to 0.04, while that of the ZDDP tribofilm increased from 0.11 to 0.12 [10].

9.2.2 MICROSTRUCTURES

Recent analytical studies of the sliding surface have suggested that the ZDDP tribofilm forms a graded composite, multilayer structure on a steel surface that is usually very thin, with a thickness of less than 100 nm [16–23]. Figure 9.3 shows the typical structure of the ZDDP tribofilm, which consists of at least four to five layers from the surface to the steel base. Amorphous polyphosphate base layers are presumed to dominate the bulk mechanical properties of the ZDDP tribofilm, and a sulfide layer near the surface controls the friction property of the film.

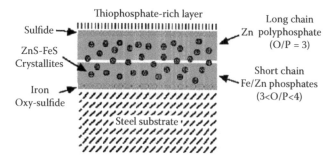

FIGURE 9.3 Schematic of the typical microstructure of the ZDDP tribofilm based on previous studies [16–23].

9.2.3 TWO MAIN PRESUMPTIONS

There are two main presumptions regarding the mechanism of the friction reduction attributed to the MoDTC additive [11–15]. One assumes a difference in tribofilm materials, and the other concerns morphological differences such as surface roughness. As for the former presumption, many previous studies reported that MoS_2 derived from the MoDTC additive was detected in the MoDTC/ZDDP tribofilm and was also observed in some wear fragments collected from the residual oil present on the wear scar [11–14]. Because MoS_2 has a layered structure bonded by weak van der Waals forces, it is thought to act as a lamellar solid lubricant that lowers the friction force. With regard to the latter presumption, the MoDTC additive was reported to have an effect on smoothing the sliding surface [15]. This suggests that it may reduce friction by promoting the formation of a hydrodynamic or elastohydrodynamic film.

These presumptions are based on the results of chemical composition analyses and structural determinations by beam-based analytical methods. To verify the mechanisms involved, it is essential to clarify the frictional and mechanical properties of each layer in the multilayered structure of the tribofilm. Most importantly, with respect to the uppermost surface and the underlying area, which are thought to control the macroscale friction behavior, it is indispensable to determine differences not only in surface roughness, but also in shear strength and frictional behavior relative to depth on a nanometer scale. It is also important to make clear the distribution of the chemical composition relative to the tribofilm depth and to make this chemical distribution consistent with the nanometer-scale frictional and mechanical properties. However, estimating these properties experimentally on a nanometer scale is a difficult task, although several attempts have been made [24–28].

9.3 NANOMETER-SCALE CONTROLLING FACTORS

During this past decade, SPM has been widely used as a technique for determining nanometer-scale mechanical properties. However, only a few applications of this method have so far been reported for studying tribofilms. Norton and coworkers investigated the use of an interfacial force microscope (IFM) for examining the difference in modulus and shear strength among various ZDDP tribofilms based on force-displacement curves [24–25]. Bec and coworkers used a surface-force apparatus (SFA) to measure the mechanical properties of ZDDP tribofilms before and after washing them with n-heptane and proposed that mechanical properties were related to the tribofilm structure [26–27]. Aktary and coworkers also measured the modulus and hardness of ZDDP tribofilms for different rubbing times and suggested that the large variations observed in modulus and hardness were likely due to the heterogeneity in contact pressure at the pin–sample interface [28]. However, these results varied considerably with relatively large error levels because the measurements apparently did not take into account nanomechanical property differences along the depth of the tribofilms or the heterogeneous thickness distribution of the films on the steel substrate. The reasons for the differences in friction behavior between the MoDTC/ZDDP and ZDDP tribofilms are still not clear [29].

In contrast, over these past several years, Ye and coworkers demonstrated that the nanoprobing SPM technique combined with beam-based analytical methods could determine the nanometer-scale differences in tribological properties between the MoDTC/ZDDP and ZDDP tribofilms [1–3, 30–34]. They systematically measured and analyzed the differences in the nanometer-scale mechanical properties and the nanostructural properties between these two tribofilms, and the results revealed the real reason for the friction reduction due to the MoDTC additive. In this section, Ye's approach is described. For the probing techniques, nanoindentation and nanoscratch measurements were used to quantify mechanical properties such as modulus, hardness, and friction coefficient relative to the tribofilm depth as well as shear strength near the tribofilm surface [1–3, 30–32]. AFM height image and AFM phase image techniques were used to detect surface roughness and surface nanostructure differences [33–34]. Meanwhile, AES and XPS were employed as the beam-based analytical methods to obtain depth profiles of the chemical composition and surface chemical state, respectively [3, 34]. The nanometer-scale properties thus obtained in terms of surface roughness, nanostructure, nanomechanical properties, nanofriction coefficient, as well as nanocomposition and chemical state were shown to be the nanometer-scale controlling factors for elucidating the mechanism of the friction reduction, as described below.

9.3.1 TOPOGRAPHIC IMAGES AND SURFACE ROUGHNESS

AFM topographic images of the tribofilms were obtained in the tapping mode using a Si cantilever with a spring constant of around 50 N/m and a resonance frequency of around 350 kHz [1–3]. Figure 9.4 shows typical AFM topographic images of the MoDTC/ZDDP and ZDDP tribofilms. Some sliding grooves were observed on the sliding surfaces. The sliding grooves on the MoDTC/ZDDP tribofilm were much deeper than those on the ZDDP tribofilm. This indicates that the MoDTC/ZDDP film was obviously much rougher than the ZDDP tribofilm. The mean roughness was R_a 21 nm for the ZDDP film, but 78 nm for the MoDTC/ZDDP film, although both sliding surfaces became rougher than their initial values. Moreover, the regions other than the sliding grooves showed no significant difference in the mean roughness between the MoDTC/ZDDP and ZDDP films, with the mean roughness being around R_a 20 nm. Thus, no evidence was found to show that the MoDTC/ZDDP

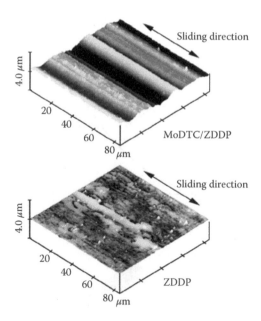

FIGURE 9.4 Typical AFM topographic images of the MoDTC/ZDDP and ZDDP tribofilms.

additives promoted the formation of a hydrodynamic or elastohydrodynamic film that would smooth the sliding surface to reduce friction.

9.3.2 NANOMECHANICAL PROPERTIES AND NANOFRICTION COEFFICIENT

9.3.2.1 Nanoindentation and Nanoscratch Measurements

Nanoindentation and nanoscratch measurements were taken to determine hardness and modulus distributions with respect to depth as well as the shear strength and friction coefficient of the tribofilms on a nanometer scale [1, 32]. Both measurements were combined with in situ AFM observations to find a flat film area less influenced by roughness and to confirm the indent or scratch made on the tribofilms from topographic images after the test. In these observations, the same diamond tip was used as the stylus for indentation or scratching as well as the AFM probe for obtaining in situ AFM images.

A Berkovich diamond tip with a total included angle of 142.3° and a radius of around 150 nm was used for the nanoindentation measurements [1–2]. Indentation load–displacement curves were obtained by applying loads ranging from 1 μN to 1 mN. The hardness and reduced elastic modulus of the tribofilms were determined with Oliver's method [35, 36], where fused silica with a Young's modulus of 69.7 GPa was used as a standard sample for tip-shape calibration to determine the function of the contact area with respect to the contact depth in a range of 1.5–50 nm. Figure 9.5 shows indentation load–displacement curves obtained for the MoDTC/ZDDP and ZDDP tribofilms at a maximum load of 600 μN and in situ AFM images of the residual indent. A plastic pileup was clearly observed around the indent on both the MoDTC/ZDDP and ZDDP tribofilms.

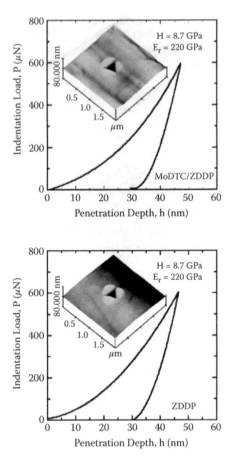

FIGURE 9.5 Indentation load–displacement curves of the MoDTC/ZDDP and ZDDP tribo-films at a maximum load of 600 µN and in situ AFM images of the residual indent. No significant differences in hardness and modulus between the MoDTC/ZDDP and ZDDP tribofilms were observed under this loading condition.

Nanoscratch measurements were obtained in a constant-force nanoscratching process using a conical diamond tip as a stylus for scratching [2, 32]. The nanoscratching process consisted in applying a constant normal force by pressing the diamond stylus on the sample surface and then displacing the stylus to induce a lateral scratch on the surface. By detecting the lateral force and normal displacement, the nanofriction coefficient with respect to the scratch depth was determined on the basis of the ratio of the lateral force to the normal force. In this scratching process, a constant normal force ranging from 10 to 1000 µN and a scratch displacement ranging from 1 to 3 µm within 60 s were applied. Figure 9.6 shows nanoscratch data for the MoDTC/ZDDP tribofilm at a constant normal force of 1000 µN and an in situ AFM image of the residual scratch. The scratching conditions, consisting of the introduced normal force P_n and lateral scratch displacement d_s, are plotted in fig. 9.6a as a function of time t. The obtained lateral force and normal displacement (so-called

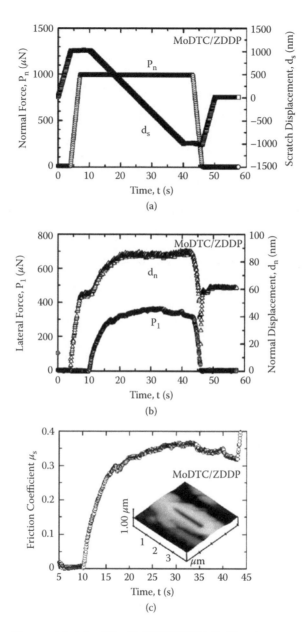

FIGURE 9.6 Nanoscratch data of the MoDTC/ZDDP tribofilm at a constant normal force of 1000 μN. (a) Normal force vs. time and scratch displacement vs. time, (b) lateral force vs. time and normal displacement vs. time, and (c) friction coefficient vs. time and in situ AFM image of the residual scratch groove.

scratch depth) with respect to time t are shown in fig. 9.6b, and the nanofriction coefficient μ_s is shown in fig. 9.6c. Since small deviations in lateral force, normal displacement, and nanofriction coefficient were observed in a time range from 25 to 40 s, a friction coefficient of 0.353 at a scratch depth of 85 nm was determined for the MoDTC/ZDDP tribofilms from their average values. Thus, the nanofriction coefficient distribution relative to the scratch depth was determined by scratching the sample surface at various levels of constant normal force.

9.3.2.2 Hardness/Modulus and Nanofriction Coefficient vs. Depth

Figure 9.7 shows the hardness and modulus distributions of the MoDTC/ZDDP and ZDDP tribofilms, carburized steel disc, and fused silica as a function of the contact depth [1]. Since the fused silica as the standard sample for tip-shape calibration retained constant hardness and modulus values in a contact depth range from 5 nm, the hardness and modulus of the tribofilms were examined at contact depths greater than 5 nm. It was found that both kinds of tribofilms had the same hardness and modulus distributions relative to the contact depth. When the contact depth was greater than 30–40 nm, the mechanical properties were constant, with hardness staying at the fused silica level of 10 GPa and the modulus at the carburized disc level of 215 GPa. However, when the contact depth was reduced from 30 nm to 5 nm near the film surface, film hardness decreased from the fused silica level of 10 GPa to 6 GPa, and the modulus decreased from the carburized steel level of 215 GPa to 150 GPa.

As shown in fig. 9.8, both the MoDTC/ZDDP and ZDDP tribofilms displayed the same friction coefficient distributions relative to the scratch depth [32]. The friction coefficient retained a high constant value of 0.35 when the scratch depth was greater than 60 nm. However, when the scratch depth was around 10 nm near the film surface,

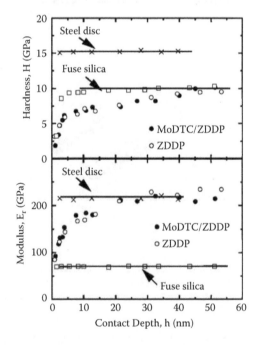

FIGURE 9.7 Depth distributions of hardness and modulus in the MoDTC/ZDDP and ZDDP tribofilms, carburized disc, and fused silica.

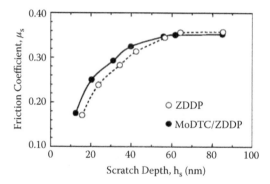

FIGURE 9.8 Depth distributions of nanofriction coefficient in the MoDTC/ZDDP and ZDDP tribofilms.

the friction coefficient decreased sharply to 0.16. Thus, the friction coefficient distribution relative to depth agreed with the hardness and modulus distributions as a function of the film depth.

These results indicated that the MoDTC/ZDDP and ZDDP tribofilms possessed the same mechanical and frictional property distributions relative to the film depth except near the tribofilm surface. Their hardness, modulus, and friction coefficient became smaller with decreasing film depth. Most importantly, there was no evidence that the friction reduction due to the MoDTC additive could be attributed to these properties at a depth greater than 10 nm.

9.3.2.3 Shear Yield Stress and Friction Behavior near the Surface

For the purpose of detecting differences in elastoplasticity in the film surface region, indentation measurements were made at ultralow loads [1]. Figure 9.9 shows the indentation load–displacement curves of the MoDTC/ZDDP and ZDDP tribofilms, where the dark and gray circles denote the loading and unloading data, respectively. The maximum load P_{max} was applied in the range of 4 to 11 μN for both types of tribofilms. The MoDTC/ZDDP tribofilm exhibited plastic deformation at a P_{max} value of 6 μN. In contrast, significant plastic deformation began to appear at $P_{max} = 11$ μN for the ZDDP tribofilm. Thus, the initial P_{max} level causing plastic deformation obviously differed between the MoDTC/ZDDP and ZDDP tribofilms.

For further analysis, all the loading and unloading data presented in fig. 9.9 were traced in $P \approx h^{3/2}$ and $P \approx (h - h_p)^{3/2}$ by the least squares method, where P, h, and h_p are the indentation load, displacement, and plastic depth, respectively. Reliability factors for all the approximated curves were in the range of 0.95–0.99. The proportional exponent 1.5 confirmed that the indentation contact area was in the spherical regime before the contact area in the pyramidal regime at high indentation loads [37]. By applying the Hertz contact theory under the Tresca yield criterion and assuming a Poisson's ratio of $v = 0.3$ [37–39], the maximum shear yield stress was estimated to be $\tau_{max} = 0.46\, p_m$, and it occurred at a depth of $h_\tau = 0.47\, R$ from the contact surface, where p_m and R refer, respectively, to the initial

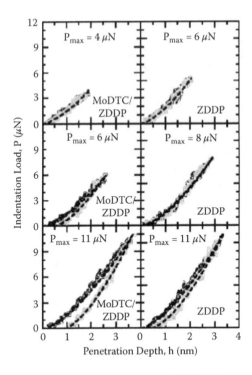

FIGURE 9.9 Load–displacement curves of the MoDTC/ZDDP and ZDDP tribofilms at low-indentation loads.

maximum indentation pressure causing plastic deformation and the radius of the contact area. In this calculation, a tip radius of 150 nm was used. As a result, the MoDTC/ZDDP tribofilm was calculated to possess a shear yield stress of $\tau_{max} = 2.3$ GPa at a depth of $h_\tau = 9.3$ nm, while the ZDDP tribofilm had $\tau_{max} = 3.3$ GPa at $h_\tau = 10.4$ nm.

With respect to the nanofriction coefficient in the surface region at a depth of less than 10 nm, as shown in fig. 9.10, both tribofilms showed a reduction in the initial friction coefficient to its minimum value and exhibited a valley-shaped nanofriction coefficient distribution [32]. However, the minimum nanofriction coefficient of the MoDTC/ZDDP tribofilm was 0.084 at a scratch depth of 4.2 nm, which was lower than the corresponding minimum value of 0.104 for the ZDDP tribofilm at a depth of 6.0 nm. Because plastic deformation was observed when the tribofilm surface was indented under the same load conditions as the nanoscratching process, the initial reduction of the friction coefficient was confirmed to be due to yielding in the surface region. The different frictional behaviors of the MoDTC/ZDDP and ZDDP tribofilms in the surface region agreed with the above-mentioned ultralow-load nanoindentation measurements. That is, the lower yield strength near the surface of the MoDTC/ZDDP tribofilm caused the nanofriction coefficient to decline to a lower value than in the case of the ZDDP tribofilm.

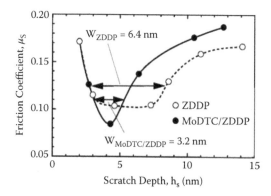

FIGURE 9.10 Depth distribution of nanofriction coefficient in the surface region at a depth of less than 15 nm. The half-value width of the MoDTC/ZDDP tribofilm was about 3.2 nm and was much narrower than the corresponding value of 6.4 nm for the ZDDP tribofilm.

9.3.3 NANOSTRUCTURAL AND LOCAL MECHANICAL PROPERTIES IN THE SURFACE REGION

9.3.3.1 AFM Phase Images

The local mechanical properties of different components and their mechanical property differences relative to depth in the surface area of the tribofilms were determined by an AFM phase image technique [33–34]. This technique is performed in the tapping mode, in which the sample surface is tapped by an oscillating AFM tip. The tip–sample interaction force can be adjusted with the amplitude A_0 of the freely oscillating cantilever and the amplitude A_f of the interacting cantilever [40–41]. In light tapping with a small A_0 and an A_0/A_f ratio close to 1.0, as shown in fig. 9.11, the height image reproduces the surface topography and the phase image obtained shows individual components on the uppermost surface. In hard tapping with a large A_0 and an A_0/A_f ratio deviating from 1.0, the material surface may be mechanically deformed, causing a change in the surface topography in the height image and revealing the individual components with different mechanical properties below the surface. In this experiment, AFM phase images were also observed using a Si cantilever at a resonance frequency of around 280 kHz. The cantilever used had a spring constant of approximately 50 N/m, a tip radius of less than 5 nm, and a half-cone angle of less than 10°. The tip–sample interaction force was adjusted by increasing A_0 from 660 to 850 mV at a constant A_0/A_f of 0.95, where A_0 was expressed in millivolts instead of nanometers.

9.3.3.2 Nanostructure and Local Elasticity

Figure 9.12 shows height and phase images recorded for the MoDTC/ZDDP tribofilm under light and hard tapping conditions in the elastic deformation range [33–34]. These images were obtained at $A_0 = 660$ mV and $A_0 = 850$ mV, respectively, while keeping A_0/A_f constant at 0.95. At $A_0 = 660$ mV, the height image revealed a smooth topography and the phase image did not exhibit well-defined features. However, the

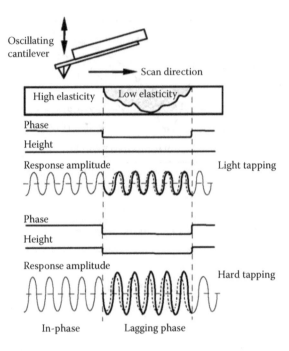

FIGURE 9.11 AFM phase imaging technique for determining the nanostructures and mechanical properties on a nanometer scale by adjusting the tip–sample interaction force.

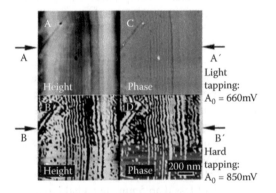

FIGURE 9.12 AFM height and phase images of the MoDTC/ZDDP tribofilm obtained under light and hard tapping conditions: (A) height and (C) phase at $A_0 = 660$ mV and $A_0/A_f = 0.95$; (B) height and (D) phase at $A_0 = 850$ mV and $A_0/A_f = 0.95$. The arrows A-A′ and B-B′ indicate the line scans in fig. 9.13.

height and phase images dramatically changed as a result of increasing A_0. Many nanostrips oriented in the sliding direction of the pin-on-disc test were observed with dark contrast in the height image and with bright contrast in the phase image at $A_0 = 850$ mV. Figure 9.13 shows the height and phase profiles at the sections denoted by the arrows A-A′ and B-B′ in fig. 9.12. In hard tapping at $A_0 = 850$ mV, the nanostrips were tapped at a height of approximately 10 nm below the uppermost surface and

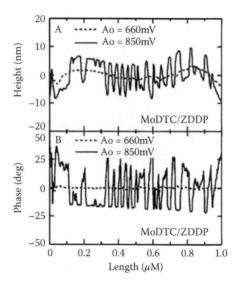

FIGURE 9.13   AFM height and phase line scans of the MoDTC/ZDDP tribofilm at the sections indicated in fig. 9.12.

exhibited a width of approximately 20 nm while the phase difference increased to approximately 40°, as compared with the height and phase images obtained in light tapping at A_0 = 660 mV. These observations indicated that nanostrips were present near the tribofilm surface and possessed material properties different from the surface material around the nanostrips. The nanostrips were more easily mechanically deformed, and their modulus was lower than that of the surface material.

The height and phase images of the ZDDP tribofilm obtained at light and hard tapping conditions in the elastic deformation range are shown in fig. 9.14 [33–34]. Line scans denoted by the arrows C-C′ and D-D′ in fig. 9.14 are shown in fig. 9.15. Since the AFM interacting cantilever vibrated and became unstable when A_0 was increased to values larger than 760 mV at a constant A_0/A_f of 0.95, the hard tapping condition was raised to A_0 = 760 mV, while the light tapping condition was kept at A_0 = 660 mV, the same as that for the MoDTC/ZDDP tribofilm. When the tapping condition was changed from light to hard, as shown in figs. 9.14 and 9.15, no significant difference in the height and phase images or profiles of the ZDDP tribofilm was observed as compared with those seen for the MoDTC/ZDDP tribofilm. It was found that some rings and spots with bright contrast in the height images and with low contrast in the phase images were homogeneously dispersed in the surface material. The sizes of the spots and rings ranged from a few nanometers to 180 nm in diameter. These results indicated that the mechanical properties did not differ very much between the top-surface and the near-surface materials.

9.3.3.3 Nanostructure and Local Elastoplasticity

For the purpose of investigating the mechanical properties of the nanostrips in more detail, the tip–sample interaction force was adjusted by further increasing A_0, causing the surface material of the MoDTC/ZDDP tribofilm to change from elastic to

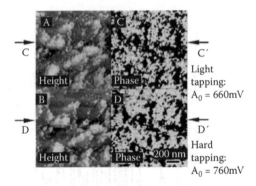

FIGURE 9.14 AFM height and phase images of the ZDDP tribofilm obtained under light and hard tapping conditions: (A) height and (C) phase at $A_0 = 660$ mV and $A_0/A_f = 0.95$; (B) height and (D) phase at $A_0 = 760$ mV and $A_0/A_f = 0.95$. The arrows C-C′ and D-D′ indicate the line scans in fig. 9.15.

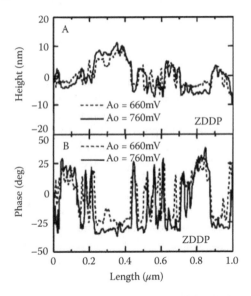

FIGURE 9.15 AFM height and phase profiles of the ZDDP tribofilm at the sections indicated in fig. 9.14.

plastic deformation [33]. Figure 9.16 shows height and phase images for the MoDTC/ZDDP tribofilm when A_0 was increased from 660 mV to 900 mV and then decreased to 850 mV at a constant A_{sp} of 0.95. The corresponding line scans denoted by 1-1′ to 5-5′ in fig. 9.16 are shown in fig. 9.17. At $A_0 = 660$ mV, as stated above, no significant features were observed in either the height or phase images (figs. 9.16A and 9.16F; figs. 9.17A and 9.17F). At $A_0 = 760$ mV, several oriented nanostrips appeared with high contrast in the phase image (fig. 9.16G) at a lower location than in the height image (fig. 9.16B), and the corresponding signals are indicated by the numbers 1 and 8 in fig. 9.17. The number of observable nanostrips increased simultaneously with the contrast with increasing A_0, and both reached their maximum values at $A_0 = 850$ mV

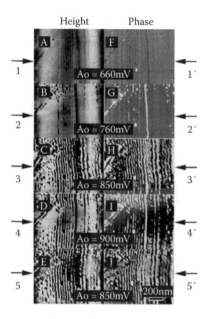

FIGURE 9.16 AFM height and phase images of the MoDTC/ZDDP tribofilm when A_0 was increased from 660 mV to 900 mV and then decreased to 850 mV at constant $A_{sp} = 0.95$. The arrows from 1-1' to 5-5' indicate the line scans in fig. 9.17.

(figs. 9.16C and 9.16H; figs. 9.17C and 9.17H). However, the contrast in the height and phase signals from most of the nanostrips decreased and even disappeared when A_0 was continuously increased to $A_0 = 900$ mV (figs. 9.16D and 9.16I; figs. 9.17D and 9.17I). The average compressive displacement was 6–7 nm at $A_0 = 900$ mV, which was smaller than the value of 10 nm observed at $A_0 = 850$ mV. These observations confirmed that nanostrips with low modulus were present at a location not on the uppermost surface or far from the surface but just near the top surface.

Furthermore, when the same area was scanned again with A_0 returned to 850 mV, some permanent deformations appeared in both the height and phase images (figs. 9.16E and 9.16J; figs. 9.17E and 9.17J). It was observed that the height of the surface material around the nanostrips was elevated, but the nanostrips remained with irreversible deformations (figs. 9.16E and 9.17E). Meanwhile, the phase difference between the nanostrips and surrounding surface material, as shown in figs. 9.16J and 9.17J, became slightly less distinct and relatively smaller as compared with the phase image and profile in figs. 9.16H and 9.17H. However, when A_0 was continuously decreased to less than $A_0 = 850$ mV, the AFM cantilever no longer tapped the sample surface and no AFM image was obtained. This indicated that the surface mechanical deformation compressed by the AFM cantilever could not be recovered completely with a decrease of A_0 due to the irreversible deformation of the nanostrips. Therefore, these height and phase images at the different tip–sample interactive conditions revealed that the nanostrips exhibited plastic deformation at $A_0 = 900$ mV, although the surrounding surface material still showed elastic deformation at the same A_0.

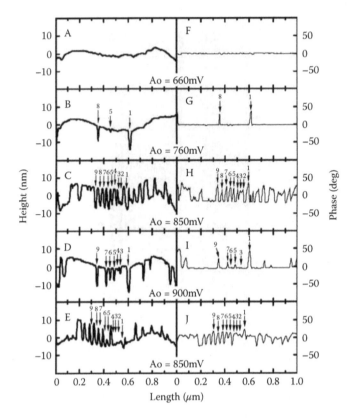

FIGURE 9.17 AFM height and phase line scans of the MoDTC/ZDDP tribofilm as indicated in fig. 9.16. The numerals indicate the height and phase signals from the nanostrips.

The AFM phase images revealed that the nanostrips were just under the tribofilm surface and were more easily mechanically deformed because of their lower hardness than that of the surrounding surface material. Because the height and phase contrasts from most of the nanostrips decreased and even disappeared when A_0 was continuously increased to $A_0 = 900$ mV, the nanostrips were found to be present at a depth of around 10 nm from the top surface. Moreover, because the nanostrips exhibited irreversible deformation, they were concluded to have lower yield strength than that of the surface material around the nanostrips.

9.3.4 COMPOSITION AND CHEMICAL STATE

9.3.4.1 AES and XPS Analyses

Compositions relative to the depth of the tribofilms were analyzed by AES, in which secondary electrons were used for imaging the tribofilm surface to search for a flat test area [34]. In this experiment, an electron gun with a minimum electron beam diameter of 100 nm was operated at a beam energy of 5 keV. The electron beam scanning area for analysis was 1×2 µm. An Ar ion beam with 1 keV of energy was used for sputtering the sample surface to obtain a depth profile of the composition.

Meanwhile, the chemical binding state of the nanostrips near the surface of the tri-bofilms was examined by XPS analysis. Photoelectrons were excited by monochro-matic Al-Kα (1486.6 eV) and detected with a concentric hemispherical analyzer. The x-ray tube was operated at 40 W. The pass energy of Concentric Hemispherical Analyzer (CHA) was 58.7 eV for scan spectra. The analyzed area was 20 × 20 μm and was located in the center of the irradiated area. The take-off angle was 45°. The binding energies of the spectra were calibrated by setting the C1s peak at 284.6 eV.

9.3.4.2 Depth-Related Composition and Surface Chemical State

Figure 9.18 shows AES sputter-depth profiles of the MoDTC/ZDDP and ZDDP tribo-films [34]. Oil additive elements such as Zn, Mo, O, S, P, and Ca, and other elements such as Fe from the steel disk, were observed in these two tribofilms. Both tribofilms tended to have higher concentrations of Ca, O, P, and S in the earlier sputtering period, and then these elements began to decrease and disappeared in the latter sputtering period, while the Fe concentration increased from an ultralow level to the level of the steel disk. Most importantly, Mo was not detected from the ZDDP tribofilm but only from the MoDTC/ZDDP tribofilm. It was found that Mo was not distributed over a deep range but only in the surface region. In this AES analysis, secondary electron images showed that the surface material was immediately melted and damaged, as the elec-tron beam was incident on the tribofilm surface. Thus, because of the electron-beam damage, in addition to a limited escape depth of 3 nm for Auger electrons, no differ-ences in composition were observed between the uppermost surface and the underly-ing area within a depth of several nanometers. Figure 9.19 shows an XPS spectrum of

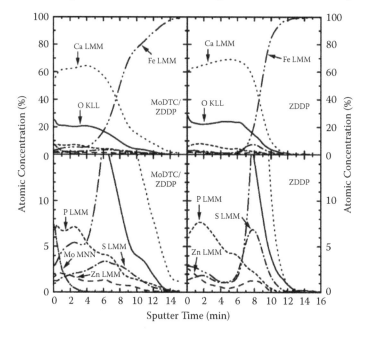

FIGURE 9.18   AES sputter-depth profiles of the MoDTC/ZDDP and ZDDP tribofilms.

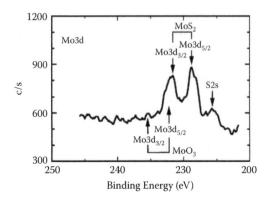

FIGURE 9.19   XPS spectrum of Mo3d from the MoDTC/ZDDP tribofilms.

Mo3d from the MoDTC/ZDDP tribofilms [34]. It was observed that the Mo3d3/2 and Mo3d5/2 peaks were far from the binding energy of MoO$_3$, but subsequently shifted and reached the binding energy level of MoS$_2$. Thus, Mo near the surface was identified to be combined with S to form a compound of MoS$_2$.

9.4 FUNCTIONAL STRUCTURE AND FRICTION-REDUCTION MECHANISM

The aforementioned nanoprobe and beam-based analyses results made clear the differences in surface roughness, mechanical/frictional properties, surface nanostructure, chemical composition, and surface chemical state on a nanometer scale between the MoDTC/ZDDP and ZDDP tribofilms. These results demonstrated that the nanostructural, nanometer-scale mechanical and chemical properties acted as the controlling factors in the macroscale friction behavior of the tribofilms. From these results, a microstructure model of the tribofilms was developed to explain the mechanism of friction reduction, as shown in fig. 9.20 [3, 34].

9.4.1 FUNCTIONALLY GRADED MATERIAL

The nanoindentation and nanoscratch measurements revealed that both the MoDTC/ZDDP and ZDDP tribofilms possessed the same hardness, modulus, and friction coefficient distributions relative to depth except in the near-surface area. The decrease in hardness, modulus, and friction coefficient with a shallower depth indicated that the tribofilms were heterogeneous and friction-functionally graded materials. It is thought that the base layer on the steel disk, having a higher constant hardness, modulus, and friction coefficient, serves to decrease mechanical property differences at the interface and provides high shear strength and adhesion strength to prevent flaking or chipping of the tribofilm from the steel disk during sliding. Meanwhile, the graded middle layer with a depth ranging from 10 to 30 nm and possessing lower values of hardness, modulus, and friction coefficient gives low friction performance to the tribofilm in the boundary lubrication regime. The AES analysis revealed that the graded middle layer had a depth range corresponding to the earlier sputtering period in the

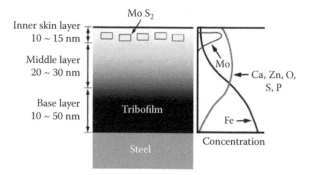

FIGURE 9.20   Model of the functionally graded structure of the MoDTC/ZDDP tribofilm. A low-friction inner-skin layer formed by MoS_2 nanostrips has low yield strength and acts as a lamellar solid lubricant in lowering the friction coefficient of the MoDTC/ZDDP tribofilm. High concentrations of Ca, O, P, and S oil additive elements in the middle layer decreased and disappeared from the base layer, while the Fe concentration increased from an ultralow level in the surface layer to the same level as the steel disk at the tribofilm–disk interface.

AES sputter-depth profile and that the base layer had a depth range corresponding to the latter sputtering period. Thus, the graded middle layer contained main concentrations of Ca, Zn, O, P, and S, while the base layer had a main concentration of Fe, although the compositions and mechanical properties exhibited continuous gradient changes in the depth range from the middle layer to the base layer.

9.4.2 Inner-Skin Layer with Low-Yield Stress

As for the surface region of the tribofilms, the nanoindentation measurements revealed that the MoDTC/ZDDP tribofilm possessed a lower yield stress than the ZDDP tribofilm at a depth of around 10 nm. This agreed with the nanoscratch measurements, i.e., the MoDTC/ZDDP tribofilm exhibited a lower friction coefficient in a scratch depth range from 3.0 to 5.2 nm than the ZDDP tribofilm. The AFM phase observations also confirmed the lower shear strength exhibited by the MoDTC/ZDDP tribofilm. It was found that nanostrips were present just near the surface in a depth range from a few nanometers to about 10 nm. These nanostrips possessed lower shear strength than that of the surrounding surface material. In contrast, the nanostrip structure was never found in the ZDDP tribofilm; no significant differences in nanostructure and mechanical properties were observed between the top-surface materials and the near-surface materials. Moreover, the AES depth profile measurement combined with the XPS analysis revealed that MoS_2 was not present in the ZDDP tribofilm but only in the surface region of the MoDTC/ZDDP tribofilm. The thickness of the surface region was estimated to be 10–15 nm when the sputtering time was reduced to the depth with the sputter etch rate of thermally oxidized SiO_2 films. Therefore, the nanostrips just near the top surface of the MoDTC/ZDDP tribofilm were identified to be a compound layer of MoS_2, although the AES depth profiles also showed Mo in the top-surface area at a depth of less than a few nanometers due to the problem of electron beam damage as well as the limited escape depth for Auger electrons.

Consequently, the nanometer-scale measurements demonstrated that the MoDTC/ZDDP tribofilm possessed a lower shear strength near the surface than the ZDDP tribofilm. The MoDTC/ZDDP tribofilm had an inner-skin layer just below the tribofilm surface at a depth from a few nanometers to around 10 nm. The inner-skin layer was composed of MoS_2 nanostrips, which were oriented in the sliding direction and possessed a lower shear strength.

9.4.3 Low-Friction Mechanism

No evidence was found to indicate that the MoDTC/ZDDP additive promoted smoothing of the sliding surface to reduce friction. Nor were there any significant differences in the mechanical/frictional properties or in composition between the MoDTC/ZDDP and ZDDP tribofilms with respect to the tribofilm depth except in the near-surface area, whereas differences in the shear yield stress and friction coefficient were observed only near the surface. Accordingly, the reduction in friction due to the MoDTC additive is concluded to originate from the inner-skin layer just below the surface of the MoDTC/ZDDP tribofilm. The inner-skin layer consists of MoS_2 nanostrips oriented in the sliding direction. These nanostrips exhibit low-yield strength and act as a lamellar solid lubricant in lowering the friction coefficient of the MoDTC/ZDDP tribofilm.

9.5 CONCLUDING REMARKS

Nanometer-scale differences in the morphological, mechanical, frictional, structural, and chemical properties between MoDTC/ZDDP and ZDDP tribofilms—and correlations among these controlling factors—were successfully evaluated by using several nanoprobe and beam-based analytical methods, including nanoindentation and nano-scratch measurements, AFM and phase-imaging techniques, and AES/XPS analysis. In the surface range in particular, from the top-surface to the near-surface area at a depth of less than 10 nm, very important data such as yield stress, friction coefficient, nanostructure, and composition distributions were directly obtained. From these results, the macroscale friction reduction due to the MoDTC additive was elucidated in relation to the nanoscale tribological properties and nanoscale phenomena.

Both tribofilms examined were found to be heterogeneous and friction-functionally graded materials. They possessed the same mechanical and frictional distributions relative to depth except in the near-surface area. The base layer on the steel disk with higher constant hardness, modulus, and friction coefficient is thought to provide high shear strength and adhesion strength; the graded middle layer—at a depth ranging from 10 to 30 nm and possessing lower hardness, modulus, and friction coefficient—is believed to be responsible for low-friction performance. In contrast, in the surface area, the MoDTC/ZDDP tribofilm possessed a lower yield stress and a lower friction coefficient near the surface than the ZDDP tribofilm. The friction reduction due to the MoDTC additive is concluded to originate from the inner-skin layer just below the surface of the MoDTC/ZDDP tribofilm. This inner-skin layer consists of MoS_2 nanostrips oriented in the sliding direction. These nanostrips have low-yield strength and act as a solid lubricant in lowering the friction coefficient of the MoDTC/ZDDP tribofilm.

This chapter has demonstrated that macroscopic effects can be clarified from the perspective of nanoproperties and microscopic phenomena by evaluating quantitatively and qualitatively the mechanical properties on a nanometer scale. The nanoprobing SPM technique is a powerful tool for deriving simple and clear analysis results for apparently complicated phenomena.

ACKNOWLEDGMENT

This work was based on collaboration between Nissan Motor Co., Ltd., and Nissan ARC, Ltd.

REFERENCES

1. Ye, J., Kano, M., and Yasuda, Y. 2002. *Tribol. Lett.* 13: 41.
2. Ye, J., Kano, M., and Yasuda, Y. 2003. *Jpn. Soc. Tribologists* 48: 60.
3. Ye, J., Kano, M., and Yasuda, Y. 2006. *Jpn. Soc. Tribologists* 51: 627.
4. Isoyama, H., and Sakurai, T. 1974. *Tribol. Int.* 7: 151.
5. Retzloff, J. B., Davis, B. T., and Pietras, J. M. 1979. *Lubrication Eng.* 35: 568.
6. Zheng, P., Han, X., and Wang, R. 1986. *STLE Trans.* 31: 22.
7. Yamamto, Y., and Gondo, S. 1986. *Wear* 112: 79.
8. Yamamoto, Y., and Gondo, S. 1989. *Tribol. Trans.* 32: 251.
9. Kubo, K., Kibukawa, M., and Shimakawa, Y. 1985. In *Proc. Institution of Mechanical Engineers Conference.* C 68/85. London.
10. Kano, M., Yasuda, Y., and Ye, J. 2001. *Proc. 2nd World Tribology Congress.* Vienna.
11. Martin, J. M., Mansot, J. L., Berbezier, I., and Belin, M. 1996. *Wear* 197: 335.
12. Grossiord, C., Varlot, K., Martin, J. M., Le Mogne, T., Esnouf, C., and Inoue, K. 1998. *Tribol. Int.* 31: 737.
13. Croggiord, C., Martin, J. M., Le Mogne, T., Inoue, K., and Igarashi, J. 1999. *J. Vac. Sci. Technol.* A17: 884.
14. Martin, J. M., Donnet, C., and Le Mogne, Th. 1993. *Phys. Rev.* B48: 10583.
15. Tohyama, M., Ohmori, T., Shimura, Y., Akiyama, K., Ashida, T., and Kojima, N. 1995. *Proc. Int. Tribology Conference.* Yokohama.
16. Bird, B. J., and Galvin, G. D. 1976. *Wear* 37: 143.
17. Martin, J. M., Belin, M., and Mansot, J. M. 1986. *ASLE Trans.* 29: 523.
18. Belin, M., Martin, J. M., and Mansot, J. M. 1989. *Tribol. Trans.* 32: 410.
19. Belin, M., Martin, J. M., Tourillon, G., Constns, B., and Bernasconi, C. 1995. *Lubrication Sci.* 8: 3.
20. Yin, Z., Kasrai, M., Bancroft, G. M., Laycock, K. F., and Tan, K. H. 1993. *Tribol. Int.* 26: 383.
21. Fulier, M., Yin, Z., Kasrai, M., Bancroft, G. M., Yamaguchi, E. S., Ryason, P. R., Willermet, P. A., and Tan, K. H. 1997. *Tribol. Int.* 30: 305.
22. Willermet, P. A., Daily, B. P., Carter III, R. O., Schmits, P. J., and Zhu, W. 1995. *Tribol. Int.* 28: 177.
23. Yamaguchi, E. S., Ryason, P. R., Yeh, S. W., and Hansen, T. P. 1998. *Tribol. Trans.* 41: 262.
24. Warren, O. L., Graham, J. F., Norton, P. R., Houston, J. E., and Milchaske, T. A. 1998. *Tribol. Lett.* 4: 189.
25. Graham, J. F., McCague, C., and Norton, P. R. 1999. *Tribol. Lett.* 6: 149.

26. Bec, S., Tonck, A., Georges, J. M., Coy, R. C., Bell, J. C., and Roper, G. W. 1999. *Proc. R. Soc. Lond.* A455: 4181.
27. Bec, S., and Tonck, A. 1995. *Proc. 22nd Leed-Lyon Symposium.* Lyon, France.
28. Aktary, M., McDermott, M. T., and McAlpine, G. A. 2002. *Tribol. Lett.* 12: 155.
29. Topolovec-Miklozic, K., Graham, J., and Spikes, H. A. 2001. *Tribol. Lett.* 11: 71.
30. Ye, J., Kano, M., and Yasuda, Y. 2002. *Tribotest* 9: 13.
31. Ye, J., Kano, M., and Yasuda, Y. 2002. *Tribol. Lett.* 13: 41.
32. Ye, J., Kano, M., and Yasuda, Y. 2004. *Tribol. Lett.* 16: 107.
33. Ye, J., Kano, M., and Yasuda, Y. 2003. *J. Appl. Phys.* 93: 5113.
34. Ye, J., Araki, S., Kano, M., and Yasuda, Y. 2005. *Jpn. J. Appl. Phys.* 44: 5358.
35. Oliver, W. C., and Pharr, G. M. 1992. *J. Mater. Res.* 7: 1564.
36. Doerner, M. F., and Nix, W. D., *J. Mater. Res.* 1: 601.
37. Hertz, H. 1881. *J. reine und angewandte Math.* 92: 156.
38. Timoshenko, S. P., and Goodier, J. N. 1970. *Theory of elasticity.* 3rd ed. New York: McGraw-Hill.
39. Tabor, D. 1951. *Hardness of metals.* Oxford: Clarendon Press.
40. Hoper, R., Gesang, T., Possart, W., Hennemann, O.-D., and Boseck, S. 1995. *Ultramicroscopy* 60: 17.
41. Mogonov, S. N., Elings, V., and Whangbo, M.-H. 1997. *Surf. Sci. Lett.* 375: L385.

10 Synthesis, Characterization, and Tribological Properties of Sodium Stearate–Coated Copper Nanoparticles

Lei Sun, Zhishen Wu, and Zhijun Zhang

ABSTRACT

Sodium stearate coated copper nanoparticles were synthesized in water medium through the chemical reduction method. The morphology, structure and thermal properties of coated Cu nanoparticles were investigated with transmission electron microscopy, x-ray powder diffraction, Fourier transform infrared spectroscopy, thermogravimetric analysis, and differential thermal analysis. Their tribological behaviors were evaluated on a four-ball test machine. The results show that the sodium stearate coated Cu nanoparticles exhibit a uniform particle size distribution and have a size in the range of 5–10 nm, and the coated Cu nanoparticles as oil additives can improve the load-carrying capacity and tribological property of the base oil.

10.1 INTRODUCTION

The properties of metal nanoparticles are quite different from those of bulk metals or individual molecules. Therefore, research of metal nanoparticles has been carried out in many fields of science and technology [1–5]. In recent years, nanoparticles as oil additives have received increased attention in tribological studies. However, incompatibility between inorganic particles and oil has been the main factor preventing their use. This above problem can be resolved by using dispersing agents or employing surface-modification techniques. If the surface-modification agents are high molecular weight hydrocarbons, the nanoparticles thus prepared will have good dispersibility in organic solvents. The tribological properties of chalcogenides [6, 7], oxides [8], rare earth compounds [9, 10], and polyoxometalate compounds [11] used as oil additives have been investigated, and the results have indicated that nanoparticles may have the potential to be used as additives of lubricating oils in the near future.

213

Although the use of metal nanoparticles as oil additives is seldom reported, they have been widely used as magnetic materials [12, 13], catalysts [14, 15], etc.

It was reported that Cu nanoparticles modified by dialkyldithiophosphate (DDP) exhibited good tribological properties [16], but the modification agent contained S and P, which may cause some problems such as toxicity, pollution, etc. So other additives have been explored [17].

In this chapter, we report on the synthesis of Cu nanoparticles modified with an aliphatic acid salt. The characterization of the as-prepared product was performed with a variety of methods, including x-ray diffraction (XRD), transmission electron microscopy (TEM), Fourier transform infrared (FT-IR) spectroscopy, thermogravimetric analysis (TGA), and differential thermal analysis (DTA), and their tribological properties were investigated on a four-ball test machine.

10.2 EXPERIMENTAL

10.2.1 PREPARATION OF CU NANOPARTICLES

Copper nitrate trihydrate ($Cu(NO_3)_2 \cdot 3H_2O$) and hydrazine hydrate ($N_2H_4 \cdot H_2O$) were both analytical reagent (A.R.) grade, and sodium stearate ($CH_3(CH_2)_{16}COONa$) was a chemically pure (C.P.) reagent. These were purchased and used without further treatments. Distilled water was used as the solvent.

The synthesis of Cu nanoparticles was similar to that described by Zhou et al. [16]. A series of samples was synthesized. The preparation methods for all nanoparticles are essentially the same, the difference being in the ratio of $Cu(NO_3)_2 \cdot 3H_2O$ to sodium stearate used. A typical procedure for $[Cu^{2+}]:[CH_3(CH_2)_{16}COO^-] = 4:1$ was as follows: 0.383 g (1.25 mmol) of sodium stearate was dissolved in 50 ml of distilled water by heating, transferred to a 250-ml flask, and then 5.0 ml (100 mmol) of $N_2H_4 \cdot H_2O$ was added dropwise into the flask. Subsequently, 1.21 g (5.0 mol) of $Cu(NO_3)_2 \cdot 3H_2O$ dissolved in 45 ml distilled water was added to the flask. The color of the reaction mixture quickly changed from blue to brown, and the solution was opaque. A large number of bubbles could be seen on the solution surface. The reaction was allowed to continue for 2 h at 60°C under magnetic stirring. As the reaction proceeded, the color of the solution turned to brick red. At the end of the reaction, the solution was allowed to stand for 12 h at room temperature and was then vacuum filtered. The precipitate was rinsed several times with distilled water and dried in a degassed desiccator at ambient temperature for 2 d. Finally, the target product, a red powder of Cu nanoparticles coated with sodium stearate, was obtained. Uncoated Cu particles were also prepared using the same procedure without adding sodium stearate.

10.2.2 CHARACTERIZATION

The TEM morphology of sodium stearate–coated Cu nanoparticles was investigated on a JEOL JEM-2010 transmission electron microscope at an acceleration voltage of 200 kV. The prepared powders were dispersed in chloroform under ultrasonic agitation for 20 min and then deposited on a copper grid covered with a perforated carbon film. The resulting specimen was then subjected to TEM analysis.

XRD patterns were recorded on a Philips X'pert pro x-ray powder diffractometer using Cu K_α radiation (λ = 1.5418 Å). The operating voltage and current were 40 kV and 40 mA, respectively. The scan rate was 10°/min.

FT-IR spectra were taken on a Nicolet AVATAR 360 Fourier transform infrared spectrometer, which covered from 4000 to 400 cm^{-1}, to characterize the surface structure of Cu nanoparticles. The prepared Cu nanoparticles were mixed with KBr powder and pressed into a pellet for measurement. Background correction was made using a reference blank KBr pellet.

TGA and DTA analyses were conducted in nitrogen on a Seiko EXSTAR 6000 thermal analyzer at a scan rate of 10°C/min.

10.2.3 Friction and Wear Properties

The tribological properties of Cu nanoparticles were determined on an MRS-10A four-ball test machine at 1450 rpm in ambient conditions. The 12.7-mm-diameter balls used in the test were made of bearing steel (composition: 0.95%–1.05% C, 0.15%–0.35% Si, 0.24%–0.40% Mn, <0.027% P, <0.020% S, 1.30%–1.67% Cr, <0.30% Ni, and <0.025% Cu) with a Rockwell hardness (Rc) of 61–64. The base oil was chemically pure liquid paraffin (LP), which has a distillation range of 180°C–250°C and density of 0.835–0.855 g/cm^3. Before each test, the balls and specimen holders were ultrasonically cleaned in petroleum ether (normal alkane with a boiling point of 60°C–90°C) and then dried in hot air. At the end of each test, the wear-scar diameters (WSD) of the three lower balls were measured on a digital-reading microscope to an accuracy of 0.01 mm. Then the average wear-scar diameter from the three balls was calculated.

10.3 RESULTS AND DISCUSSION

10.3.1 Particle Dispersibility

The sodium stearate–coated Cu nanoparticles can disperse in various organic solvents, such as chloroform, benzene, and liquid paraffin, with the assistance of ultrasonic agitation. The formed homogeneous suspensions were stable for several hours without particle deposition. As a comparison, noncoated Cu particles cannot disperse in organic solvents, even after ultrasonication. The results showed that after surface modification with the sodium stearate surfactant, the dispersibility of Cu nanoparticles in oil was improved. It can be concluded that the surfaces of Cu nanoparticles were coated with sodium stearate.

10.3.2 Morphology and Structure of Nanoparticles

Figure 10.1 shows the XRD pattern of sodium stearate–coated Cu nanoparticles. It is seen that the peaks at 2θ = 43.2°, 50.3°, 73.9°, and 89.9° are assigned to diffractions from the (111), (200), (220), and (311) lattice planes of copper, respectively, which are in agreement with the face-centered cubic (fcc) copper phase (Joint Committee on Powder Diffraction Standards, File No. 03-1005). The weak peak at 2θ = 36.5° corresponds to the (111) lattice plane diffraction of CuO (Joint Committee on Powder

Diffraction Standards, File No. 78-0428). It can also be seen from fig. 10.1 that there are some relatively weak and broad peaks before $2\theta = 30°$ that arise from the diffraction of modification layer on the coated nanoparticles. Thus, it is concluded that the Cu nanoparticles with fcc structure were successfully prepared and that there are some oxides on its surface.

Figure 10.2 presents the TEM morphologies of (a) sodium stearate–coated Cu nanoparticles and (b) noncoated Cu nanoparticles. It is seen from fig. 10.2b that noncoated Cu nanoparticles tend to agglomerate owing to their high surface energy, and the mean size of the particles is about 100 nm. In contrast, the TEM image of sodium stearate–coated Cu nanoparticles shows relatively small size (average size ≈10 nm), and no obvious aggregation occurs. This is because the surfactant sodium stearate is adsorbed on the surface of nanoparticles, thereby reducing the surface energy of nanoparticles and preventing their aggregation. The results show that the existence

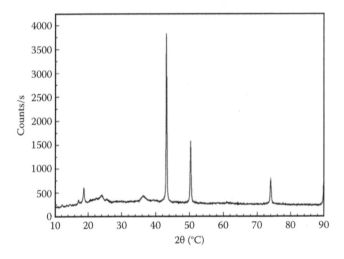

FIGURE 10.1   XRD pattern of sodium stearate–coated Cu nanoparticles.

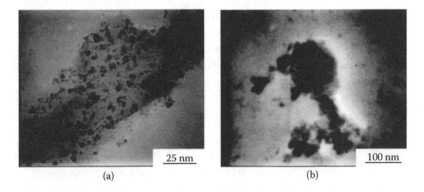

FIGURE 10.2   TEM images of (a) sodium stearate–coated Cu nanoparticles and (b) uncoated particles.

of a surfactant layer can control the size of particles and thus improve the dispersion of the nanoparticles.

10.3.3 FT-IR RESULTS

Figure 10.3 shows the FT-IR spectrum of sodium stearate–coated Cu nanoparticles. The bands at 1456 and 1374 cm^{-1} are attributed to the asymmetric and symmetric bending vibrations of -CH$_3$, respectively. The bands at 2917 and 2852 cm^{-1} correspond to the asymmetric and symmetric stretching vibrations of -CH$_2$-, and the vibration in the 720-cm^{-1} region is a characteristic of a minimum of four methyl groups, (CH$_2$)$_4$, in a row and assigned to the methylene rocking vibration. The bands at 1539 and 1411 cm^{-1} correspond to the asymmetric and symmetric stretching vibrations of COO$^-$, respectively. Thus it can be concluded that the long alkyl acid radical does exist in the coated Cu nanoparticles. The band at 3446 cm^{-1} is attributed to the stretching vibrations of -OH, which arise from the adsorbed water on/in the coated nanoparticles.

10.3.4 THERMAL ANALYSIS OF NANOPARTICLES

Figure 10.4 shows the TGA and DTA curves of sodium stearate–coated Cu nanoparticles. It can be seen from the TGA curve that, from 70°C to 100°C, the mass loss is about 5%, which corresponds to the vaporization of adsorbed water. The mass loss from 200°C to 400°C is about 52%, which corresponds to the decomposition of the coated layer. According to the stoichiometric relation of Cu^{2+} and CH$_3$(CH$_2$)$_{16}$COO$^-$, i.e., [Cu^{2+}]:[CH$_3$(CH$_2$)$_{16}$COO$^-$] = 4:1, the amount of sodium stearate coated on Cu nanoparticles should be 52.5%, which is consistent with the TGA result. This indicates that the modification reaction proceeded completely. It can be seen from fig. 10.4 that there is an exothermic region at 98°C, which corresponds to

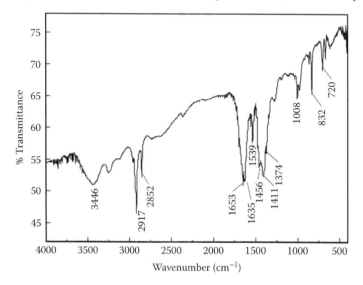

FIGURE 10.3   FT-IR spectrum of sodium stearate–coated Cu nanoparticles.

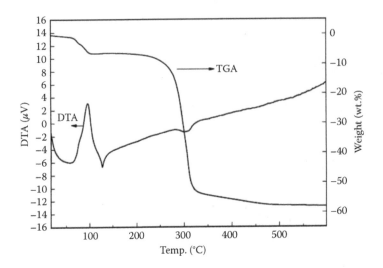

FIGURE 10.4   TGA and DTA curves of sodium stearate–coated Cu nanoparticles.

the dehydration, while the endothermic region at 308°C corresponds to the melting and decomposition of the coated layer.

According to the above results, we propose the formation mechanism of sodium stearate–coated Cu nanoparticles as follows: Before the addition of copper nitrate, hydrazine hydrate did not react with sodium stearate. As copper nitrate was added, two reactions proceeded, one being the formation of copper stearate, and the other being the reduction of Cu^{2+} to Cu. The former is a reversible reaction, with copper stearate decomposing into copper ions and a stearic radical as reduction proceeded, and the fresh copper ions being reduced by the excess hydrazine hydrate. It is well known that the initial nanocrystalline is active [18], i.e., there are a large number of defect sites and dangling bonds on its surface, so the initial Cu nanocrystalline will react with the stearic radical to form a surface-modified layer that deters the growth and condensation of Cu. Since the reaction solution had alkalescence, some of the copper ions formed copper oxide, although in the presence of a modification agent.

10.3.5 Tribological Properties of Nanoparticles

Figure 10.5 shows the tribological properties of sodium stearate–coated Cu nanoparticles at different additive contents in LP and in pure LP (additive content: 0%). The test conditions were as follows: rotating rate 1450 rpm, load 300 N, test duration 30 min. The results show that sodium stearate–coated Cu nanoparticles can improve the antiwear properties (reduce the values of WSD) of the base oil appreciably, even at relatively low content (below 0.12%). With an increase in additive content, the values of WSD increase slightly, and when the content exceeds 0.15%, the WSD decreases and its curve goes smooth. The nanoparticles show the optimum antiwear property when the additive content is 0.50%, and at this content, the WSD of LP can be reduced to about one-half. The friction-reducing properties of the sodium

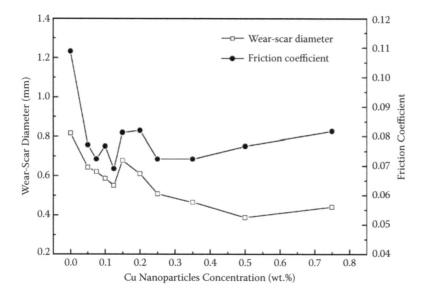

FIGURE 10.5 Effect of concentration of sodium stearate–coated Cu nanoparticles in liquid paraffin on tribological properties (four-ball test machine; applied load: 300 N; speed: 1450 rpm; test duration: 30 min).

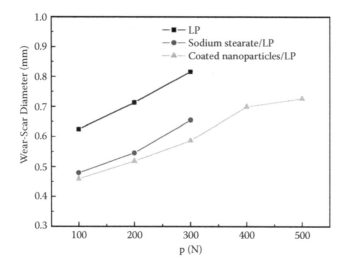

FIGURE 10.6 (**See color insert following page 80.**) Effect of load on wear-scar diameter (four-ball test machine; additive concentration in liquid paraffin (LP): 0.1%; speed: 1450 rpm; test duration: 30 min).

stearate–coated Cu nanoparticles show similar behavior. This result indicates that the sodium stearate–coated Cu nanoparticles exhibit good tribological properties.

Figure 10.6 presents the tribological properties of sodium stearate, sodium stearate–coated Cu nanoparticles, and pure LP versus applied load. (The conditions were: additives content 0.10%, rotating rate 1450 rpm, and test duration 30 min.)

It can be seen that the sodium stearate–coated Cu nanoparticles can improve the load-carrying capacity of LP from 300 to 500 N, and at the same load, the WSD of LP containing nanoparticles is smaller than that of pure LP and LP with surfactant sodium stearate. Thus it can be concluded that, at the low loads, the sodium stearate–coated Cu nanoparticles exhibit good antiwear properties and can improve the load-carrying capacity of base oil.

10.4 CONCLUSIONS

Sodium stearate–coated Cu nanoparticles were synthesized by the chemical reduction method, and their morphology and structure were characterized by TEM, XRD, FT-IR, TGA, and DTA. The results show that the coated nanoparticles could disperse in organic solvents well, with an average particle size of 10 nm. The tribological properties of nanoparticles as lubrication additives in LP were investigated, and the results showed that the as-prepared nanoparticles possessed good antiwear and friction-reduction ability.

ACKNOWLEDGMENT

The authors wish to acknowledge the financial support from the National Science Foundation of China (grant no. 50701016).

REFERENCES

1. David, I. G., Donald, B., and David, J. S. 2000. *Nature* 408: 67–69.
2. Lu, L., Wang, L. B., and Ding, B. Z. 2000. *J. Mater. Res.* 15: 270–73.
3. Petit, C., Cren, T., and Roditcher, D. 1999. *Adv. Mater.* 11: 1198–1202.
4. Link, S., Wang, Z. L., and El-Sayed, M. A. 2000. *J. Phys. Chem. B* 104: 7867–70.
5. Suryanarayanan, R., Frey, C. A., and Sastry, S. M. L. 1996. *J. Mater. Res.* 11: 439–49.
6. Zhang, Z. J., Xue, Q. J., and Zhang, J. 1997. *Wear* 209: 8–12.
7. Chen, S., Liu, W. M., and Yu, L. G. 1998. *Wear* 218: 153–58.
8. Xue, Q. J., Liu, W. M., and Zhang, Z. J. 1997. *Wear* 213: 29–32.
9. Zhou, J. F., Wu, Z. S., and Zhang, Z. J. 2001. *Wear* 249: 333–37.
10. Zhang, Z. F., Yu, L. G., and Liu, W. M. 2001. *Tribol. Int.* 34: 83–88.
11. Sun, L., Zhou, J. F., and Zhang, Z. J. 2004. *Wear* 256: 176–81.
12. Black, C. T., Murray, C. B., and Sandstrom, R. L. 2000. *Science* 290: 1131–34.
13. Pileni, M. P. 2001. *Adv. Funct. Mater.* 11: 323–36.
14. Kamat, P. V. 2002. *J. Phys. Chem. B* 106: 7729–44.
15. Araña, J., Piscina, P. R., and Llorca, J. 1998. *Chem. Mater.* 10: 1333–42.
16. Zhou, J. F., Wu, Z. S., and Zhang, Z. J. 2000. *Tribol. Lett.* 8: 213–18.
17. Huang, W. J., Dong, J. X., and Wu, G. F. 2004. *Tribol. Int.* 37: 71–76.
18. Hoffmann, A. J., Mills, G., and Yee, H. 1992. *J. Phys. Chem.* 96: 5540–46.

Part IV

Polymeric and Bio-Based Surfactants

11 Surface Friction and Lubrication of Polymer Gels

Miao Du and Jian Ping Gong

ABSTRACT

Biological connective tissues, such as cartilage and corneal stroma, are essentially hydrogels consisting of fibrous collagen and proteoglycans. Few facts are known of the surface properties of the hydrogels, although we observe fascinating tribological behavior of biological soft tissues, such as the extremely low friction between animal cartilages. We consider that the role of a solvated polymer network existing in extracellular matrix as a gel state is critically important in the specific frictional behavior of cartilages. In order to elucidate the general tribological features of solvated polymer matrix, friction of various kinds of hydrogels have been investigated, and very rich and complex frictional behaviors are observed. The friction force and its dependencies on the load are quite different depending on the chemical structures of the gels, surface properties of the opposing substrates, and the measurement condition, which are totally different from those of solids. Most importantly, the coefficient of friction of gels, μ, changes in a wide range and exhibits very low values ($\mu \approx 10^{-3}$–10^{-4}), which cannot be obtained from the friction between two solid materials. A repulsion–adsorption model has been proposed to explain the gel friction, which says that the friction is due to lubrication of a hydrated layer of polymer chain when the polymer chain of the gel is non-adhesive (repulsive) to the substrate, and the friction is due to elastic deformation of the adsorbed polymer chain when it is adhesive to the substrate.

11.1 INTRODUCTION

Sliding friction is one of the oldest problems in physics and certainly one of the most important subjects from a practical point of view. Many different terms and approaches have been used to describe solid friction [1–3]. For solid friction with lubricating oil, the frictional behavior is usually divided into three regimes. In the boundary lubrication (BL) regime, which occurs at low sliding velocity when there is negligible fluid entrainment into the contact zone, the load is carried by the contacting asperities, and friction is dependent on the surface and interfacial film properties at the molecular scale. In the hydrodynamic lubrication regime, a film of lubricant, whose thickness depends on the viscosity and entrainment velocity, is entrained to fully separate the solid surfaces. The friction now depends on the rheological properties of the lubricant film in the contact zone at the high-shear-rate condition that prevails there. Sophisticated solutions of the Reynolds lubrication equation have evolved to explain friction on the basis of the continuum flow of liquids. The fundamental view here is that liquids resist deformation against force that increases with velocity [3]. The dependence of force on velocity is linear for the simplest case (said to be "Newtonian"). Even modern treatments consider that when velocity is increased, the force required to accomplish deformation increases with a positive slope. In the mixed regime, termed here as the friction transition regime, which lies between boundary lubrication and hydrodynamic lubrication, both the boundary film and bulk lubricant play crucial roles in determining friction.

Some biological surfaces display fascinating low-friction properties [4–13]. For example, cartilages of animal joints have a friction coefficient in the range 0.001–0.03, remarkably lower than even for hydrodynamically lubricated journal bearings [4, 5]. It is not well understood why the cartilage friction of the joints is so low, even in such conditions as when the pressure between the bone surfaces reaches as high as 3–18 MPa and the sliding velocity never exceeds more than a few centimeters per second [4]. Under such conditions, the lubricating liquid layer cannot be sustained between two solid surfaces, and the hydrodynamic lubrication does not work.

The authors consider that these fascinating tribological properties of the biological systems originate from the soft and wet nature of tissues and organs. That is, the role of a solvated polymer network existing in an extracelluar matrix (ECM) as a *gel state* is critically important in the specific frictional behavior of the biological systems.

Hydrogels consist of an elastic cross-linked macromolecular network, with a liquid filling the interstitial space of the network. Such hydrogels display properties characteristic of both solids and liquids. Like solids, they deform with stress and recover their initial shape after the stress is removed. Like liquids, they can support fluid convection and diffusion of solutes that are smaller than the mesh size of the network. Therefore, studying the frictional behavior of hydrogels is helpful for understanding the low-friction mechanism observed in biological systems, and may be useful in finding novel approaches in the design of low-friction artificial organs.

In order to elucidate the general tribological features of a solvated polymer matrix, frictions of various kinds of hydrogels have been investigated in the last several years, and very rich and complex frictional behaviors have been observed [14–30]. To describe the frictional behavior of a gel sliding on a smooth substrate, we have proposed a thermodynamic model from the viewpoint of a polymer–solid interfacial interaction [15].

Recently, we also studied the effect of polymer solutions and surfactant solutions on the friction of an adhesive gel on a glass substrate. For example, we have studied the friction behaviors of polyvinyl alcohol (PVA) gel sliding against a glass surface in a dilute polyethylene oxide (PEO) aqueous solution with various molecular weights, M_w, and concentrations [31]. We have found that in the low-sliding-velocity region, where adsorption of PVA gel on glass plays the dominant role in friction, the PEO polymer chain screens the adsorption of PVA gel to the glass surface, thereby reducing the friction dramatically. Similar screening phenomena were also observed in the ionic surfactant solution.

This chapter consists of five sections that summarize the experimental and theoretical results on the surface-sliding friction of hydrogels [14–22]. In section 11.2, the general experimental features of gel friction are described. In section 11.3, a thermodynamic model from the viewpoint of a polymer–solid interfacial interaction is described to understand the frictional behavior of the polymer gels. In section 11.4, the theoretical predictions are compared with the experimental observations, and the essential feature of the polymer repulsion–adsorption is discussed. In section 11.5, attempts to apply robust gels with low friction as substitutes for biological tissues, such as artificial cartilages, are discussed.

11.2 EXPERIMENTAL RESULTS

11.2.1 EXPERIMENTAL DETAILS

11.2.1.1 Sample Preparation

The poly(2-acrylamido-2-methyl-1-propanesulfonic acid) (PAMPS) gel was synthesized by radical polymerization. An aqueous solution of 1.0 M 2-acrylamido-2-methyl-1-propanesulfonic acid (AMPS), 5.0 mol% cross-linking agent N,N'-methylenebisacrylamide (MBAA), and 0.1 mol% initiator 2-oxoglutaric acid was purged in a reaction cell with nitrogen gas for 30–45 min and then irradiated with ultraviolet (UV) light for 5 h at 20°C. The reaction was carried out between a pair of glass substrates, separated with a 2-mm-thick spacer. After polymerization, the gel was immersed in a large amount of water for 1 week to equilibrate and to wash away the residual chemicals. The surface of gel prepared on the glass was mirrorlike and was used for the measurements. Other samples were prepared in the same manner; the details are described in the literature [16–22].

11.2.1.2 Measurements

11.2.1.2.1 Friction Measurement Using a Tribometer
The friction of gels against a piece of solid substrate, such as a piece of glass, was measured in air or in water using a commercial tribometer (Heidon 14S/14DR, SHINTO Scientific Co. Ltd., Japan), as shown in fig. 11.1. A gel sample with a prescribed size was embedded in a square frame of adjustable size attached to the upper board and pressed against a piece of glass plate, which was fixed on the lower board and was driven to move horizontally and repeatedly at a prescribed velocity over a distance of 90 mm. The normal load (or sliding velocity) dependence was studied by using a sample for which the load (or velocity) was changed in a stepwise manner from lower

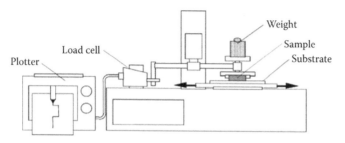

FIGURE 11.1 Illustration of the tribometer used to measure the gel friction. (Reproduced with permission from Gong, J. P., et al. 1999. *J. Phys. Chem. B*: 103: 6001. Copyright 1999, Am. Chem. Soc.)

to higher values without separating the two surfaces during the interval of measurement. A detailed description of the measurements is given in the literature [16].

11.2.1.2.2 Friction Measurement Using a Rheometer

A commercial rheometer (3-ARES-17A, Rheometric Scientific Inc.), as shown in fig. 11.2, was also used for measuring the friction of a gel against a solid surface or a gel in water or in aqueous NaCl solutions at prescribed temperatures. Samples were glued to the upper surface of a coaxial disk-shaped platen. As the opposing substrate, a gel or glass plate a little larger than the upper gel was fixed on the lower platen. The interface was immersed in water or in aqueous salt solutions.

The difference in the measurements between using a tribometer and a rheometer is that, with the former, the friction tests are performed under a constant normal compressive stress, but the latter runs in a constant compressive-strain mode. In both cases, prior to the measurement, the gel sample was loaded with a normal load (or normal stress) and reached an equilibrium state. After achieving relaxation equilibrium, a displacement with a velocity, $v(\omega)$, was applied to the lower platen of the tribometer (rheometer) to generate the frictional force (torque). By using a rheometer, the total frictional force, F, can be related to the torque by assuring that the unknown frictional shear at a radius r changes with the sliding velocity in a power law as $F \propto (\omega r)^{\alpha}$, where α is a constant. A detailed description of the measurement is given in the literature [17].

11.2.2 COMPLEX FRICTIONAL BEHAVIOR OF GELS

Amonton's law says that the frictional force F between two solids is proportional to the load W, forcing them together, i.e., $F = \mu W$ [32], and does not depend on the apparent contact area A of the two solid surfaces nor on the sliding velocity v. At the same time, the proportional coefficient μ, known as the frictional coefficient, depends not on the sliding velocity or on the apparent contact area of the two surfaces but, rather, only on the moving materials according to this law. The proportionality constant μ usually lies within a range of 0.5–1.0 [33]. However, the gel friction does not simply obey

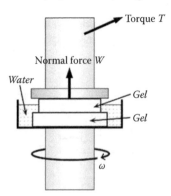

FIGURE 11.2 Schematic illustration of the rheometer used to measure the gel friction. (Reproduced with permission from Gong, J. P., Kagata, G., and Osada, Y. 1999. *J. Phys. Chem. B*: 103: 6007. Copyright 1999, Am. Chem. Soc.)

Amonton's law, and it shows rich and complex features. The frictional forces of the gels obtained show only a slight dependence on the load W in the range investigated, but they strongly depend on the sliding velocity v, contact area A, and the surface properties of the opposing substrates. Most importantly, these gels have frictional coefficients μ of $\approx 10^{-3}$, which are much lower than that observed in friction between solids.

11.2.2.1 Load Dependence

Figure 11.3 shows the friction behavior of hydrogels with different chemical structures: poly(vinyl alcohol) (PVA), gellan gum (a kind of polysaccharide),

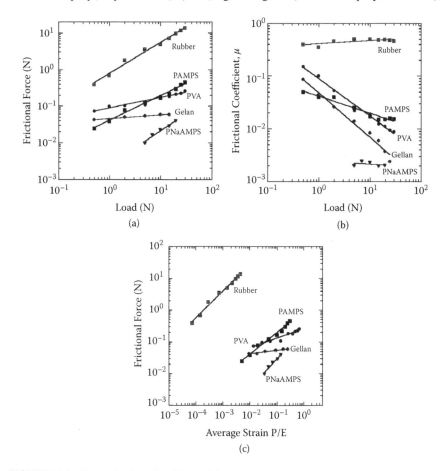

(a)

(b)

(c)

FIGURE 11.3 Dependencies of (a) frictional force F vs. load W, (b) frictional coefficient vs. load, and (c) frictional force vs. average strain P/E (P = normal pressure and E = compressive modulus of gel) for various kinds of hydrogels on a glass substrate. Sliding velocity: 7 mm/min; sample sizes for PVA, gellan, and rubber: 3×3 cm; sample sizes for PAMPS and PNaAMPS: 2×2 cm; compressive modulus E for PVA: 0.014 MPa, for gellan: 0.06 MPa, for PAMPS: 0.25 MPa, for PNaAMPS: 0.35 MPa, and for rubber: 7.5 MPa; degree of swelling $q(g/g)$ for PVA: 17, for gellan: 33, for PAMPS: 21, and for PNaAMPS: 15. Measurements were performed with a tribometer in air. (Reproduced from Gong, J. P., et al. 1999. *J. Phys. Chem. B*: 103: 6001. With permission.)

poly(2-acrylamido-2-methylpropanesulfonic acid) (PAMPS) and its sodium salt PNaAMPS gels, and rubber [16]. These hydrogels were slid on a smooth glass substrate at a sliding velocity of 7 mm/min using a tribometer. Figure 11.3a shows the relationship between the normal load and the frictional force. The relationship between the frictional force (F) and the normal load (W) obeys a power law, $F \propto W^{\alpha}$, where the scaling exponent α lies in a range of 0–1.0, depending on the chemical structure of the gels. The frictional coefficient μ, which is defined as the ratio of the frictional force to the applied load, is shown in fig. 11.3b. The μ of these gels accordingly shows a unique load dependency that is quite different from those of solids. The μ values for PVA, gellan, and PAMPS gels all decrease with an increase of load. On the other hand, the μ value of the PNaAMPS gel is constant over the range of the load, similar to that of rubber, but its μ is as low as 0.002, two orders of magnitude lower than those of solids. PAMPS and PNaAMPS gels are different only in their counterions, but they show a striking difference in frictional behavior on the glass surface.

The above results indicate that the gel friction strongly depends on the properties of the gel as determined by its chemical structure, such as hydrophilicity, charge density, cross-linking density, water content, and elasticity. The frictional force of the gels is two or three orders of magnitude lower than that of a rubber. This behavior can be continuously observed for a few hours. Some physically cross-linked gels even show negative load dependencies, as shown in fig. 11.4. When the average normal pressure exceeds a critical value, the frictional forces of gellan gel and κ-carrageenan gel (both are physically cross-linked polysaccharide gels) show a pronounced negative load dependency. The observed specific phenomenon is attributed to the physical cross-linking nature of the polysaccharide gel. A high normal pressure leads to the loosening and/or dissolution, in part, of the cross-linkages and results in an increased viscous layer of the linear polymer at the friction interface that serves as a good lubricator under the high load [18].

11.2.2.2 Sliding Velocity Dependence

We have found that the frictional force increases with an increase in the sliding velocity, v. The double logarithmic plots of μ against v showed approximately linear relations in a range of 7–500 mm/min. For all gel samples measured, the slopes of the lines change from 0.21 to 0.67, depending on the gel species (fig. 11.5). This kind of behavior of gels is also quite different from that of a solid material, for which μ is independent of the sliding velocity, as prescribed by Amonton's law. Obviously, the specific behavior of the gel friction should be associated with the water absorbed in the gel. Under the load, the gel deforms, and a portion of the water might be squeezed out from the bulk gel to serve as a lubricator, leading to a boundary lubrication or even to a hydrodynamic lubrication. The strong dependence of the gel friction on the pressure P (fig. 11.3) and velocity v (fig. 11.5) suggests a hydrodynamic lubrication mechanism. However, one should note that the hydrodynamic lubrication is usually sustained only when two solids rotate at a very high speed with lubricating oil. When two solid surfaces are allowed to slide under controlled conditions, the lubricant layers

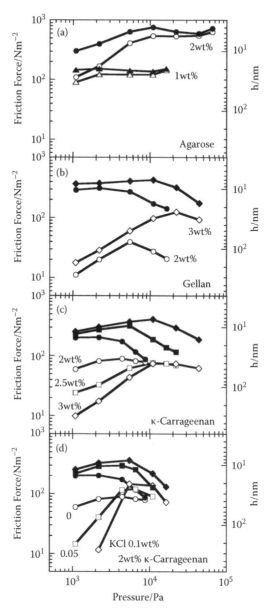

FIGURE 11.4 Frictional force per unit area and the hydrodynamic water layer thickness, h, as a function of normal pressure measured in water (open symbols) and in air (closed symbols). (a) Agarose gel concentration: ($\triangle\blacktriangle$) 1 wt%, ($\bigcirc\bullet$) 2 wt%; (b) gellan gel concentration: ($\bigcirc\bullet$) 2 wt%, ($\diamond\blacklozenge$) 3 wt%; (c) κ-carrageenan gel concentration: ($\square\blacksquare$) 2 wt%, ($\square\blacksquare$) 2.5 wt%, ($\diamond\blacklozenge$) 3 wt%; (d) 2 wt% κ-carrageenan gel at various levels of KCl concentration: ($\bigcirc\bullet$) 0 wt%, ($\square\blacksquare$) 0.05 wt%, ($\diamond\blacklozenge$) 0.1 wt%. Sliding velocity, 3 mm/s; gel surface area, 3 × 3 cm²; temperature, 25°C. (Reproduced from Gong, J. P., Iwasaki, Y., and Osada, Y. 2000. *J. Phys. Chem. B*: 104: 3423. With permission.)

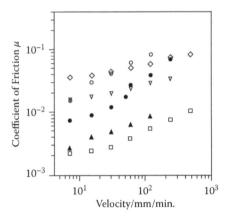

FIGURE 11.5 Coefficient of friction μ as a function of sliding velocity v for various gels sliding against a glass. For one set of an experimental run, the sliding velocity measurement was carried out using one sample, starting from a lower velocity and increasing the velocity continuously without separating the two sliding surfaces during the interval of measurement. The sliding distances used in the measurement were adjusted from 14 to 30 mm, depending on the velocity. (●) gellan gel, $P = 1.1 \times 10^3$ Pa; (□) κ-carrageenan gel, $P = 1.3 \times 10^3$ Pa; (▲) PVA gel, $P = 2.2 \times 10^3$ Pa; (◇) konjak gel (the konjak is made from the konjak potato, called "devil's tongue" in English.), $P = 2.2 \times 10^3$ Pa; (▽) PAMPS gel, $P = 1.3 \times 10^3$ Pa; (○) PNaAMPS gel, $P = 1.3 \times 10^3$ Pa. (Reproduced from Gong, J. P., et al. 1997. *J. Phys. Chem. B* 101: 5487. With permission.)

are squeezed out quickly, and the hydrodynamic lubrication cannot be sustained. This is why a water-soaked sponge conforms to Amonton's law and does not exhibit hydrodynamic lubrication. The very strong hydration ability of the gel probably makes it possible to sustain the supposed hydrodynamic lubrication even at a very low sliding velocity and under a high pressure.

11.2.2.3 Sample Area Dependence

To elucidate the features of the interface contact between the gel and the opposing plate, the frictional forces of various kinds of gels were measured by varying the contact area of the gel A under a constant load W [16]. It was found that the F also shows a power law with A, which can be denoted as $F \propto A^\beta$. Combining the results of $F \propto W^\alpha$ and $F \propto A^\beta$, one has $F \propto W^\alpha A^\beta$. As shown in fig. 11.6, a correlation between α and β was found as $\beta \approx 1 - \alpha$. Therefore,

$$F \propto A P^\alpha \tag{11.1}$$

where $P = W/A$ is the average normal pressure and $\alpha = 0 - 1$, depending on the chemical structure of the gel.

This result demonstrates that the frictional force per unit area (frictional stress) is related to the normal pressure P instead of the load W by the power law. For solid friction, eq. (11.1) is also valid, with $\alpha = 1$. A polymer gel is easily deformable, with a typical elasticity ranging from 1 to 1000 kPa, owing to the presence

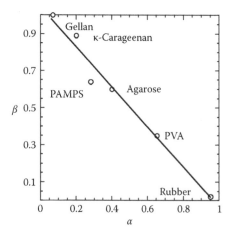

FIGURE 11.6 Relation between α and β, where α and β are the exponents of the relation $F \propto W^{\alpha}A^{\beta}$. The α was measured at $A = 9$ cm^2, and the β was measured at $W = 0.98$ N. Sliding velocity: 180 mm·min^{-1}. Degree of swelling $q(g/g)$: PVA, 20; gellan, 33; κ-carrageenan, 33; agarose, 50; PAMPS, 17; rubber, 1. Measurements were performed with a tribometer in air. (Reproduced from Gong, J. P., et al. 1999. *J. Phys. Chem. B*: 103: 6001. With permission.)

of a large amount of water. A small pressure would be sufficient to cause a large deformation of a gel. This favors interfacial contact with the opposing surface. As shown in fig. 11.3c, the average strains, $\lambda = P/E$, of gels under the experimental load range are more than several percent higher than that of a rubber and, needless to say, much higher than that of a solid. Here, E is the compressive elastic modulus, or the stiffness of the gel. Since the surface roughness of gels synthesized on smooth surfaces such as a glass plate or a silicon wafer is several nanometers, on the order of the network mesh size of a gel, the whole gel surface should contact with the smooth glass substrate on the mesh size scale. All the gel samples were measured under a similar strain range; nevertheless, they showed quite different pressure dependencies.

11.2.2.4 Substrate Effect

The frictional behavior of gels also depends on the opposing substrates. When the nonionic PVA gel is allowed to slide on a poly(tetrafluoreoethylene) (Teflon) plate, for example, the behavior is the same as that on a glass surface. However, the behavior of a strong anionic PAMPS gel on Teflon greatly changes and becomes similar to that of PVA on glass (fig. 11.7). When a pair of polyelectrolyte gels carrying the same charges, for example, two like-charged PNaAMPS gels with PNaAMPS gel, were slid against each other, very low frictional force was observed [16]. On the other hand, when two polyelectrolyte gels carrying opposite charges were slid against each other, the adhesion between the two gels was so high that the gels were broken during the measurement [17]. This phenomenon indicates that the interfacial interaction between the gel surface and the opposing substrate is crucial in gel friction.

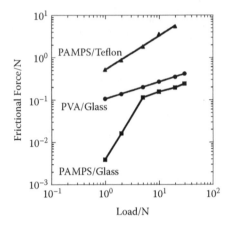

FIGURE 11.7 Load dependence of gel friction on various substrates measured in water. Sliding velocity: 90 mm/min. Gel size: PVA, $3 \times 3 \times 0.8$ cm³; PAMPS, $3 \times 3 \times 0.6$ cm³. Degree of swelling q(g/g): PVA, 10; PAMPS, 15. Elastic modulus: PAMPS, 1.0 MPa; PVA, 0.024 MPa. (Reproduced from Gong, J. P., et al. 1999. *J. Phys. Chem. B*: 103: 6001. With permission.)

11.3 PROPOSED GEL FRICTION MECHANISM: REPULSION–ADSORPTION MODEL

Based on the above complex frictional behavior of polymer gels, Gong and coworkers have proposed a repulsion–adsorption model from the viewpoint of polymer–solid interfacial interactions to describe the friction behavior of gels on a smooth substrate (fig. 11.8) [15].

A gel has many common features with other soft cross-linked polymers in terms of its cross-linked polymer structure and viscoelastic properties. On the other hand, different from a rubber, a gel contains a large amount of a low-molecular-weight component, water. The water content in biological gels, such as cartilage and other soft tissues, is ca. 70–80 wt%, and in synthetic gels the water content can be as high as 99.9 wt%. Water in gels is strongly solvated to the polymer network and cannot be squeezed out easily like a sponge. Due to the presence of a large amount of water, the internal friction of a gel is much less than that of a cross-linked polymer melt or rubber. For a typical gel, tan δ, a characteristic parameter of the viscoelastic properties of soft materials, is on the order of 0.01, while that of a rubber is never less than 0.1.

The simplest molecular picture of a polymer gel is given by the c* gel in the scaling theory [34]. The scaling theory describes a gel as a collection of adjacent blobs with a radius R_F that has a characteristic relaxation time $\tau_f \approx \eta R_F^3/T$, where T is the absolute temperature in energy units, and η is the viscosity of solvent. Each blob is associated with one partial polymer chain (the polymer chain between two next-neighboring cross-linking points). The scaling theory relates this molecular structure of the gel to its elastic modulus E by $E \approx T/R_F^3$.

The main argument of the adsorption–repulsion model is as follows. By considering a c* gel in contact with a solid wall in water, in analogy to a polymer solution, the polymer network on the surface of the gel will be repelled from the solid

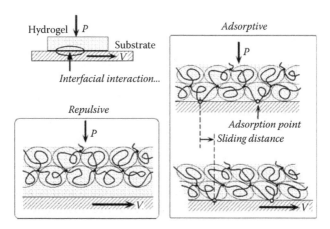

FIGURE 11.8 Schematic illustration of the repulsion–adsorption model for gel friction on a solid substrate.

surface if the interfacial interaction is repulsive, and it will be adsorbed to the solid surface if it is attractive. In the repulsive case, the friction is due to the lubrication of the hydrated water layer of the polymer network at the interface, which predicts that the frictional stress f should be proportional to the sliding velocity. In the attractive case, the friction of a gel is from two contributions: (1) elastic deformation of the adsorbed polymer chain and (2) lubrication of the hydrated layer of the polymer network, where the first contribution is the same as the adhesive friction, as proposed by Schallamach [35–39].

According to Schallamach's model for rubber [35], the frictional stress from the first contribution, f_{el}, arising from the elastic stretching of the polymer chains adsorbed on the substrate, is expressed as the product of the adsorbing site number per unit area, m, the stretching velocity, v, and the lifetime of adsorption, τ_b, i.e., $f_{el} \propto mv\tau_b$. It takes a time τ_f for a desorbed polymer to be readsorbed on the surface. Taking into consideration the thermal agitation, $\tau_b^{-1} = \tau_f^{-1}\exp[(-F_{ads} - F_{el})/T]$. Here, F_{ads} is the adsorption energy per polymer chain, and F_{el} is the elastic energy due to stretching of the adsorbed chain. The elastic energy accumulated by stretching disfavors the adsorption. Therefore, τ_b decreases with an increase in chain stretching.

By applying the scaling relation to Schallamach's model, one obtains a characteristic velocity [15],

$$v_f \approx R_F/\tau_f = T/\eta R_F^2 \tag{11.2}$$

which can be expressed in terms of elastic modulus E of the gel as

$$v_f \approx T^{1/3}E^{2/3}/\eta \tag{11.3}$$

and f_{el} has a maximum at $v\tau_f/R_F \approx 1$. When the sliding velocity is slow enough so that $v\tau_f/R_F \ll 1$, the friction force is due to the elastic force of stretched polymer chain, and it increases with the sliding velocity. When the sliding velocity is high, such that $v\tau_f/R_F \gg 1$, the polymer does not have enough time to form an adsorbing site, and

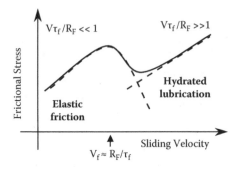

FIGURE 11.9 Schematic curve for the friction of a gel that is adhesive to the substrate in liquid. The friction is the sum of elastic force due to polymer adsorption and viscous force due to hydration of the polymer. At $v \ll v_f$, the first component is dominant. At $v \gg v_f$, the second component is dominant. Transition from elastic friction to lubrication occurs at the sliding velocity characterized by the polymer chain dynamics $v_f \approx R_F/\tau_f = T/\eta R_F^2$. (Reproduced from Kurokawa, T., et al. 2005. *Langmuir* 21: 8643. With permission.)

the friction decreases with the increase of the velocity in this velocity region. This characteristic peak of elastic friction is related to the viscoelastic G'' peak of the gel, since the latter is also determined by the characteristic relaxation time of the network τ_f and mesh size R_F.

The friction from the second contribution, i.e., the viscous friction, increases with the velocity monotonously. Therefore, the first contribution is dominant when $v\tau_f/R_F \ll 1$, and the second contribution is dominant when $v\tau_f/R_F \gg 1$. Around $v\tau_f/R_F \approx 1$, a transition from elastic friction to hydrated layer lubrication occurs, as shown schematically in fig. 11.9.

The repulsion–adsorption model also predicts that: (1) when the substrate is repulsive, the frictional force is lower than the attractive case, and it linearly increases with the normal pressure and velocity when the compressive strain (normal load) is not very high, and (2) when the substrate is attractive, the frictional force increases with the attraction strength. For weak attraction, the pressure dependence of the frictional force is much weaker than in the repulsive case. It becomes stronger when the attraction strength increases [15].

11.4 COMPARISON OF MODEL WITH OBSERVATION

11.4.1 ADHESIVE GEL ON A SMOOTH GLASS

The model provided in eqs. (11.2) and (11.3) implies that, in the attractive case, the transition velocity v_f depends on (a) stiffness of the gel, E or R_F, the latter being related to the partial chain length between the neighboring cross-linking points as well as on the solvent quality; (b) viscosity of solvent; and (c) temperature. As an attraction system, the velocity dependence of friction of a poly(vinyl alcohol) (PVA) gel against a glass substrate in water has been experimentally studied [20]. Figure 11.10 shows the frictional force when a physically cross-linked PVA gel is slid against a glass surface in water at room temperature.

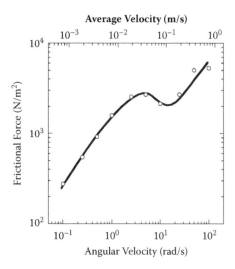

FIGURE 11.10 Velocity dependence of the frictional force for ring-shaped PVA gel rotated on a piece of glass surface in pure water at 25°C as measured with a rheometer. Sample size: inner radius, 5 mm; outer radius, 10 mm; thickness, 3 mm. Degree of swelling q(g/g): 11; normal strain: 30%; normal stress: 8.5 kPa. (Reproduced from Kagata, G., Gong, J. P., and Osada, Y. 2002. *J. Phys. Chem. B*: 106: 4596. With permission.)

The friction increases with the sliding velocity and reaches a maximum at an average sliding velocity of 3.8×10^{-2} m/s, which agrees with the theoretical prediction [15]. From the velocity at which the frictional force attains a maximum, the radius of the partial polymer chain, R_F, was estimated as 16 nm by using eq. (11.2). On the other hand, the Flory radius of the polymer chain R_F was about 3 nm, estimated from the swelling degree $q = 11$ using the c* gel theory. Considering that this order estimation was made using scaling relations, in which all the numerical factors were neglected, these two results are in good agreement.

The temperature effect on the gel friction is shown in fig. 11.11 [20]. An increase in temperature from 5°C to 45°C leads to both a decrease in the frictional force and an increase in the velocity where the friction force shows the maximum. This temperature dependence agrees well with the surface adhesion mechanism. An increase in temperature should result in a decreased friction force due to an increase in the thermal agitation that favors desorption. At the same time, the v_f of the polymer chain increases with an increase in temperature, originating partly from the increased thermal energy and partly from the decreased viscosity of the solvent, as shown by eq. (11.2). As shown in fig. 11.11, when the temperature is raised from 5°C to 45°C, the velocity at which the friction shows a maximum, v_f, increases five times. The theory expressed in eq. (11.2) predicts about a three-times increase in v_f, which roughly agrees with the experimental results.

As suggested by eq. (11.2), the solvent viscosity directly influences the dynamics of the partial chain in a gel. Solvent viscosity was systematically varied over a wide range of $\eta/\eta_0 = 1\text{–}10^3$ (η and η_0 are viscosities of various solvents and water, respectively), and its effect on friction was studied using a tough PAMPS/PAAm (polyacrylamide)

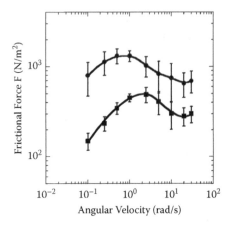

FIGURE 11.11 Frictional forces as a function of the angular velocity for ring-shaped PVA gel rotated against a piece of glass surface in pure water at 5°C (●) and 45°C (■) as measured with a rheometer. Sample size: inner radius, 5 mm; outer radius, 10 mm; thickness, 3 mm. Degree of swelling q(g/g): 11. The error ranges in the figure are standard deviations of the mean values over four samples. (Reproduced from Kagata, G., Gong, J. P., and Osada, Y. 2002. *J. Phys. Chem. B*: 106: 4596. With permission.)

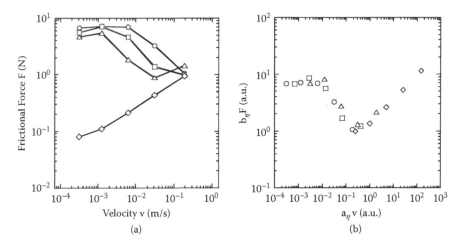

FIGURE 11.12 (a) Sliding velocity v dependence of frictional force F for PAMPS/PAAm DN gels swollen with various solvents and (b) master curve of frictional force $b_\eta F$ against the relative sliding velocity $a_\eta v$. Here, $b\eta$ is an arbitrary factor and a_η is a factor equal to the relative viscosity of the solvent to water. Relative viscosity to water: (○), 1 (pure water); (□), 3.6 (water–ethylene glycol mixture of 1:1 by volume); (△), 16.9 (ethylene glycol); (◇), 945 (glycerol). Normal load: 20 N. Sample size: 20 × 20 × 3 mm. (Reproduced from Kurokawa, T., et al. 2005. *Langmuir* 21: 8643. With permission.)

double-network (DN) gel that is able to sustain a normal pressure as high as several MPa [28]. This DN gel, with PAAm on its surface, is strongly adhesive to a glass surface. Figure 11.12a shows the velocity dependencies of the frictional force, F, for PAMPS/PAAm gels swollen with water, a water–ethylene glycol mixture, ethylene glycol, and glycerol, respectively. The data in fig. 11.12a can be reconstructed into a master curve by shifting the plots of various viscosities by a factor of $a_\eta = \eta/\eta_0$ along the horizontal axis and by an arbitrary factor along the vertical axis. Figure 11.12b shows the master curve thus obtained. All the data fall on a single curve, which is similar to the schematic friction curve, as shown in fig. 11.9. This indicates that $v\tau_f/R_F$ or $\eta v/T^{1/3}E^{2/3}$ is a characteristic parameter for describing the elastic friction of a gel. This velocity–viscosity superposition principle is similar to the v–T superposition principles like, for example, Williams-Landel-Ferry (WLF) [40]. It is found that this velocity–viscosity superposition principle is valid, even when the normal pressure on the gel is as high as the megapascal (MPa) range, a pressure sustained by human articular cartilages [28]. The repulsion–adsorption model estimated the hydration layer as being on the order of the polymer partial-chain radius (or, in other terms, the mesh size) in the region where lubrication becomes dominant after the transition. However, a recent study shows that the thickness of the lubricant solvent layer is also influenced by the hydrodynamic effect, i.e., $\mu \propto (\eta v/P)^{1/2}$ at $v\tau_f/R_F \gg 1$, which is the same as that for hydrodynamic lubrication of solids [12, 28].

11.4.2 Repulsive Gel on a Smooth Glass

As a repulsive case, the friction of negatively charged gel PNaAMPS sliding against a glass substrate was studied in water. Figure 11.13 shows the relationship between the sliding velocity, v, and the frictional force per unit area, f, for PNaAMPS gel slid against a glass surface in water at 25°C under various normal compressive strains measured with a rheometer [20]. As shown in fig. 11.13, f increases with an increase

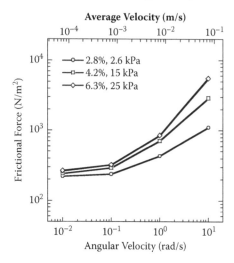

FIGURE 11.13 Velocity dependence of the frictional force for ring-shaped PNaAMPS gel rotated on a piece of glass surface in pure water at 25°C as measured with a rheometer. The numbers in the figure are the values of the normal compressive strains and normal stresses applied during the measurement. Sample size: inner radius, 5 mm; outer radius, 10 mm; thickness, 3 mm. Degree of swelling, 27. (Reproduced from Kagata, G., Gong, J. P., and Osada, Y. 2002. *J. Phys. Chem. B*: 106: 4596. With permission.)

in velocity, where the profiles depend on the applied normal compressive strain and, therefore, on normal stress. At low normal strain (2.8%), f is almost constant for low velocities and then gradually increases at higher velocities. At higher normal strains, f increases only moderately in the low-velocity range but increases distinctly at higher velocities. In other words, the velocity dependence is less notable at low velocities (normal strains) and becomes stronger at higher velocities (normal strains).

The results demonstrate that the frictional behavior of repulsive gels is quite different from that of adhesive gels. The mechanism of gel friction in the repulsive case, however, cannot be explained only in terms of the hydration effect. Actually, an extensive static friction has been observed when two like-charged PNaAMPS gels were slid against each other in water [21]. The two like-charged gel surfaces could not slip against each other at the initial shearing until the shear stress acting on the interface exceeded a certain critical value. The critical yield shear stress (static friction) and strain did not show a distinct dependence on the shearing rate, but they did decrease with increasing temperature (fig. 11.14). The value of static friction also increased with the increase in the normal pressure, and the effect of pressure became more substantial at a higher temperature. Furthermore, the static friction decreased three to four times when the counterions of the polyelectrolyte gels were changed from Na^+ to Cs^+. The observation of the static friction indicates that the proposed repulsion–adsorption model, which predicts a hydrated lubrication mechanism with no static friction for two like-charged (repulsive) hydrogels in pure water, requires modification.

11.4.3 FRICTION REDUCTION BY REPULSIVE DANGLING CHAINS

The effect of solid surface modification with polymer brushes on sliding friction has been investigated using surface force apparatus (SFA) measurements [31–43].

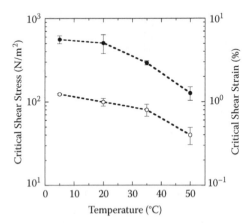

FIGURE 11.14 Temperature dependencies of the critical shear stress (●) and strain (○) for two pieces of disc-shaped PNaAMPS gel (negatively charged) that begin to slide passed each other as measured with a rheometer. The error ranges in the figure are standard deviations of the mean values over three samples. Sample size: radius, 7.5 mm; thickness, 2 mm. Normal pressure: 14 kPa. Angular velocity: 10^{-3} rad/s. (Reproduced from Kagata, G., Gong, J. P., and Osada, Y. 2003. *J. Phys. Chem. B*: 107: 10221. With permission.)

For example, Klein and coworkers reported a massive lubrication between mica surfaces modified by repulsive polyelectrolyte brushes in water [43]. To quantitatively investigate the relationships between the friction and the polymer-brush properties, such as polymer chain length and density, hydrogels of poly(2-hydroxyethyl methacrylate) (PHEMA) with well-defined polyelectrolyte brushes of poly(sodium 4-styrenesulfonate) (PNaSS) of various lengths (molecular weights) were synthesized, keeping the distance between the polymer brushes constant at ca. 20 nm [25]. The effect of polyelectrolyte brush length on the sliding friction against a glass plate, an electrostatic repulsive solid substrate, was investigated in water in a velocity range of 7.5×10^{-5} to 7.5×10^{-2} m/s, as shown in fig. 11.15. Behaviors of interpenetrated network (IPN) gels from PNaSS and PHEMA with a molar ratio R and swelling degree q similar to NaSS/HEMA were also studied as comparisons. It was found that the presence of polymer brushes could dramatically reduce the friction when the

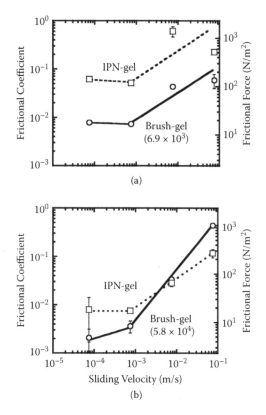

FIGURE 11.15 Velocity dependencies of the frictional coefficient for the gels sliding on a piece of glass surface in pure water at 25°C under a normal pressure of 2.3 kPa. ($\square$) Brush-gels consist of neutral polymer PHEMA as network and negatively charged PNaSS as brushes; ($\bigcirc$) IPN-gels consist of PHEMA and PNaSS with interpenetrated network structure. Numbers in the figure are molecular weights of polymer brushes. (a) $R = 1.2\%$; brush-gel $q = 10.5$, IPN-gel $q = 11$. (b) $R = 15\%$; brush-gel $q = 44$, IPN-gel $q = 52$. (R is NaSS/HEMA molar ratio, and q(g/g) is swelling degree of gels.) (Reproduced from Ohsedo, Y., et al. 2004. *Langmuir* 20: 6549. With permission.)

polymer brushes were not very long. With an increase in the length of the polymer brush, this drag-reduction effect only works at a low sliding velocity, and a gel with long polymer brushes shows an even higher friction than that of a normal network gel at a high sliding velocity.

The strong polymer-length and sliding-velocity dependencies indicate a dynamic mechanism of the polymer-brush effect. Supposing that the hydrated water layer behaves as a lubricant, the apparent lubricating layer thickness L_T can be estimated from the frictional force f by $f = \eta v/L_T$. Figure 11.16 shows that the lubricating layer thickness L_T thus obtained monotonically increases with the sliding velocity from ca. 5 nm to ca. 200 nm for a brush-gel with a length of 8.4 nm, indicating the shear-thinning effect. For the brush-gel with a brush length of 70.2 nm, longer than the next-neighboring distance (20 nm), L_T shows a maximum with the sliding velocity, while the IPN-gel with a similar R and q shows a monotonic increase in L_T with v (fig. 11.16). Figure 11.16 also shows that for the same PNaSS content, the brush-gel shows a thicker L_T than that for the IPN-gel when the brush is short (8.4 nm) or when the sliding velocity is not very high for a long brush (70.2 nm), indicating the drag-reduction effect of the brush. What is interesting is that the change in L_T with sliding velocity is about 100 nm, much larger than the brush length (fig. 11.16a) or the mesh size of the network (fig. 11.16b).

Why does a gel possessing branched dangling polymer chains on its surface exhibit such low friction? Two possible reasons for this might be as follows. First is the enhanced thickness of the solvent layer between branched dangling polymer chains and the sliding substrate. When the interfacial interaction is repulsive, a water layer is retained at the interface, even under a large normal load, to give a low friction. Under the same pressure, the static solvent layer thickness should be the same for both the chemically cross-linked gel and the gel having branched dangling polymer

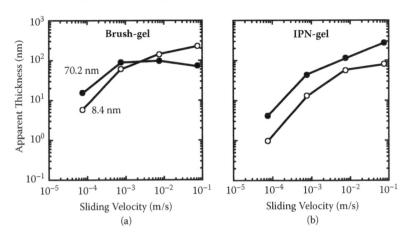

FIGURE 11.16 Sliding velocity dependences of the apparent hydrated layer thickness for (a) brush-gels and (b) IPN-gels of similar degree of swelling q or NaSS/HEMA molar ratio R. The numbers in the figure are contour lengths of PNaSS brushes. (○): $q = 19$; brush-gel $R = 5.9\%$, IPN-gel $R = 13\%$. (●): $R = 15\%$; brush-gel $q = 44$, IPN-gel $q = 52$. (Reproduced from Ohsedo, Y., et al. 2004. *Langmuir* 20: 6549. With permission.)

chains on its surface. However, during sliding, the polymer brushes are deformed more easily than the cross-linked network, and this would increase the effective thickness of the lubricating layer and reduce the shear resistance. The second reason may be the decreased coherence length of the surface layer of the gel. A branched surface would have a shorter coherence length than a cross-linked network surface due to the free polymer chain ends. This ensures independent deformation and chain relaxation on the scale of one macromolecule. This, in turn, suggests a possible bulk contribution to energy dissipation for networked surfaces, which we did not take into consideration in the previous discussion.

11.5 APPLICATION OF ROBUST GELS WITH LOW FRICTION AS SUBSTITUTES FOR BIOLOGICAL TISSUES

Most biological tissues are in the gel state, for example, an articular cartilage is a natural fiber-reinforced hydrogel composed of proteoglycans, type II collagen, and approximately 70% water. The normal cartilage tissue highly contributes to the joint functions, involving ultralow friction, distribution of loads, and absorption of impact energy. When normal cartilage tissues are damaged, it is extremely difficult to regenerate these tissues with currently available therapeutic treatments. Therefore, it is important to develop substitutes for the normal cartilage tissue as a potential therapeutic option. The important characteristic of a material to be considered for load-bearing tissue replacement is that it have mechanical properties comparable with the native tissue. For example, the severe loading conditions imposed on human articular cartilage is as high as 1.9–14.4 MPa [5].

11.5.1 ROBUST GELS WITH LOW FRICTION: AN EXCELLENT CANDIDATE AS ARTIFICIAL CARTILAGE

It has been reported that a hydrogel with branched dangling polymer chains on its surface can effectively reduce the surface sliding frictional coefficient to a value as low as 10^{-4} [19]. Unfortunately, conventional hydrogels are mechanically too weak to be practically used in any stress- or strain-bearing applications, which hinders the extensive application of hydrogels as industrial and biomedical materials. Design and production of hydrogels with a low surface friction and high mechanical strength are crucially important in the biomedical applications of hydrogels as contact lenses, catheters, artificial articular cartilage, and artificial esophagus [44].

In previous work, we reported on a novel method to obtain hydrogels with extremely high mechanical toughness [45, 46]. The hydrogel is composed of two kinds of hydrophilic polymers (double-network [DN] structure) that, despite containing 90% water, have 0.4–0.9-MPa elastic modulus and exhibit a compressive fracture strength as high as several tens of MPa. The point of this discovery is that a DN is composed of a combination of a stiff and brittle first network and a soft and ductile second network. The molar composition and the cross-linking ratio of the first and second networks are also important. A DN hydrogel is completely different in concept from the common interpenetrating polymer networks or a fiber-reinforced hydrogel, which is only a linear combination of two component networks.

Based on this previous research, the new soft and wet materials, with both lower friction and high strength, are synthesized by introducing a weakly cross-linked PAMPS network (to form a triple-network, or TN, gel) or a non-cross-linked linear polymer chain (to form a DN-L gel) as a third component into the optimally tough PAMPS/PAAm DN gel [27]. The TN and DN-L gels were synthesized by UV irradiation after immersing the DN gels in a large amount of a third solution of 1M AMPS and 0.1 mol% 2-oxoglutaric acid with (TN) and without the presence of 0.1 mol% MBAA (DN-L). The mechanical properties of the gels are summarized in table 11.1. After introducing cross-linked or linear PAMPS to DN gel, the fracture strength of TN and DN-L gels remains on the order of megapascals (MPa), and the elasticity of the gels is higher than that of DN gels (≈ 2 MPa). In addition, the fracture strength of DN-L remarkably increases because PAMPS linear chains can effectively dissipate the fracture energy.

Figures 11.17a and 11.17b show the frictional forces (F) and frictional coefficients (μ) of the three kinds of gels as a function of normal pressure (P). The gels were slid against the glass plate in water. The results clearly show that the frictional coefficient decreases in the order DN > TN > DN-L, indicating that introducing PAMPS, especially linear PAMPS, as the third network component obviously reduces the frictional coefficient of the gels.

The DN gel has a relatively large value of frictional coefficient ($\approx 10^{-1}$), since the second network, nonionic PAAm, dominates the surface of the DN gel, which is adsorptive to the glass substrate. However, when a PAMPS network is introduced to the DN gel as the third component, the frictional coefficient of the TN gel decreases to $\approx 10^{-2}$, which is two orders of magnitude lower than that of the DN gel, since the surface of the TN gel is dominated by PAMPS, resulting in repulsive interaction with the glass substrate and thus reducing the frictional force. Furthermore, when linear PAMPS chains are introduced to the surface of a DN gel, the frictional coefficient significantly reduces to $\approx 10^{-4}$, which is one to three orders of magnitude less than that of the TN and DN gels. This demonstrates that the linear PAMPS chains on the gel surface reduce the frictional force due to further repulsive interaction with the glass substrate [19].

It should be emphasized that the lower friction coefficient of DN-L gel can be observed under a pressure range of 10^{-3}–10^5 Pa, which is close to the pressure exerted on articular cartilage in synovial joints. The results demonstrate that the

TABLE 11.1
Mechanical Properties of DN, TN, and DN-L Gels

Gel	Water content (wt%)	Elasticity (MPa)	Fracture stress (max (MPa)	Fracture strain (max (%)
DN	84.8	0.84	4.6	65
TN	82.5	2.0	4.8	57
DT-L	84.8	2.1	9.2	70

Source: Reproduced with permission from Kaneko, D., et al. 2004. *Adv. Mater.* 17: 535.

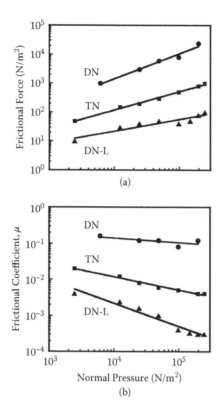

FIGURE 11.17 Normal pressure dependencies of (a) frictional force and (b) frictional coefficient of hydrogels against a glass plate in pure water. Sliding velocity: 1.7×10^{-3} m/s. Symbols denote DN (●), TN (■), and DN-L(▲) gels. (Reproduced from Kaneko, D., et al. 2005. *Adv. Mater.* 17: 535. With permission.)

linear polyelectrolyte chains are still effective in retaining lubrication, even under an extremely high normal pressure.

11.5.2 WEAR PROPERTIES OF ROBUST DN GELS

For application of DN gels to artificial articular cartilage, it is critical to evaluate the gel's wear property, because articular joints are subjected to rapid shear force of high magnitude for millions of cycles over a lifetime. However, there are no established methods for evaluating the wear property of a gel. The pin-on-flat wear testing that has been used to evaluate the wear property of ultrahigh-molecular-weight polyethylene (UHMWPE), which is the only established rigid and hard biomaterial used in artificial joints, was used to evaluate the wear property of DN gels.

The wear properties of four kinds of DN gels, which were composed of synthetic or natural polymers, were evaluated [47]. The first gel was a PAMPS/PAAm DN gel, which consists of poly(2-acrylamide-2-methyl-propane sulfonic acid) and polyacrylamide. The second gel was a PAMPS/PDMAAm DN gel, which consists of poly(2-acrylamide-2-methyl-propane sulfonic acid) and poly(N,N′-dimethyl acrylamide).

The third gel was a cellulose/PDMAAm DN gel, which consists of bacterial cellulose (BC) and poly-dimethyl-acrylamide. The fourth gel was a cellulose/gelatin DN gel, which consists of bacterial cellulose (BC) and gelatin. These four unique gel materials have great potential for application as an artificial cartilage.

Under 1 million times of cyclic friction, which was equivalent to 50 km of friction (50 mm $\times$ 10^6), the maximum wear depths of the PAMPS/PAAm, PAMPS/PDMAAm, BC/PDMAAm, and BC/gelatin gels were 9.5, 3.2, 7.8, and 1302.4 μm, respectively. It is amazing that the maximum wear depth of the PAMPS/PDMAAm DN gel is similar to the value of UHMWPE (3.33 μm). In addition, although the maximum wear depths of the PAMPS/PAAm DN gel and the BC/PDMAAm DN gel were about two to three times higher than that of UHMWPE, these gels could bear the 1 million friction cycles. The results demonstrate that PAMPS/PAAm, PAMPS/PDMAAm, and BC/PDMAAm DN gels are resistant to wear to a greater degree than conventional hydrogels, and the PAMPS/PDMAAm DN gel can potentially be used as a replacement material for artificial cartilage. On the other hand, BC/gelatin DN gel, which is composed of natural materials, shows extremely poor wear properties compared with the other DN gels. The lower wear property of the BC/gelatin DN gel is attributed to its relatively low water content, its higher friction coefficient, and its tendency to become rough by abrasion.

When designing materials for potential use as artificial cartilage, one must consider several requirements, including suitable viscoelasticity, high mechanical strength, durability to repetitive stress, low friction, high resistance to wear, and resistance to biodegradation within the living body. It is difficult to develop a gel material that satisfies even two of these requirements at the same time. However, a recent breakthrough in synthesizing mechanically strong hydrogels changes the conventional gel concept, thereby opening a new era of soft and wet materials as substitutes for articular cartilage and other tissues.

11.6 SUMMARY

The frictional behavior of polymer gels is more complex than that of solids, being dependent on the chemical structure and properties of the gels, the surface properties of sliding substrates, and the measurement conditions. The surface friction of gels on a smooth substrate can be divided into two categories, i.e., adhesive gels and repulsive gels, as determined by the specific combinations of the gels and the opposing surfaces. For adhesive gels, friction is from two contributions, namely, surface adhesion and hydrated lubrication. The former is dominant at low sliding velocities, and the latter is dominant at high velocities. The friction shows an S-shaped curve with the sliding velocity. A transition in friction occurs at the sliding velocity, and this transition is characterized not only by the mesh size and polymer chain relaxation time of the gel, but also by the mechanical properties of the gel. For a repulsive gel, the friction is believed to be due to lubrication of the hydrated water layer. The friction shows a monotonic increase with the sliding velocity due to the nature of hydrodynamic lubrication. The presence of repulsive dangling chains on gel surfaces dramatically reduces the friction, with brush gels showing a friction coefficient as low as 10^{-4}, which is comparable with that found for animal joints.

Looking for materials with a low surface friction has been one of the classical and everlasting research topics for material scientists and engineers. Despite many efforts, it has been shown that surface modification or adding lubricants is not a very effective way to reduce the steady-state sliding friction between two solids, which show a frictional coefficient $\mu = 10^{-1}$ even in the presence of a lubricant. The discovery of an extremely low-friction gel should enable hydrogels to find wide application in many fields where low friction is required.

REFERENCES

1. Dowson, D. 1979. *History of tribology*. London: Longman.
2. Suh, N. P. 1986. *Tribophysics*. Englewood Cliffs, NJ: Prentice-Hall.
3. Dhinojwala, A., Cai, L., and Granick, S. 1996. *Langmuir* 12: 4537.
4. McCutchen, C. W. 1962. *Wear* 5: 1.
5. McCutchen, C. W. 1978. *Lubrication of joints: The joints and synovial fluid*. New York: Academic Press.
6. Dowson, D., Unsworth, A., and Wright, V. 1970. *J. Mech. Eng. Sci.* 12: 364.
7. Ateshian, G. A., Wang, H. Q., and Lai, W. M. 1998. *J. Tribology* 120: 241.
8. Hodge, W. A., Fijian, R. S., Carlson, K. L., Burgess, R. G., Harris, W. H., and Mann, R. W. 1986. *Proc. Natl. Acad. Sci. U.S.A.* 83: 2879.
9. Grodzinsky, A. J. 1985. *CRC Crit. Rev. Anal. Chem.* 9: 133.
10. Buschmann, M. D., and Grodzinsky, A. J. 1995. *J. Biomech. Eng.* 117: 179.
11. Wojtys, E. M., and Chan, D. B. 2005. *Instr. Course Lect.* 54: 323.
12. Presson, B. N. J. 1998. *Sliding friction: Physical principles and applications*. In *Nanoscience and technology series*. 2nd ed. Berlin: Springer-Verlag.
13. Fung, Y. C. 1993. *Biomechanics: Mechanical properties of living tissues*. 2nd ed. New York: Springer-Verlag.
14. Gong, J. P., Higa, M., Iwasaki, Y., Katsuyama, Y., and Osada, Y. 1997. *J. Phys. Chem. B* 101: 5487.
15. Gong, J. P., and Osada, Y. 1998. *J. Chem. Phys.* 109: 8062.
16. Gong, J. P., Iwasaki, Y., Osada, Y., Kurihara, K., and Hamai, Y. 1999. *J. Phys. Chem. B*: 103: 6001.
17. Gong, J. P., Kagata, G., and Osada, Y. 1999. *J. Phys. Chem. B*: 103: 6007.
18. Gong, J. P., Iwasaki, Y., and Osada, Y. 2000. *J. Phys. Chem. B*: 104: 3423.
19. Gong, J. P., Kurokawa, T., Narita, T., Kagata, K., Osada, Y., Nishimura, G., and Kinjo, M. 2001. *J. Am. Chem. Soc.* 123: 5582.
20. Kagata, G., Gong, J. P., and Osada, Y. 2002. *J. Phys. Chem. B*: 106: 4596.
21. Kagata, G., Gong, J. P., and Osada, Y. 2003. *J. Phys. Chem. B*: 107: 10221.
22. Kurokawa, T., Gong, J. P., and Osada, Y. 2002. *Macromolecules* 35: 8161.
23. Baumberger, T., Caroli, C., and Ronsin, O. 2002. *Phys. Rev. Lett.*: 88: 75509.
24. Baumberger, T., Caroli, C., and Ronsin, O. 2003. *Eur. Phys. J. E* 11: 85.
25. Ohsedo, Y., Takashina, R., Gong, J. P., and Osada, Y. 2004. *Langmuir* 20: 6549.
26. Tada, T., Kaneko, D., Gong, J. P., Kaneko, T., and Osada, Y. 2004. *Tribol. Lett.* 17: 505.
27. Kaneko, D., Tada, T., Kurokawa, T., Gong, J. P., and Osada, Y. 2004. *Adv. Mater.* 17: 535.
28. Kurokawa, T., Tominaga, T., Katsuyama, Y., Kuwabara, R., Furukawa, H., Osada, Y., and Gong, J. P. 2005. *Langmuir* 21: 8643.
29. Nitta, Y., Haga, H., and Kawabata, K. 2002. *J. Phys. IV* 12: 319.
30. Jiang, Z. T., Tominaga, T., Kamata, K., Osada, Y., and Gong, J. P. 2006. *Colloids Surf. A: Physicochem. Eng. Aspects* 284: 56.

31. Du, M., Maki, Y., Tominaga, T., Furukawa, H., Gong, J. P., Osada, Y., and Zheng, Q. 2007. *Macromolecules* 40: 4313.
32. Amontons, M. 1699. *Mem. Acad. R. Sci.* 206.
33. Adamson, A. W. 1990. *Physical chemistry of surfaces.* New York: Wiley.
34. de Gennes, P. G. 1979. *Scaling concept in polymer physics.* Ithaca, NY: Cornell University Press.
35. Schallamach, A. 1963. *Wear* 6: 375.
36. Chernyak, Y. B., and Leonov, A. I. 1986. *Wear* 108: 105.
37. Savkoor, A. R. 1965. *Wear* 8: 222.
38. Ludema, K. C., and Tabor, D. 1966. *Wear* 9: 329.
39. Vorvolakos, K., and Chaudhury, M. K. 2003. *Langmuir* 19: 6778.
40. Ferry, J. D. 1980. *Viscoelastic properties of polymers.* 3rd ed. New York: Wiley.
41. Klein, J., Kumacheva, E., Mahalu, D., Perahia, D., and Fetters, L. 1994. *Nature* 370: 634.
42. Grest, G. S. 1999. *Adv. Polym. Sci.* 138: 149.
43. Raviv, U., Giasson, S., Kampf, N., Gohy, J. F., Jerome, R., and Klein, J. *Nature* 425: 163.
44. Freeman, M. E., Furey, M. J., Love, B. J., and Hampton, J. M. 2000. *Wear* 241: 129.
45. Gong, J. P., Katsuyama, Y., Kurokawa, T., and Osada, Y. 2003. *Adv. Mater.* 15: 1155.
46. Na, Y.-H., Kurokawa, T., Katsuyama, Y., Tsukeshiba, H., Gong, J. P., Osada, Y., Okabe, S., and Shibayama, M. 2004. *Macromolecules* 37: 5370.
47. Yasuda, K., Gong, J. P., Katsuyama, Y., Nakayama, A., Tanabe, Y., Kondo, E., Ueno, M., and Osada, Y. 2005. *Biomaterials* 26: 4469.

12 Friction of Elastomers against Hydrophobic and Hydrophilic Surfaces

Sophie Bistac

ABSTRACT

Self-assembled monolayers (SAMs), formed by chemical adsorption of surfactant molecules on solid surfaces, can be used as boundary lubricants. The grafting of monolayers on a rigid substrate allows to control the surface chemistry and consequently the surface energy and reactivity without varying the roughness and the mechanical properties. The objective of this work was to illustrate the influence of the nature of interfacial interactions on the sliding friction of polymers. Two different substrates were used: a hydrophilic substrate (hydroxylated silicon wafer) and a hydrophobic one (silicon wafer grafted with a CH_3 terminated silane). Both substrates exhibit identical stiffness and roughness and differ only by their surface chemistry. Friction of model elastomers (crosslinked poly(dimethylsiloxane), PDMS) against both types of substrates was quantified, by using a translation tribometer for different normal forces and friction speeds. Experimental results indicate that the friction coefficients of PDMS against hydrophilic and hydrophobic substrates are different for low friction speed (with a greater friction coefficient for hydrophilic wafer). However, at higher speed, friction coefficients obtained with both types of substrates become identical. This work evidences the complex competition between interfacial interactions and polymer surface rheological behaviour.

12.1 INTRODUCTION

Polymers generally exhibit complex tribological behaviors due to different energy dissipation mechanisms, notably those induced by internal friction (chain movement), which is dependent on both time and temperature. Polymer friction is then governed by interfacial interactions and viscoelastic dissipation mechanisms that are operative in the interfacial region and also in the bulk, especially in the case of soft materials. Friction of a polymer can be closely linked to its molecular structure. The role of chain mobility has been studied in the case of elastomers, based on dissipation phenomena during adhesion and friction processes of the elastomer in contact with a silicon wafer covered by a grafted layer [1–5].

Self-assembled monolayers (SAMs), formed by chemical adsorption of molecules on a solid surface [6, 7], can be used as boundary lubricants [8–11]. The grafting of monolayers of surfactant molecules onto a rigid substrate allows control of the surface chemistry and, consequently, the surface energy and reactivity without altering the roughness and the mechanical properties. Friction of these thin films on a nanoscale has often been studied using atomic force microscopy (AFM). Moreover, computer-simulation techniques for studying these phenomena at the atomic scale have allowed a better understanding of the molecular mechanisms of thin-film tribology [10, 11]. Studies of polymer tribology are generally focused on the effects of molecular weight [12, 13], sliding velocity, or normal load [14–16]; however, the role of interface is rarely analyzed.

Another difficulty in polymer tribology is due to the fact that the rheology of a polymer surface can be completely different from that of the bulk [17, 18]. Moreover, the surface rheology cannot be easily known by an extrapolation of polymer bulk rheology. Surface force apparatus facilitate investigation of the rheology and tribology of thin films between shearing surfaces through the measurement of parameters such as the real area of contact, the local load, and the sheared film thickness. For very thin films, the viscosity is greatly increased from that expected from continuum behavior. This effect is due to the close approach of surfaces that modify the liquid structure between the surfaces [19, 20].

The previous scientific studies have, therefore, shown that the friction of polymers is greatly influenced by the adhesion phenomena. The nature and number of interfacial interactions between the polymer and the substrate can indeed directly affect the friction coefficient. The objective of the work described in this chapter was to illustrate the complex role of interface. Polymer friction has been investigated in contact with two types of smooth and rigid substrates: a hydrophilic silicon wafer (hydroxylated by a piranha treatment) and a hydrophobic silicon wafer, obtained by chemical grafting of a CH_3-terminated silane (corresponding to chemical adsorption of a surfactant layer onto the wafer). Both substrates exhibit identical stiffness and roughness and differ only in their surface chemical compositions. Two cross-linked poly(dimethylsiloxane) (PDMS) elastomers, varying in their cross-linking degrees, were rubbed against both types of substrates. Friction experiments were performed with a translation tribometer at various normal forces and speeds. The evolution of the friction coefficient of both PDMS samples as a function of the substrate chemistry, speed, and normal force was analyzed.

12.2 MATERIALS AND METHODS

Two vinyl-terminated poly(dimethylsiloxane) (PDMS) samples were used (provided by Gelest, Morrisville, PA). The two samples differed by their initial molecular weights M_w: $M_w = 6,000$ g/mol (PDMS A) and $M_w = 17,200$ g/mol (PDMS B). PDMS samples were cross-linked with tetrakis(dimethylsiloxy)silane (four functional sites), using a platinum catalyst at room temperature. The stoichiometry ratio (cross-linker/ PDMS) was equal to 1.1 (10% excess of the cross-linker) to ensure a complete reaction. Increasing the initial molecular weight of PDMS induces major consequences in the form of a lower cross-link density (higher chain length between chemical cross-links) and, consequently, a lower modulus. The initial molecular weight M_w, mean molecular weight between cross-links M_c (determined by the swelling method), glass transition temperature T_g (measured by differential scanning calorimetry at a scan

TABLE 12.1
PDMS Characteristics: Initial Molecular Weight before Cross-Linking (M_w), Mean Molecular Weight between Cross-Links (M_c), Glass Transition Temperature (T_g), and Young's Modulus (E) of Cross-Linked PDMS A and B

Polymer	M_w (g/mol)	M_c (g/mol)	T_g (°C)	E (MPa)
PDMS A	6,000	7,500	−123	1.40
PDMS B	17,200	18,500	−123	0.42

rate of 10°C/min), and Young's modulus E (determined by tensile test at 10 mm/min) of PDMS A and B are presented in table 12.1. PDMS hemispheres (diameter = 16 mm) were used for the friction experiments.

The surface energy of PDMS films was determined by wettability measurements. Equilibrium contact angles of drops of water (a polar liquid) and diiodomethane (a nonpolar liquid) were measured with an automated Kruss apparatus. Contact-angle hysteresis (difference between receding and advancing angles) was measured for water and diiodomethane by a tensiometry technique (speed of 10 mm/min, immersion length equal to 7 mm).

Two kinds of model substrates were used: a hydrophilic silicon wafer and a hydrophobic grafted wafer. Substrates measuring 1.5 × 2 cm were slices of polished silicon wafers Si(100) (from Mat. Technology, Morangis, France).

The hydrophilic surface was obtained by a treatment with a piranha solution (H_2SO_4/H_2O_2). This treatment was aimed to produce a high-surface silanol density.

The hydrophobic substrate was obtained by grafting a methyl-terminated silane (hexadecyltrichlorosilane) on a silicon wafer (previously piranha treated). The terminal group of the grafted layer is then a -CH_3 function that exhibits a hydrophobic character. After the cleaning and the hydroxylation (by piranha solution) steps, the wafers were immersed in a solution of 0.1 vol% hexadecyltrichlorosilane (HTS) in carbon tetrachloride (CCL_4) for 8 h at room temperature. The substrates were then rinsed in CCl_4 solution to remove any excess physisorbed silanols, rinsed with deionized water, and dried under a nitrogen flow.

Friction properties were measured using a translation tribometer (fig. 12.1). The PDMS hemisphere was brought into contact with the substrate under a given normal force. The substrate was then moved at a given speed, and the tangential force, which corresponds to the friction force, was measured. At least five friction measurements were performed for each experimental condition to obtain a mean friction coefficient value. The friction coefficient μ is defined as the quotient of the tangential (or frictional) force F_T divided by the applied normal force F_N, i.e., $\mu = F_T/F_N$. All data were collected at ambient conditions (25°C).

Friction experiments were carried out for both PDMS samples in contact with both hydrophobic and hydrophilic substrates at different friction speeds (25, 50, and 120 mm/min) and normal forces (1, 2, and 3 N). Only a single passage was performed for each sample.

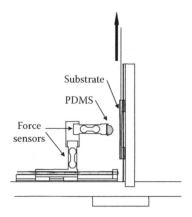

FIGURE 12.1   Schematic of the translation tribometer.

12.3 RESULTS AND DISCUSSION

Surface energy γ_s was calculated using the Fowkes method [21] to determine the dispersion and polar (or nondispersion) component of γ_s. Wettability measurements indicate a surface energy close to 20 mJ/m² for the hydrophobic wafer (CH₃ grafted surface), with a polar component equal to zero. This result confirms the hydrophobic character of the grafted wafer. A surface energy equal to 79 mJ/m² was obtained for the hydrophilic substrate (polar component = 42 mJ/m²).

Surface energy and contact-angle hysteresis of both PDMS samples are presented in table 12.2. Quite identical low values of surface energy (determined from equilibrium contact angles) are observed for both samples, with a nondispersion (or polar) component equal to zero, which shows that the ability of PDMS to exchange polar or specific interactions (e.g., hydrogen bonding) can be considered as negligible. This low surface energy (close to 27 mJ/m²) is mainly due to the orientation of the CH₃ groups. (Moreover, PDMS chains are initially vinyl terminated, and do not contain any OH function.)

Contact-angle hysteresis measurements (performed with water) indicate, however, significant differences between PDMS samples, with an increase of hysteresis for PDMS B. Contact-angle hysteresis can be induced by a chemical or topological heterogeneity. However, the chemical composition of both PDMS samples is identical, and the surface roughness is low and also identical for both samples (close to

---

TABLE 12.2
Surface-Energy Values and Contact-Angle Hysteresis of PDMS: Dispersion (γ_s^d) and Nondispersion (γ_s^{nd}) Components of Surface Energy γ_s

	Surface energy			Hysteresis with water (°)
	γ_s^d (±1 mJ/m²)	γ_s^{nd} (±1 mJ/m²)	γ_s (±2 mJ/m²)	
PDMS A	27	0	27	12
PDMS B	26	0	26	17

10 nm, determined by atomic force microscopy in the tapping mode). The origin of this hysteresis is probably linked to energy-dissipation phenomena associated with interfacial chain movement. The longer free and pendant chains of PDMS B will dissipate more energy during the interfacial shear induced by the liquid flow, explaining a higher value of hysteresis.

The elastomers differ in their cross-linking degree, which induces a great difference in their mechanical properties (modulus). For this reason, the contact area between the PDMS hemisphere and the substrate will be different for the two samples. The contact area of the PDMS hemisphere in contact with a smooth glass substrate (whose transparency allows the measurement) was experimentally determined using a video camera equipped with a microscope. The values obtained for different normal forces are given in table 12.3. Contact areas increase as a function of normal force. Greater contact areas are also obtained for PDMS B, in correlation with its mechanical properties.

Other structural differences are also induced by varying the cross-linking degree. The cross-linking reaction is indeed usually incomplete, leading to an imperfect network. Increasing the initial molecular weight of PDMS induces globally two consequences: (a) a lower cross-link density (higher chain length between chemical links) and (b) a greater quantity of free chains (not linked to the network) and pendant chains (chemically linked to the network by only one chain end). Even though they are chemically identical, the PDMS samples differ in their molecular structures.

Friction experiments were carried out for both PDMS samples in contact with both hydrophobic and hydrophilic substrates at various friction speeds (25–120 mm/min) and normal forces (1–3 N) using a translation tribometer. Only a single pass was performed, which means that the PDMS was always rubbed on a fresh wafer surface.

Table 12.4 presents friction coefficient values of PDMS A and B for various normal forces and speeds on both substrates. The influence of each parameter (cross-link degree of PDMS, type of substrate, normal force, and friction speed) will be analyzed and discussed below.

Table 12.4 indicates that the friction coefficients of both PDMS samples are quite similar, irrespective of the experimental conditions. However, the two elastomers exhibit different stiffness and, consequently, the corresponding friction stress (friction force divided by contact area) will be different, with a greater friction

---

TABLE 12.3

Contact Area between PDMS Hemispheres and a Glass Substrate, for Different Normal Forces

| Normal force (N) | Contact area (mm²) | |
	PDMS A	PDMS B
1	7	14
2	11	18
3	15	24

TABLE 12.4
Friction Coefficient (μ) Values of PDMS A and B, in Contact with Hydrophobic and Hydrophilic Substrates for Different Friction Speeds and a Force Equal to 1 and 3 N (Δμ = ±0.02)

Normal force (N)		1 N		3 N	
Speed (mm/min)		Hydrophobic substrate	Hydrophilic substrate	Hydrophobic substrate	Hydrophilic substrate
25	PDMS A	1.04	1.58	0.81	0.88
	PDMS B	1.05	1.47	0.76	1.00
50	PDMS A	1.29	1.53	0.95	0.86
	PDMS B	1.10	1.42	0.80	1.02
120	PDMS A	1.35	1.36	1.04	0.93
	PDMS B	1.27	1.23	0.92	0.91

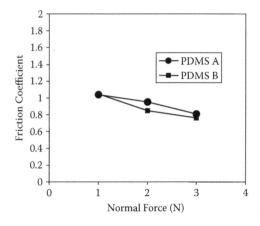

FIGURE 12.2 Friction coefficient of PDMS A and B as a function of normal force in contact with hydrophobic wafer (speed = 25 mm/min).

stress for PDMS A. Explanations based on the respective roles of elastic and adhesive contact have been previously given for these friction stress differences [12, 22–24].

The influence of cross-link degree on the friction coefficient is negligible, as illustrated in fig. 12.2, which shows the evolution of the friction coefficient of PDMS A and B as a function of normal force.

Table 12.4 indicates that the friction coefficient increases when the normal force is decreased for all PDMS samples and wafer substrates. This strong influence of normal force is illustrated in fig. 12.3, which shows the evolution of the friction coefficient of PDMS A as a function of normal force for both substrates (speed = 25 mm/min).

It is interesting to note that the difference in coefficient of friction between the two substrates becomes smaller at higher normal force. This effect could be due to the higher confinement of the interface under a greater normal load. The elastic contact becomes predominant, and the effect of interface chemistry is then masked. The same evolution, i.e., an increase of friction coefficient at low normal force, is also observed for higher friction speed, as shown in fig. 12.4 for PDMS B (speed = 120 mm/min).

For both PDMSs, for both substrates, and for all friction speeds, a great effect of normal force is observed. The higher friction coefficient observed at low normal force could be explained by the role of adhesion, which is magnified at low load (where the bulk contribution is lower). The contribution of interfacial interactions (or adhesive contact) is then magnified. These interfacial interactions will activate viscoelastic dissipation mechanisms, increasing the friction resistance.

Moreover, fig. 12.4 (corresponding to a high friction speed) also indicates that the friction coefficients of both hydrophobic and hydrophilic wafers are similar whatever the normal force, contrary to the results illustrated in fig. 12.3, obtained at

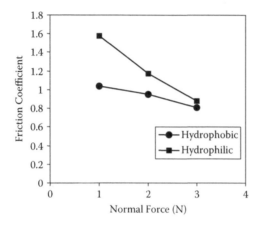

FIGURE 12.3 Friction coefficient of PDMS A as a function of normal force (friction speed = 25 mm/min).

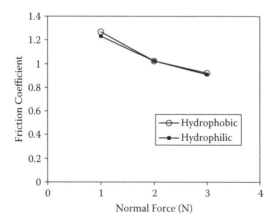

FIGURE 12.4 Friction coefficient of PDMS B as a function of normal force (friction speed = 120 mm/min).

a lower friction speed, for which a great difference between the two substrates can be observed, especially for lower normal forces.

The coupled effect of both substrate surface chemistry and friction speed is analyzed below.

Significant differences between hydrophobic and hydrophilic substrates are observed at lower speed, with a low friction for hydrophobic wafer at low normal load. For example, as illustrated in fig. 12.5 (presenting the friction coefficient of PDMS A versus speed), the friction coefficient of PDMS A (at 25 mm/min and 1 N) is equal to 1.04 for the hydrophobic wafer and 1.58 for the hydrophilic wafer (factor equal to 1.5 between the two coefficients). However, when the friction speed is increased, the difference between the friction coefficients of hydrophilic and hydrophobic systems becomes lower, and both coefficients are identical for high speed, with a value close to 1.3 (for a normal force of 1 N).

The friction coefficient is increased as a function of speed for hydrophobic substrate and is decreased for hydrophilic substrate, as illustrated in figs. 12.5 and 12.6, which present the evolution of the friction coefficient as a function of speed for both substrates for PDMS A and B, respectively (at a normal force equal to 1 N). This

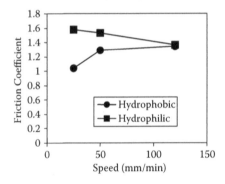

FIGURE 12.5 Evolution of the friction coefficient of PDMS A as a function of speed, in contact with hydrophilic and hydrophobic wafers ($F_N = 1$ N).

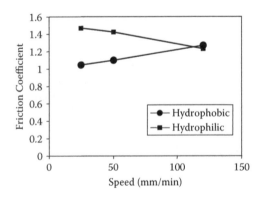

FIGURE 12.6 Evolution of the friction coefficient of PDMS B as a function of speed, in contact with hydrophilic and hydrophobic wafers ($F_N = 1$ N).

result means that the speed dependence of the friction coefficient is strongly dependent on the nature of the substrate, and vice versa.

These results indicate that at higher speeds, friction coefficients of both hydrophobic and hydrophilic systems are equal, i.e., the effect of the interface becomes negligible. The difference between the two substrates is also lower for high normal force. These more severe conditions (higher speed or normal force) are able to strongly confine the PDMS surface. The specific rheological behavior of this confined interfacial layer could explain these complex speed dependences. A competition between interfacial interactions and polymer cohesion can be envisaged. At low speeds (and low normal loads), interfacial interactions will control the friction, and the role of the substrate surface chemistry is then significant. At higher speeds, the friction is governed by the intrinsic rheological behavior of the constrained polymer surface, and the influence of the substrate surface becomes negligible.

Higher shear rate and stress are indeed able to induce a chain orientation at the PDMS surface [24]. Such an orientation will probably modify the rheological behavior of the polymer interface, and the anisotropy of the confined interfacial layer is able to induce a lower shear resistance. Moreover, it is also necessary to interpret the speed dependence of the friction for both substrates, with an increase of the friction coefficient with speed for the hydrophobic wafer and a decrease for the hydrophilic one.

At low speeds, strong interactions between PDMS and the hydrophilic substrate activate a dissipation mechanism (chain deformation), leading to a high friction coefficient. However, when the speed is increased, the chain orientation at the interface during sliding can hamper the mechanical stress transmission from the surface to the bulk polymer, thereby reducing the elastomer deformation and consequently the friction resistance. At high speeds, the interfacial shear of this oriented layer will then consume less energy (less dissipative), explaining the decrease of the friction coefficient with speed observed for the hydrophilic wafer.

For hydrophobic substrates, the interfacial interactions are weak, leading to a low friction coefficient at low speed. However, when the speed (or normal load) is increased, the same chain confinement and orientation will occur at the interface during sliding, but in that case, the shear of the oriented layer will be more dissipative than the effect of interfacial interactions, which are very low. This competition between interfacial interactions and surface rheological behavior is able to explain the increase of the friction coefficient with speed for the hydrophobic substrate.

In summary, a high shear rate is able to confine and orient the chains of the PDMS surface. The cohesion, or shear resistance, of this layer will be lower (i.e., less dissipative) than hydrophilic–wafer/PDMS interactions, explaining the decrease of the friction with speed for the hydrophilic substrate, but greater (more dissipative) than hydrophobic–wafer/PDMS interactions, explaining the increase of the friction with speed for the hydrophilic substrate. Finally, at high speed, shear should occur preferentially within this confined layer (and consequently not exactly at the polymer–substrate interface), leading to similar friction for both substrates. Analysis of the transfer layer (observed by atomic force microscopy) allows a better understanding of the involved mechanisms [25–27].

This complex behavior cannot be generalized to other polymers. The low stiffness of elastomers is able to generate a low-cohesion interfacial layer, and then to make possible a balanced competition between interfacial interaction and shear resistance of the confined layer.

For other types of polymers, especially glassy polymers that exhibit a high stiffness at room temperature, the effect of the substrate chemistry on the tribological behavior can be different. For example, previous studies have been performed on polystyrene (PS) samples sliding against both hydrophobic and hydrophilic wafers [28, 29].

First, the influence of the substrate chemistry was more significant compared with the case of PDMS, with PS friction coefficients close to 0.40 and 0.15, respectively, for hydrophilic and hydrophobic wafers. The friction coefficient of the hydrophilic wafer was always much greater than that of the hydrophobic one, whatever the experimental conditions (speed, normal force). Moreover, the effect of friction speed was totally different in the case of polystyrene compared to PDMS. A slight increase of the friction coefficient with speed could indeed be observed for both types of substrates, attributed to viscoelastic effects. It was also shown that the normal force had a negligible influence.

Polystyrene exhibits a greater cohesion. For polystyrene, the competition between polymer cohesion and interfacial interactions is not balanced. Friction is still governed by interfacial interactions, whatever the experimental conditions (for the studied ranges of normal forces and speed).

The tribological behavior of this glassy polymer is thus completely different compared with PDMS elastomers. The influence of the substrate chemistry on the friction of PDMS and PS is very different. The effect of interface chemistry on friction is thus not so evident. The friction coefficient of hydrophilic and hydrophobic substrates can be identical or different, depending on the experimental conditions. Friction speed also plays a major role through its influence on polymer interfacial rheology, especially in the case of soft polymers.

12.4 CONCLUSION

The aim of this study was to illustrate the complex role of the interface through the study of PDMS friction in contact with both hydrophilic and hydrophobic silicon wafers. Experimental results have shown that the friction coefficient of hydrophilic substrates can be either greater than or similar to that of hydrophobic wafers, depending on the friction speed (and normal load). At low speed, a significant difference between the two substrates is observed, but the influence of the substrate chemistry becomes negligible at higher friction speed.

The presence of a confined interfacial layer, with specific rheological behavior, is proposed to explain this complex behavior. The low stiffness of PDMS allows a competition between the (low) cohesion and the confined chain layer at the PDMS surface and the adhesion level (interfacial interactions between PDMS and substrates). At low speeds, interfacial interactions have a significant effect and partly govern the friction, and at high speeds the influence of the substrate surface becomes negligible and friction is then governed by the polymer's intrinsic viscoelastic behavior. Experimental results underline the subtle competition between interfacial interactions and polymer rheological properties, especially for PDMS samples. Comparison with a glassy polymer (polystyrene) indicates that the influence of surface chemistry of the substrate on friction is subtle. Friction speed appears to be a major parameter, especially through its influence on polymer interfacial properties (orientation, rheology, etc.).

REFERENCES

1. Ghatak, A., Vorvolakos, C., She, H., Malotky, D. L., and Chaudhury, M. K. 2000. *J. Phys. Chem. B* 104: 4018.
2. She, H., Malotky, D. L., and Chaudhury, M. K. 1998. *Langmuir* 14: 3090.
3. Chaudhury, M. K., and Owen, M. J. 1993. *J. Phys. Chem.* 97: 5722.
4. Deruelle, M., Léger, L., and Tirrell, M. 1995. *Macromolecules* 28: 7419.
5. Amouroux, A., and Léger, L. 2003. *Langmuir* 19: 1396.
6. Ulman, A. 1991. In *An introduction to ultrathin organic films: From Langmuir–Blodgett to self-assembly.* San Diego: Academic Press.
7. Sellers, H., Ulman, A., Shnidman, Y., and Eilers, J. 1993. *J. Am. Chem. Soc.* 115: 9389.
8. Zhang, Q., and Archer, L. A. 2005. *Langmuir* 21: 5405.
9. Chandross, M., Grest, G. D., and Stevens, M. J. 2002. *Langmuir 18*: 8392.
10. Bhushan, B., Israelachvili, J. N., and Landman, U. 1995. *Nature* 374: 607.
11. Bhushan, B., Kulkarni, A. V., Koinkar, V. N., Boehm, M., Odoni, L., Martelet, C., and Belin, M. 1995. *Langmuir* 11: 3189.
12. Galliano, A., Bistac, S., and Schultz, J. 2003. *J. Colloid Interface Sci.* 265: 372.
13. Lee, S. W., Yoon, J., Kim, H. C., Lee, B., Chang, T., and Ree, M. 2003. *Macromolecules* 36: 9905.
14. Gasco, M. C., Rodriguez, F., and Long, T. 1998. *J. Appl. Polym. Sci.* 67: 1831.
15. Zhang, S., and Lan, H. 2002. *Tribol. Int.* 35: 321.
16. Rubinstein, S. M., Cohen, G., and Fineberg, J. 2004. *Nature* 430: 1005.
17. Wallace, W. E., Fischer, D., Efimenko, K., Wu, W. L., and Genzer, J. 2001. *Macromolecules* 34: 5081.
18. Luengo, G., Schmitt, F. J., Hill, R., and Israelachvili, J. 1997. *Macromolecules* 30: 2482.
19. Luengo, G., Israelachvili, J., and Granick, S. 1996. *Wear* 200: 328.
20. Israelachvili, J., and Kott, S. J. 1989. *J. Colloid Interface Sci.* 129: 461.
21. Fowkes, F. M. 1968. *J. Colloid Interface Sci.* 28: 493.
22. Galliano, A., Bistac, S., and Schultz, J. 2003. *J. Adhesion* 79: 973.
23. Bistac, S., and Galliano, A. 2005. *Tribol. Lett.* 18: 21.
24. Bistac, S., and Galliano, A. 2005. In *Adhesion: Current research and application.* Ed. W. Possart. Weinheim: Wiley-VCH.
25. Elzein, T., Galliano, A., and Bistac, S. 2004. *J. Polym. Sci. B* 42: 2348.
26. Elzein, T., Kreim, V., Ghorbal, A., and Bistac, S. 2006. *J. Polym. Sci.* 44: 3272.
27. Elzein, T., Kreim, V., and Bistac, S. 2006. *J. Polym. Sci.* 44: 1268.
28. Ghorbal, A., Schmitt, M., and Bistac, S. 2006. *J. Polym. Sci.* 44: 2449.
29. Bistac, S., Ghorbal, A., and Schmitt, M. 2006. *Prog. Organic Coatings* 55: 345.

13 Tribological Properties of Ag-Based Amphiphiles

Girma Biresaw

ABSTRACT

Most ag-based materials are amphiphilic because they comprise polar and non-polar groups within the same molecule. One of the major categories of amphiphilic ag-based materials are seed oils, which have been actively investigated as substitutes for petroleum in a wide variety of consumer and industrial applications. Due to their amphiphilicity, seed oils adsorb on surfaces and alter various surface and interfacial properties. The adsorption properties of seed oils at metal-metal, starch-metal, hexadecane-water interfaces were investigated as a function of seed oil chemical structure, seed oil concentration, and substrate surface properties. The effect of vegetable oil adsorption on surface properties was monitored using interfacial tension and boundary friction measurements. The data were then analyzed using various adsorption models to estimate the free energies of adsorption of the vegetable oils as a function of vegetable oil and substrate characteristics. The result showed that the estimated free energies of adsorptions were independent of the method used to probe the adsorption of the vegetable oil at the interface (interfacial tension vs. boundary friction). However, the estimated free energies of adsorption were found to be functions of vegetable oil chemical structure, substrate surface properties, and the adsorption model used to analyze the adsorption data.

13.1 INTRODUCTION

The dwindling supply of petroleum has intensified the search for alternative sources of raw materials for various applications, including lubrication [1–5]. Of particular interest are raw materials from renewable sources. Farm products meet these requirements and are the subject of intense investigation for material development. In addition to being renewable, farm products are biodegradable, thus making them the preferred environmentally friendly raw materials for the production, use, and disposal of commercial and consumer products. Farm products are also nontoxic to humans and other living organisms. As a result, the processing, use, and disposal of products made from farm-based materials present little or no risk to workers, consumers, or the environment. Another factor that favors farm products as the preferred raw materials is the current surplus supply of farm products over market demand. Thus, new uses are required to use up the current excess farm products in the market, and thereby ensure the profitability of farming.

Plant-based raw materials that are currently the focus of intense research include starches, proteins, gums, cellulose, fibers, oils, and waxes. These are investigated for various consumer and industrial applications in such sectors as automotive, aerospace, housing, medical devices, pharmaceuticals, agriculture, apparel, cosmetics, etc. [1–10]. In most instances, these efforts are aimed at replacing current petroleum-based products with plant-based biopolymer products. However, successful replacement of petroleum-based products requires overcoming certain inherent weaknesses of plant-based products in these applications. Examples of such weaknesses include poor water resistance, poor thermal and oxidative stability, and poor bioresistance. Various approaches are being pursued to overcome these shortcomings, including modification of biopolymers through chemical, thermal, enzymatic, and other methods; blending with synthetic biodegradable polymers; etc. [1–10].

One class of plant-based products that have a great deal of potential as substitutes for petroleum are vegetable oils. Vegetable oils are important raw materials for a variety of applications currently under development, including biofuels, biopolymers, soaps and detergents, lubricants, cosmetics, pharmaceuticals, adhesives, and coatings [5–6, 11–14]. Besides being renewable, biodegradable, safe, and in abundant supply, vegetable oils have important chemical and physical properties that make them attractive raw materials for the various applications. First, most vegetable oils are liquid at room temperature and, thus, are easy to handle and process. Second, vegetable oils have functional groups and unsaturations that serve as reactive sites for converting them into various materials. Thus, through the application of chemical, enzymatic, heat, or other methods on the reactive sites, vegetable oils can be modified to improve their properties (e.g., oxidative stability) or converted into a variety of products (e.g., biopolymers). The ability to modify or convert vegetable oils to other products is particularly important when one considers the various weaknesses that must be overcome to successfully replace petroleum-based products with vegetable oils.

Vegetable oils have the following two important properties that are particularly desirable for lubricant application. They are generally liquid at room temperature and they are also amphiphilic, i.e., they have distinctively separated polar and nonpolar groups within the same molecule. Most lubrication processes occur in one of the following lubrication regimes [15]:

1. *Boundary regime*: characterized by processes occurring at low speed and high load, involving contact between the rubbing surfaces due to the absence of lubricant film separating them, resulting in *high friction*
2. *Hydrodynamic regime*: characterized by a process occurring at high speed and low load, involving no contact between the friction surfaces due to the presence of thick lubricant film separating them, resulting in *low friction* between the rubbing surfaces
3. *Mixed-film regime*: characterized by processes with speeds and loads that are intermediate between those of boundary and hydrodynamic regimes, involving areas of contact and noncontact between the rubbing surfaces, resulting in an *intermediate coefficient of friction*

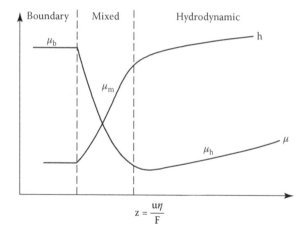

FIGURE 13.1 Schematic of lubrication regimes: effect of process parameter (z) on film thickness (h) and coefficient of friction (μ) in various lubrication regimes.

A schematic depicting these lubrication regimes is illustrated in fig. 13.1, where μ is the coefficient of friction (COF) and h is film thickness. In fig. 13.1, μ and h are plotted as functions of the lubrication process parameter, z, defined as follows:

$$z = \frac{u\eta}{F} \tag{13.1}$$

where u is speed, η is lubricant viscosity, and F is the load applied to press the friction surfaces together.

As can be seen in fig. 13.1, the boundary regime is characterized by high COF (μ) and low film thickness (h), whereas the hydrodynamic regime is characterized by low COF and high film thickness.

The fact that vegetable oils are generally liquid or fluid at room temperature allows them to be readily applicable in lubrication processes where formation of thick lubricant films is necessary. As described above, formation of high lubricant film thickness is important in processes that occur in hydrodynamic and mixed-film regimes. Thus, vegetable oils are suitable for formulating lubricants that will be applied in hydrodynamic and mixed-film regimes.

As described above, vegetable oils are also amphiphilic, since they have distinctly separate polar and nonpolar groups in the same molecule. Most vegetable oils are triglycerides, and some are monoesters of long-chain fatty acids and long-chain alcohols, also known as waxy esters. A schematic of a triglyceride vegetable oil is shown in fig. 13.2. As can be seen in fig. 13.2, the triglycerides have three ester groups on one side of the molecule attached to three nonpolar long-chain hydrocarbons. As a result of such amphiphilic structure, a vegetable oil can adsorb onto polar or high-energy surfaces with its polar group, while its nonpolar hydrocarbon chain orients away from the polar surface and possibly interacts with a nonpolar medium (e.g., air, oil). As a result, the vegetable oil can act as a boundary additive and can be used in formulating lubricants that can be used in processes occurring in the boundary regime. Thus,

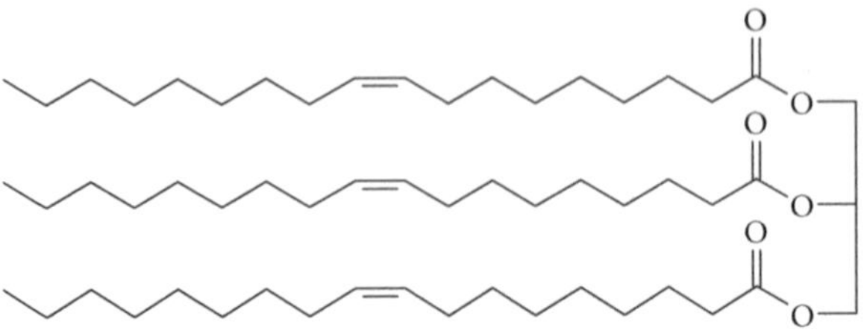

FIGURE 13.2 Schematic of a triglyceride molecule of vegetable oil.

vegetable oils, because of their fluid and amphiphilic properties, can be used in lubrication processes that occur in all lubrication regimes: boundary, hydrodynamic, and mixed-film. Because of this unique combination of properties, vegetable oils are also referred to as functional fluids.

Even though the basic chemical structure of vegetable oils is triglycerides or monoesters of long-chain fatty acids and fatty alcohols, there is a wide variation in the structure of vegetable oils. The main reason for the structural variability of triglycerides is the variability in the structure of the fatty acid component of the triglyceride. Vegetable oils display the following fatty acid structural variations: (a) chain length, which can vary from C_6 to C_{22}; (b) degree of unsaturation, i.e., the number of double bonds on the fatty acid chain, which can vary from 0 to 3; (c) position of unsaturation on the chain; (d) stereochemistry due to unsaturation, i.e., *cis* vs. *trans*; (e) presence or absence of functional groups on the hydrocarbon chain; (f) the chemistry of the functional group on the hydrocarbon chain, which can be an alcohol or epoxy group; (g) the number of functional groups on the hydrocarbon chain; and (h) the location of the functional group(s) on the hydrocarbon chain.

The structural variability in triglycerides from a specific vegetable crop is rather complex. Variabilities in the structure of fatty acids within a triglyceride molecule, as well as between triglyceride molecules of a specific crop, are commonly observed. In spite of such complex variability, the composition of fatty acids for a specific crop is unique. As a result, vegetable oils from different crops can be distinguished from each other based on their characteristic composition of fatty acids. This is illustrated in table 13.1, which shows fatty acid compositions of vegetable oils from different crops obtained using various analytical methods [16]. To illustrate the crop-to-crop variation in fatty acid composition, we compare soybean oil (SBO) vs. canola (CAN) in table 13.1. As can be seen in table 13.1, the two oils have similar composition of linolenic acid (18/3) but show big differences in their composition of oleic (18/1) and linoleic (18/2) acids. These variations in the composition of fatty acid residues between SBO and CAN are responsible for the large difference in the thermal stabilities of vegetable oils from these two crops [17].

As mentioned earlier, vegetable oils, because of their amphiphilicity, can adsorb onto surfaces and change various surface properties. Properties that can be modified due to adsorption of vegetable oils onto surfaces include surface tension of liquids, surface energy of solids, interfacial tension, boundary friction, adhesion, wettability, etc. The extent to which these properties are modified depends on a variety of factors, including the properties of the surfaces and the structure of the vegetable oil. Thus,

TABLE 13.1
Fatty Acid Composition (wt%) of Selected Vegetable Oils

	14/0	16/0	18/0	20/0	22/0	24/0	16/1	18/1	20.1	22/1	24/1	18/2	22/2	18/3
								C/db[a]						
Jojoba	...	1.8	0.2	0.1	0.6	0.3	0.4	9.0	70.0	14.0	3.0	3.0	...	...
Olive	...	13.7	2.6	0.4	0.1	0.1	1.5	67.5	0.3	...	...	12.7	...	0.6
Meadowfoam	...	...	...	...	...	...	...	2.0	62.5	14.5	...	0.5	18.0	...
Rice bran	0.1	16.2	1.7		0.2	0.3	0.2	39.4	0.7	...	...	38.4	...	1.7
Sesame	0.1	9.3	5.4	0.2	0.1	...	0.1	39.8	0.2	...	...	43.7	...	0.4
Cottonseed	0.7	21.6	2.6	0.3	0.2	...	0.6	18.6	...	...	...	54.4	...	0.7
Canola	0.1	4.1	1.8	0.7	0.3	0.2	0.3	60.9	1.0	0.7	...	21.0	...	8.8
Soybean	0.1	10.6	4.0	0.3	0.3	...	0.1	23.2	...	...	...	53.7	...	7.6

[a] C, fatty acid chain length; db, number of double bonds in fatty acid chain.

Source: Lawate, S. S., Lal, K., and Huang, C. 1997. In *Tribology data handbook.* Ed. E. R. Booser. Boca Raton, FL: CRC Press.

based on such adsorption studies, it is possible to develop structure–property correlations for vegetable oils in various applications where the above-listed properties are important. In this chapter, we explore the adsorption properties of vegetable oils of varying chemical structures using three different methods. The first method involves looking at the effect of vegetable oil adsorption on metal–metal boundary friction. The second method involves investigating the effect of vegetable oil adsorption on oil–water interfacial tension. The third method involves exploring the effect of vegetable oil adsorption on boundary friction between metal and starch–oil composites. In the following sections, each of these methods will be discussed separately. This will be followed by a section where data from these three different methods are analyzed and compared. The goal of the analysis is to quantify and compare the free energies of adsorption of the vegetable oils on the various surfaces. The ultimate goal is to understand how the variations in the chemical structures of vegetable oils affect their amphiphilic properties, which are responsible for the performance of vegetable oils in lubrication and other applications.

13.2 ADSORPTION AND FRICTION AT METAL–METAL INTERFACE

When a solution of a vegetable oil in a nonpolar hydrocarbon solvent comes in contact with a hydrophilic surface, some of the vegetable oil molecules adsorb onto the hydrophilic surface. The hydrophilic surface could be solid (e.g., clean metal) or liquid (e.g., water). The concentration of the adsorbed molecules on the surface can be determined using a variety of methods, including boundary friction on solid surfaces and surface/interfacial tension on liquid surfaces. Various factors affect the concentration of the vegetable oil molecules on the hydrophilic surface, including the chemical structure of the vegetable oil, the concentration of the vegetable oil in the nonpolar solvent, and the surface energy of the hydrophilic surface. From such investigations, one can quantitatively determine the effect of vegetable oil structure on its adsorption onto various surfaces.

13.2.1 Ball-on-Disk Tribometer

In the study described in this section, the coefficient of friction between two metals under boundary conditions was investigated as a function of vegetable oil chemical structure and concentration in hexadecane. Friction measurement was conducted under point contact conditions (point contact radius of 11.9 mm) using a ball-on-disk configuration on the Falex Multi-Specimen Friction & Wear Test instrument (Falex Corp., Sugar Grove, IL). A schematic of a ball-on-disk tribometer is shown in fig. 13.3. In this configuration, a steel ball, pressed against a stationary steel disk by a specified load, moves around the disk at a specified speed. The force opposing the motion of the ball (friction force) is measured by a load cell and automatically saved by the computer at a rate of 1 data point per second. The computer also automatically saves the following data at the same rate: changes in vertical height (wear), load, speed, lubricant temperature, and specimen temperature. The computer also calculates the coefficient of friction by dividing the measured friction force by the specified load, and then saves and displays it automatically. The specifications of the ball are: 52100 steel, 12.7-mm diameter, 64–66 Rc hardness, extreme polish. The specifications of

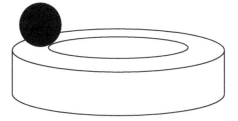

FIGURE 13.3   Schematic of a ball-on-disk tribometer.

the disk are: 1018 steel, 25.4-mm OD, 15–25 Rc hardness, 0.36–0.46-μm roughness. The ball and disk were thoroughly degreased before use.

13.2.2 Measurement of Boundary Friction on a Ball-on-Disk Tribometer

In the Falex Multi-Specimen Friction & Wear Test instrument, the ball and disk are enclosed in a cup, which is then filled with 50 mL of lubricant before the test, so as to completely immerse the ball-and-disk assembly in the lubricant. As described below, the lubricant is a solution with a known concentration of vegetable oil or methyl ester in hexadecane. Friction was measured at room temperature (25°C ± 2°C) for 15–30 min at 6.22 mm/s (5 rpm) and 181.44-kg load. The temperatures of the specimen and lubricant increased by only 1°C–2 °C at the end of each test period. Two tests were conducted with each lubricant sample using a new set of ball-and-disk specimens for each test. The COFs from the duplicate tests were then averaged to obtain the COF of the lubricant being tested. In general, the standard deviation of the COFs from the duplicate runs was less than ±5% of their mean.

The lubricants used in the test were solutions of a vegetable oil or a methyl ester of long-chain fatty acid in pure hexadecane (99+% anhydrous, Sigma-Aldrich, Milwaukee, WI). Various methyl esters and vegetable oils were obtained from commercial sources and used as supplied. For each, a series of formulations, with concentrations of vegetable oil or methyl ester ranging from 0 (pure hexadecane) to 0.6M, were prepared and used.

13.2.3 Effect of Vegetable Oil on Steel–Steel Boundary Friction

A typical data set from a test using a ball-on-disk tribometer is given in fig. 13.4. The data in fig. 13.4 are for a formulation with 0.003M soybean oil in hexadecane. As can be seen in fig. 13.4, the COF gradually increases with time until it reaches a plateau value, beyond which it remains constant for the duration of the test. The COF for the test is obtained by averaging the values in the steady-state region. The COF for the lubricant is obtained by averaging the COFs of the two consecutive tests.

The effect of vegetable oil concentration on the COF measured on the ball-on-disk tribometer is illustrated in fig. 13.5. The data in fig. 13.5 are for formulations comprising various concentrations of canola oil in hexadecane. The data in fig. 13.5 show a number of interesting features that are worth pointing out. First, the COF is very high in the absence of the vegetable oil, i.e., for pure hexadecane. COF of up to 0.5 has been measured for pure hexadecane (not shown in fig. 13.5) [18]. Second,

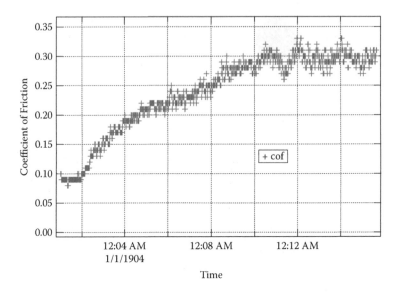

FIGURE 13.4   Typical time vs. friction data from a ball-on-disk tribometer for 0.003M soybean oil in hexadecane.

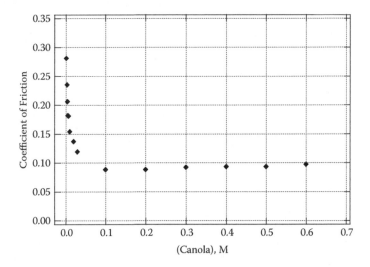

FIGURE 13.5   Effect of canola oil concentration on COF as measured with a ball-on-disk tribometer.

the incorporation of only small quantities of vegetable oil in hexadecane results in a dramatic reduction in COF. As can be seen in fig. 13.5, in the low canola concentration region ([canola] < 0.03M), the COF displayed a sharp decrease with every small incremental rise in the concentration of the vegetable oil. Third, at high concentration of vegetable oil ([canola] > 0.03M), the COF becomes more or less constant and independent of vegetable oil concentration. These three phenomena can be explained in terms of the adsorption of the amphiphilic vegetable oil on the friction surfaces. In the absence of a vegetable oil in the hexadecane, there will not be any adsorbed species on the surface to separate the rubbing surfaces. Hexadecane is nonpolar, does not adsorb, and therefore does not reduce friction between surfaces that operate in the boundary regime (low speed and high load). As a result, a high COF is expected and was observed in the absence of a polar component in the hexadecane. However, as soon as a small quantity of vegetable oil was introduced into the hexadecane, the COF displayed a dramatic reduction, indicating that adsorption of the polar vegetable oil molecules onto the friction surfaces had occurred. Addition of more vegetable oil into the hexadecane resulted in more adsorbed molecules and, hence, in further reduction of the COF. This process continues until the friction surfaces are fully covered by the adsorbed molecules. Once full surface coverage is attained, further increase in the concentration of vegetable oil in hexadecane does not result in a further reduction of COF. Thus, as shown in fig. 13.5, the COF remained constant even though the concentration of the vegetable oil in hexadecane was increased more than ten-fold.

Concentration–COF profiles similar to that shown in fig. 13.5 have been observed for all vegetable oils investigated to date, as well as for similar compounds, such as methyl esters of long-chain fatty acids [18–23].

13.3 ADSORPTION AND INTERFACIAL TENSION AT HEXADECANE–WATER INTERFACE

In principle, the amphiphilic characteristics of vegetable oils that were responsible for modifying boundary friction between metals could also result in the modification of the interfacial tension between oil and water. To effect such modification, the vegetable oil will have to adsorb on the polar water surface in a similar way as it did on the metal surface. To test this idea, a study was conducted to investigate the effect of vegetable oil dissolved in hexadecane on hexadecane–water interfacial tension. One objective of this investigation was to determine the effect of vegetable oil chemical structure on oil–water interfacial tension. Another objective was to compare the effect of vegetable oil chemical structure on changes in oil–water interfacial tension vs. metal–metal boundary coefficient of friction.

13.3.1 AXISYMMETRIC DROP SHAPE ANALYSIS (ADSA)

Interfacial tension was measured using an axisymmetric drop shape analysis (ADSA) method [24]. In this method, interfacial tension is obtained by analyzing the change in the shape of a pendant drop of one liquid suspended in a second liquid. The method is based on the Bashforth–Adams equation, which relates drop shape geometry to interfacial tension [25, 26]. A schematic of a pendant drop with the appropriate dimensions for use in the Bashforth–Adams equation is illustrated in fig. 13.6.

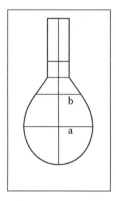

FIGURE 13.6   Schematic of a pendant drop geometry where "a" is the maximum or equatorial drop diameter and "b" is drop diameter at a height of "a" from the bottom of the drop. (From Biresaw, G. 2005. *J. Am. Oil Chem. Soc.* 82 (4): 285–92. With permission of AOCS Press, Champaign, IL.)

According to the Bashforth–Adams method, the interfacial tension between a pendant drop of a liquid and its surrounding medium is related to the drop geometry as follows [25, 26]:

$$\gamma = (\Delta \rho g a^2)/H \tag{13.2}$$

where γ is interfacial tension, $\Delta \rho = (\rho_1 - \rho_2)$ is the difference in the densities of the drop and the medium, g is gravitational acceleration, a is the maximum or equatorial diameter of the drop (see fig. 13.6), and H is the drop shape parameter.

The drop shape parameter H is a function of the drop shape factor S, which is calculated from the geometry of the drop as follows:

$$S = b/a \tag{13.3}$$

where b is the diameter of the drop at height a from the bottom of the pendant drop (fig. 13.6).

During dynamic interfacial-tension measurement, the geometry of the drop, and thus the interfacial tension between the drop and the medium, change as a function of time as more and more amphiphilic molecules diffuse from the bulk to the interface. The process continues until equilibrium geometry and, hence, equilibrium interfacial tension are attained after a relatively long period of time. The equilibrium interfacial tension is reached when the concentration of amphiphilic molecule at the interface attains its equilibrium value.

The instrument used for measuring dynamic interfacial tension was the FTA 200 automated goniometer (First Ten Angstroms, Portsmouth, VA). A schematic of the instrument is shown in fig. 13.7. The main features of the instrument include an automated pump that can be fitted with various sizes of syringes and needles to allow for control of pendant drop formation; an automated image-viewing and -capturing system with various image-capture triggering options; software for an automated

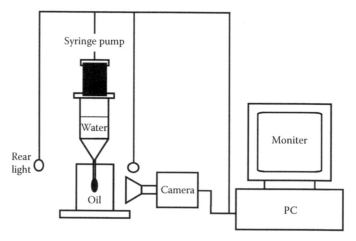

FIGURE 13.7 Schematic of an automated pendant drop goniometer. (From Biresaw, G. 2005. *J. Am. Oil Chem. Soc.* 82 (4): 285–92. With permission of AOCS Press, Champaign, IL.)

drop shape analysis of the captured drop image and for measuring the interfacial tension; and computer hardware and software for data capture, storage, analysis, and transfer. During measurement of interfacial tension using the ADSA method on the FTA 200, the image of the drop is recorded as a function of time using a high-speed camera. At the end of the experiment, image-analysis software is used to accurately measure the drop dimensions a and b from each image, and then uses these data to automatically calculate interfacial tension using eqs. (13.2) and (13.3). More details about the ADSA method can be found in the literature [24].

The vegetable oils used in this investigation were obtained from commercial sources, as explained earlier, and used as supplied. Hexadecane (99+% anhydrous) and methyl palmitate (99+%) were obtained from Sigma-Aldrich (Milwaukee, WI) and used as supplied. For each vegetable oil and methyl palmitate, a series of solutions in hexadecane with concentrations ranging from 0.00 (pure hexadecane) to 0.40M were prepared and used in measurements of interfacial tension against purified water. Water was purified by treating deionized water to a conductivity of 18.3 megohms·cm on a Barnstead EASYpure UV/UF water purification system (model D8611, Barnstead International, Dubuque, IA). The freshly treated water was further purified by filtering it on a 0.22-μL sterile disposable filter (MILLEX-GS 0.22-μL filter unit; Millipore Corp., Bedford, MA) prior to use in measurements of interfacial tension.

13.3.2 MEASUREMENT OF INTERFACIAL TENSION

All measurements of interfacial tension between water and the various solutions of vegetable oils in hexadecane were conducted at room temperature (23°C ± 2°C). Prior to conducting these measurements, the instrument was calibrated with the purified water and then checked by measuring the interfacial tension between the purified water and pure hexadecane. In a typical procedure, a 10-mL disposable syringe (Becton Dickinson & Co., Franklin Lakes, NJ), equipped with a 17-gauge (1.499-mm OD) blunt disposable needle (KDS 17-1P, Kahnetics Dispensing Systems, Bloomington, CA) was locked into place so that the end of the needle was under the surface of the hexadecane solution contained in a glass cuvette (10-mm glass spectrophotometer cell, model 22153D, A. Daigger & Company, Vernon Hills, IL). A manual trigger was then used to start the pump so that a few drops of water fell to the bottom of the cuvette. The instrument was then programmed and triggered to automatically deliver a specified volume of water at 1 µL/s and to automatically begin image capture when the pump stops. The volume of water to be automatically pumped was selected so as to generate the largest possible pendant drop that would not fall off before image acquisition was completed. All runs were programmed to acquire a total of 35 images at a rate of 0.067 s/image. At the end of the acquisition period, each image was automatically analyzed, and a plot of time vs. interfacial tension was automatically displayed. The data from each run were saved both as a spreadsheet and as a movie. The spreadsheet contained the time and interfacial tension for each image, and the movie contained each of the drop images as well as calibration information. Equilibrium values for interfacial tension were obtained by averaging the interfacial tension at very long periods, where the interfacial tension attains equilibrium and shows little or no change with time. Repeat measurements were conducted on each sample, and average values were used in data analysis.

13.3.3 EFFECT OF VEGETABLE OIL ON WATER–HEXADECANE INTERFACIAL TENSION

Typical data sets from measurements of interfacial tension on the FTA 200 automated goniometer are shown in figs. 13.8 and 13.9. Figure 13.8 shows repeat measurements of the interfacial tension of water–hexadecane following calibration with purified water. The data in fig. 13.8 indicate that the interfacial tension between water and hexadecane is constant and independent of time for the duration of the

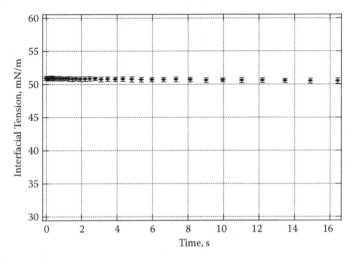

FIGURE 13.8 Average interfacial tension between purified water and pure hexadecane as a function of time, from ten repeat measurements on an FTA 200 automated goniometer. The error bar corresponds to ±1 standard deviation.

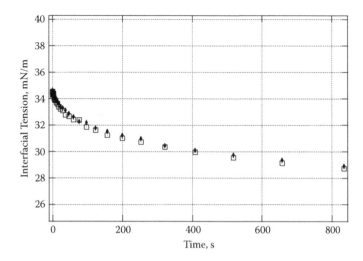

FIGURE 13.9 Typical data from repeat measurements on an FTA 200 automated goniometer of interfacial tension between purified water and hexadecane with 0.004M safflower oil.

measurement period. From the data, a value of 50.7 ± 0.3 mN/m was obtained, which is close to the literature value of 51.3 mN/m [27]. Figure 13.9 shows the interfacial tension between water and hexadecane with 0.004M safflower oil. The data in fig. 13.9 show the interfacial tension rapidly decreasing with time and leveling off after a relatively long period. This observation is consistent with the diffusion of the vegetable oil molecules from the bulk hexadecane to the hexadecane–water interface, thereby reducing the interfacial tension. In the absence of vegetable oils or other amphiphiles such as that between purified water and pure hexadecane shown in fig. 13.8, no diffusion to the interface occurs, and the interfacial tension remains constant for the entire measurement period.

Figure 13.10 shows the effect of safflower oil concentration in hexadecane on the interfacial tension between water and hexadecane. As can be seen in fig. 13.10, the concentration–interfacial-tension profile shows the familiar features exhibited due to the adsorption of the amphiphilic vegetable oils. The entire concentration–interfacial-tension profile can be divided into three regions, depending on the concentration of safflower oil in hexadecane. These are (a) zero concentration of safflower oil in hexadecane characterized by very high water–hexadecane interfacial tension; (b) intermediate concentration of safflower oil in hexadecane, up to about 0.05M, characterized by a rapid decrease in water–hexadecane interfacial tension; and (c) very high concentration of safflower oil in hexadecane, >0.1M, characterized by water–hexadecane interfacial tension that is constant and independent of safflower oil concentration in hexadecane. Similar profiles were observed for other amphiphiles dissolved in hexadecane. The profile shown in fig. 13.10 is similar to that discussed before on the effect of vegetable oil concentration on boundary friction between metals (fig 13.5).

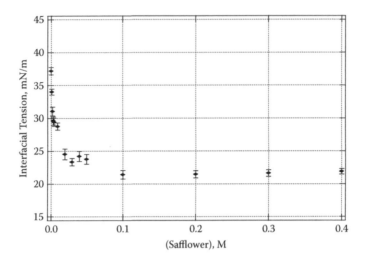

FIGURE 13.10 Effect of safflower oil concentration in hexadecane on water–hexadecane interfacial tension.

13.4 ADSORPTION AND FRICTION AT STARCH–METAL INTERFACE

13.4.1 STARCH–OIL COMPOSITES

Vegetable oils can be encapsulated into a starch matrix by a steam-jet cooking process of a slurry of oil and starch in water [28]. The resulting starch–oil composite, called Fantesk™, can be obtained in a powder form following a drying and milling process on the product from the jet cooker [29]. Fantesk is a trademark of the United States Department of Agriculture, where the steam-jet cooking process for aqueous starch and oil slurries was developed. One of the most important properties of the Fantesk starch–oil composite is that it can be reconstituted by simply redispersing the powder in water for use in the intended application. This property allows for cost-effective production, transportation, and application of Fantesk starch–oil composites. Studies have shown that encapsulation of oils through the steam-jet cooking procedure provides several benefits, including increased potency, i.e., resulting in the same effect at much lower concentrations; new functions, i.e., the ability to be effective for applications not possible with the unencapsulated oil; and a longer shelf life.

The starch component of the Fantesk starch–oil composite can vary in its chemical/structural properties (amylose, amylopectin, waxy) as well as in the type of crop source (corn, potato, rice, etc). Likewise, the oil component of the Fantesk starch–oil composite can have a variety of chemical/structural properties, compositions, and end uses. Over the years, Fantesk starch–oil composites of various combinations of starch and oil have been prepared and investigated for a variety of food and nonfood applications, including cosmetics, pharmaceuticals, polymers, coatings, and lubricants [4, 30–33].

In the area of lubrication, Fantesk starch–oil composites have been investigated for applications in oil well drilling [34] and as dry-film metalworking lubricants [35, 36]. Dry-film lubricants are those that are preapplied on sheet metal for various purposes, including the protection of the metal surface from oxidation/rusting, damage (e.g., scratching) during transportation, as well as to help in subsequent metalforming processes [15]. In the latter case, the dry-film lubricant may be the sole provider of the lubrication needs of the process or it may be used in conjunction with a compatible lubricant sprayed during the forming process. Dry-film lubricants may be applied on the sheet metal by the sheet metal manufacturer (e.g., rolling mill) or by a toll coater prior to delivery to the metal-forming facility. Successful application of dry-film lubricants in metal forming presents a number of benefits, including significant cost savings due to the elimination of facility and staff for handling, use, and control of liquid lubricants; cleaner workplace; and safer and environmentally friendly working conditions due to the elimination of vapor and mist associated with sprayed lubricants.

Successful commercialization of Fantesk starch–oil composite dry-film lubricants for sheet-metal-forming application necessitates that they meet a number of critical requirements. Among these are friction and wear properties, which can be used to infer the ability of the dry-film lubricants to allow for an efficient forming of metallic products without causing excessive tool wear. To obtain an understanding of the roles of the oil and starch in dry-film lubricants, a study of their friction and wear properties was initiated. In these studies, Fantesk starch–oil composites of varying starch and oil chemical structures and compositions, as well as starch-to-oil ratios, were investigated. These composites were deposited onto steel surfaces using various methods, and their friction properties under boundary conditions against steel were investigated.

13.4.2 Preparation of Starch–Oil Composites

Starch–oil composites of varying starch and oil chemistries and compositions were prepared by a steam-jet cooking process following the procedure detailed before [35]. A schematic of the steam-jet cooker is illustrated in fig. 13.11. The oil concentration in the composites is expressed in parts per hundred (pph) relative to dry starch. As an illustration of a typical procedure, a 45-pph soybean-oil–waxy-starch composite was prepared as follows. Waxy starch (1103 g, moisture content 10.3%)

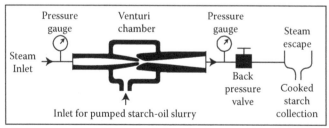

High temperature and intense turbulence in the venturi chamber causes:
(a) Complete solubilization of starch and water
(b) Intimate mixing of starch and oil phases
(c) Shear-induced lowering of starch molecular weight

FIGURE 13.11   Schematic of steam-jet cooker used for preparation of Fantesk starch–oil composite.

was added to deionized water (3000 mL) and stirred in a 4-L stainless steel Waring blender (model 37BL84; Dynamics Corporation of America, New Hartford, CT). The resulting slurry was delivered to the jet cooker utilizing a Moyno progressive cavity pump (Robbins Meyers, Springfield, OH) at a flow rate of about 1 L/min. The starch slurry and steam were combined in a Penick and Ford hydroheater (Penford Corp, Cedar Rapids, IA). Cooking temperature was 140°C using steam supplied at 448 kPa (65 psi), and the hydroheater backpressure was set at 276 kPa (40 psi). About 5000 mL of the cooked starch solution (solids content of 19.2% as determined by freeze-drying accurately weighed amounts of the dispersion in duplicate) were collected. The solids content varied because of dilution of the cooked dispersion with condensed steam. Soybean oil (178 g) was added to a portion of the starch solution (2002 g, contains 384 g starch), and the mixture was blended at high speed in a Waring blender for approximately 2 min. This oil–starch suspension was then fed through the jet cooker under the conditions described above. A center cut (approximately 2000 mL) of the white opaque dispersion was collected and drum dried on a pilot-scale double drum dryer (model 20, Drum Dryer and Flaker Company, South Bend, IN) heated with steam at 206.8 kPa (135°C) to give white composite flakes. The flakes were subsequently milled to a fine powder using a Retsch mill (model ZM-1, Brinkman Instrument Inc., Des Plaines, IL).

Vegetable oils used to prepare the starch–oil composites were obtained from commercial sources and used as supplied. Purified food-grade cornstarch (PFGS) and waxy cornstarch (Waxy No. 1, moisture content of 10.3%) were obtained from A.E. Staley Mfg. Co. (Decatur, IL) and used as supplied. Deionized water was used to prepare the starch–oil slurries for jet cooking and also to prepare dry-film-lubricant formulations from the starch–oil composites.

13.4.3 Formulation and Application of Starch– Oil Composite Dry-Film Lubricant

The Fantesk composite powders were reconstituted into dry-film lubricants by redispersing them in aqueous sucrose solutions. Sucrose was obtained from Fleming Companies, Inc. (Oklahoma City, OK) or purchased from a local supermarket (Domino, Extra Fine Granulated 100% Pure Cane Sugar, Domino Foods Inc., Yonkers, NY). Sucrose regulates the drying of the films on the metal and also improves the adhesion of the composite film to the metal surface. It was demonstrated that the concentration of sucrose had only a very minor effect on COF [35]. The dry-film-lubricant formulations were then applied onto precleaned stainless steel flat sheets using doctor-blade or spray application methods as described below.

The stainless steel sheets were type 304, 0.076 × 30.48 × 30.48-cm steel plates and were obtained from McMaster Carr Supply Co. (Elmhurst, IL) and cut into 7.6 × 15.2-cm specimens for use in friction experiments. Prior to application of the composite film, the flat sheets were cleaned using a two-step procedure. First, they were washed with soapy hot water, scrubbed with a nonscratching sponge, and rinsed with deionized water. Then they were cleaned by consecutive 5-min sonications in isopropanol and hexane or in acetone and hexane solvents. Acetone, isopropanol, and hexane (all 99.9%) used for sonication of the flat test specimen were obtained from Fisher Scientific (Fair Lawn, NJ) and used as supplied.

13.4.3.1 Doctor-Blade Application of Dry-Film Lubricants

In the doctor-blade method of dry-film-lubricant application, an aqueous sucrose solution was prepared by dissolving the required weight of sucrose in 130 g of water at 95°C. The solution was then transferred to a Waring blender, to which the appropriate amount of the Fantesk composite was added. The mixture was then blended at high speed for 5 min and the resulting dispersion applied as a thick film onto the flat steel specimen using a doctor blade with a 0.76-mm gap. The film was then allowed to dry at ambient conditions for several hours. During drying, the appearance of the film changed from opaque to transparent.

13.4.3.2 Spray Application of Dry-Film Lubricants

In the spray method of dry-film-lubricant application, an aqueous sucrose solution (7.5 g) was stirred in a Waring blender at low speed, and the starch–oil composite powder (11 g) was added slowly. Once the addition was complete, the mixture was blended at high speed for 5 min to give a homogenous dispersion. The dispersion was allowed to cool and transferred to the siphon-feed cup of a Badger model 400 detail/touch-up siphon-feed spray gun (Badger Air-Brush Co., Franklin Park, IL). The spray gun utilized a fine tip operating at 207 kPa (30 psi) and was adjusted for maximal atomization and horizontal fan spray pattern. The metal specimen was placed vertically at a slight angle in a chemical fume hood, and the aqueous composite was applied by spraying approximately 7.5 cm above the metal surface and moving the spray gun in a W-shaped fashion. This process was repeated three times, with 30-s drying time between coats.

13.4.4 BALL-ON-FLAT TRIBOMETER

The friction properties of starch–oil composite dry-film lubricants were measured in a ball-on-flat configuration using an instrument constructed by combining the SP-2000 slip/peel tester from Imass, Inc. (Accord, MA) with the model 9793A test-weight sled from Altek Co. (Torrington, CT). In this configuration, a flat-sheet metal specimen coated with the appropriate dry-film lubricant and secured onto a platen was pulled from under a sled of specified weight and connected to a load cell. The sled contacts the flat sheet with three balls installed at its bottom. The load cell, with a range of 0–2000 g, measures the friction force resisting the movement of the sled. The coefficient of friction (COF) was obtained by dividing the measured friction force by the weight of the sled. The balls were grade 100, 440-C stainless steel balls and were obtained from Altek Co. (Torrington, CT). The balls had the following specifications: diameter, 15.88 ± 0.02 mm; sphericity, 0.0254 mm; and hardness, 57–67 Rc. A schematic of the ball-on-flat tribometer is shown in fig. 13.12. A detailed description of the hardware and software components of the ball-on-flat tribometer can be found in the literature [35].

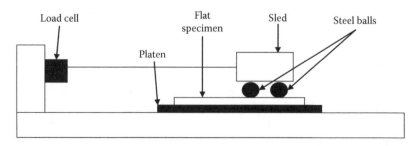

FIGURE 13.12 Schematic of a ball-on-flat tribometer. (From Biresaw, G., and Erhan, S. M. 2002. *J. Am. Oil Chem. Soc.* 79 (3): 291–96. With permission of AOCS Press, Champaign, IL.)

All friction experiments were conducted at room temperature using a 1500-g load sled at a speed of 2.54 mm/s for a total test time of 24 s. The measured friction force was automatically recorded by the microprocessor at a maximum rate of 3906 samples/s. Each dry-film lubricant was used to prepare two to three coated specimens, each of which was tested two to three times for a total of four to nine measurements. These values were averaged, and the resulting COF and standard deviation values were used in further analysis. Nitrile gloves were used to handle the test specimens, and extra care was taken to prevent contamination of the test specimen surface. After each run, the three balls were wiped with an acetone-soaked Kimwipes® and, if noticeable scratching of the ball occurred, the balls were rotated to provide undisturbed test surfaces. Between each dry-film lubricant formulation, the balls were replaced. A detailed description of the friction test procedure can be found in the literature [35].

13.4.5 Boundary Friction at Starch-Composite–Metal Interface

A typical data set from the ball-on-flat tribometer is shown in fig. 13.13. The data show time vs. friction force for a Fantesk starch–oil composite dry-film lubricant comprising soybean oil in purified food-grade starch (PFGS). The data display a number of features typically observed from such measurements. The friction force displays an initial sharp increase to a maximum, which immediately falls to a steady-state value for the remainder of the measurement period. The maximum friction force corresponds to the static friction, i.e., the resistance of the ball to the relative motion from an initial state of rest. The steady-state friction force corresponds to the kinetic friction. The COF for the test was obtained by dividing the average steady-state friction force by the normal load, which is 1500 gf.

The effect of vegetable oil concentration on the COF of the Fantesk starch–oil composite is illustrated in fig. 13.14, where the concentration of oil in the starch matrix is expressed in parts per hundred (pph), which is the ratio of oil to dry-starch weights multiplied by 100. The data in fig. 13.14 are for a composite of meadowfoam oil in PFGS and display a number of interesting features. First, the composite without vegetable oil (pure steam-jet cooked starch) displays a very high COF, on the order of 0.8. Second, incorporation of even a small quantity of vegetable oil into the composite causes a sharp reduction in the COF. The COF continues to decrease with increasing concentration of the vegetable oil in the starch–oil composite, reaching a minimum value at a vegetable oil concentration of just under 5 pph. Third, The COF of the composite remains constant and independent of vegetable oil concentration in the composite upon further increase in vegetable oil concentration. Similar results were obtained by Fantesk starch–oil composite dry-film lubricants comprising other starch and oil combinations.

These observations on the effect of vegetable oil concentration in the Fantesk starch–oil composite on the boundary COF between the starch and steel are similar

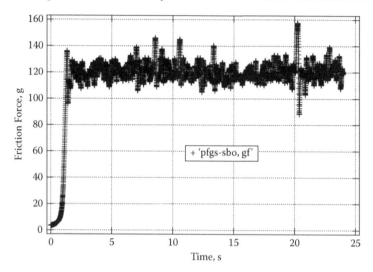

FIGURE 13.13 Typical data from friction measurements on a ball-on-flat tribometer for purified food-grade starch–soybean oil (pfgs–sbo) composite.

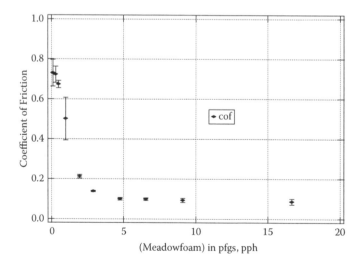

FIGURE 13.14 Effect of meadowfoam oil concentration in purified food-grade starch (pfgs) on the coefficient of friction of Fantesk starch–oil composite dry-film lubricant.

to previous observations on the effect of vegetable oil concentration on the steel–steel boundary COF as well as on the hexadecane–water interfacial tension, discussed earlier. The phenomenon in all three cases is related to the amphiphilic properties of the vegetable oils, and in particular to their adsorption properties. The common thread in all these studies is related to the effect of solubilized vegetable oil on modifying the surface properties of a solvent, thereby impacting interfacial phenomena such as friction or interfacial tension. The "solvent" can be hexadecane or starch, which, in the absence of the vegetable oil, displays a high-friction (steel–steel or steel–starch) or interfacial tension (hexadecane–water). Introduction of vegetable oil into the "solvent" results in a sudden and dramatic reduction of the interfacial property (friction or interfacial tension). This reduction in interfacial property continues with further increase in the concentration of the vegetable oil in the "solvent" until a minimum value is reached. The interfacial property then remains constant upon further increase in the concentration of the vegetable oil in the "solvent."

The effect of vegetable oil concentration on these interfacial properties is attributed to the migration of vegetable oil molecules from the bulk solvent phase to the solvent surface, thereby modifying the interfacial properties of friction and interfacial tension. In all cases, the observed interfacial property was found to be dependent on the concentration of the vegetable oil in the bulk "solvent." This observation implies the existence of some sort of equilibrium process between the bulk and surface concentrations of the vegetable oil. Since these properties are functions of the concentrations of vegetable oil molecules at the interface, such an equilibrium process can also explain the observation that the interfacial property (friction or interfacial tension) becomes constant and independent of vegetable oil bulk concentration in the bulk solvent once the minimum values are attained. Such a result is an indication of the saturation of the interface by the adsorbed vegetable oil molecules. As a result, any further increase in the concentration of the vegetable oil in the bulk "solvent" will have no effect on the concentration of the vegetable oil molecules at the interface and, hence, on the measured interfacial property (friction or interfacial tension).

The measured interfacial properties of the vegetable oils, which were attributed to their adsorption properties, can be used to quantify their free energies of adsorption, ΔG_{ads}. This requires constructing adsorption isotherms from the observed interfacial property data and analyzing the resulting isotherms using appropriate adsorption models. This analysis and quantification work is described in the next section.

13.5 ANALYSIS OF FRICTION AND INTERFACIAL-TENSION DATA

13.5.1 ADSORPTION AT INTERFACES

The use of the interfacial (friction or interfacial tension) data for the estimation of the free energy of adsorption, ΔG_{ads}, of the vegetable oils is based on the assumption that there is equilibrium between the vegetable oil in the bulk, O_b, and that on the surface, O_i. This equilibrium can be depicted as follows:

$$O_b \overset{K_{eq}}{\rightleftharpoons} O_i \qquad (13.4)$$

The concentration of the oil at the surface is expressed in terms of fractional surface coverage, θ, defined as follows:

$$\theta = \frac{O_i}{O_0} \tag{13.5}$$

where O_0 is the surface concentration of oil molecules at full surface coverage.

O_i and O_0 in eq. (13.5) can be used to represent the occupied and total number of adsorption sites on the surface. This leads to a different form of the equilibrium equation (13.4) as follows:

$$O_b + O_u \underset{}{\overset{K_0}{\rightleftharpoons}} O_i \tag{13.6}$$

or

$$K_0 = \frac{[O_i]}{[O_b][O_u]}$$

where O_u represents the unoccupied surface sites, $[O_u] + [O_i] = [O_0]$, and K_0 is the equilibrium constant.

For a simple adsorption model, such as the Langmuir model [25, 26], eq. (13.6) leads to the following relationship between surface concentration expressed in fractional surface coverage, θ, and bulk concentration in the solvent, $[O_b]$, expressed in moles/L:

$$\frac{1}{\theta} = 1 + \left(\frac{1}{\{K_0[O_b]\}} \right) \tag{13.7}$$

13.5.2 ADSORPTION ISOTHERMS

In order to apply an adsorption model such as the one given in eq. (13.7), first one has to construct an adsorption isotherm, i.e., a relationship between fractional surface coverage, θ, and bulk concentration, O_b, of vegetable oil using the measured interfacial property data. Fractional surface coverage, θ, can be obtained from the measured interfacial property data as follows:

$$\theta = \frac{(X_b - X_i)}{(X_b - X_0)} \tag{13.8}$$

where X_b is the measured interfacial property (COF or interfacial tension) in the absence of vegetable oil, i.e., of pure hexadecane or starch; X_0 is the measured interfacial property at full surface coverage; and X_i is the measured interfacial property at the vegetable oil concentration below the onset of full surface coverage.

Values for X_b and X_0 are obtained from concentration–COF or concentration–interfacial-tension data such as those shown in figs. 13.5, 13.10, and 13.14. Referring to figs. 13.5, 13.10, and 13.14, X_b is the COF or interfacial-tension value at zero veg-etable oil concentration, X_0 is the lowest COF or interfacial-tension value that remains constant and independent of vegetable oil concentration, and X_i is the COF or interfa-cial-tension value between X_b and X_0.

The effect of bulk vegetable oil concentration on fractional surface coverage obtained using eq. (13.8) for data from friction and interfacial-tension experiments are shown in figs. 13.15–13.17. The figures also show the COF and interfacial-tension data that were used to calculate the fractional surface-coverage values. The results

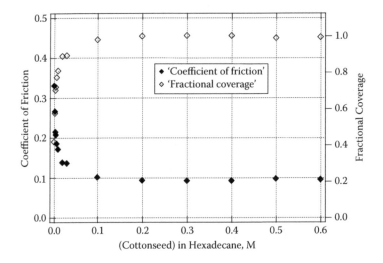

FIGURE 13.15 Adsorption isotherm for cottonseed oil in hexadecane derived from friction measurements.

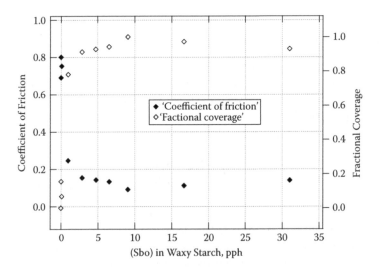

FIGURE 13.16 Adsorption isotherm for soybean oil in waxy starch–oil composite derived from friction measurements.

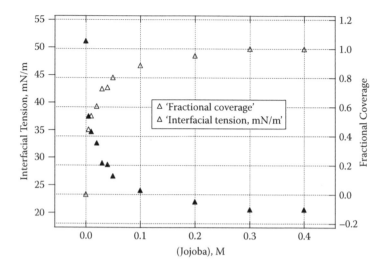

FIGURE 13.17 Adsorption isotherm for jojoba oil in hexadecane derived from measurements of hexadecane–water interfacial tension.

in figs. 13.15–13.17 indicate that the fractional surface-coverage data are a mirror image of the corresponding friction or interfacial-tension data from which they were obtained.

13.5.3 FREE ENERGY OF ADSORPTION

The free energy of adsorption, ΔG_{ads}, is obtained by analyzing the adsorption-isotherm data, such as those shown in figs. 13.15–13.17, using an appropriate adsorption model. The value for ΔG_{ads} is a measure of the net interaction between the amphiphilic adsorbate and the surface. Two types of interactions are possible [37, 38]: primary and lateral. Primary interaction is due to interactions between the polar group of the amphiphile and the polar surface (e.g., metal, water, etc.) and is the predominant contributor to ΔG_{ads}. Lateral interaction arises due to a variety of interactions between adsorbed molecules at the hydrocarbon chain or at the polar head. The net lateral interaction could be attractive, repulsive, or neutral. As a result, its contribution to ΔG_{ads} could vary as follows:

$$\Delta G_{ads} = \Delta G_0 + \alpha\theta \qquad (13.9)$$

where ΔG_0 is the free energy of adsorption due to primary interaction, and α is the free energy of adsorption due to lateral interaction.

The sign of α in eq. (13.9) is negative for a net attractive lateral interaction; it is positive for a net repulsive lateral interaction; and it is zero in the absence of lateral interaction. Thus, a net attractive lateral interaction will lead to a large negative value for ΔG_{ads}, indicating a strong adsorption of the amphiphile onto the surface.

13.5.4 LANGMUIR ADSORPTION MODEL

The simplest adsorption model is the Langmuir model [39], which assumes no lateral interactions. As a result, eq. (13.9) simplifies to

$$\Delta G_{ads} = \Delta G_0 = -RT \ln(K_0) \tag{13.10}$$

where K_0 is the equilibrium constant defined earlier in eq. (13.6) and is obtained from the analysis of the adsorption isotherm using the Langmuir model given by eq. (13.7). According to eq. (13.7), a plot of the inverse of the vegetable oil concentration in hexadecane vs. the inverse of the fractional surface coverage should result in a straight line with an intercept of 1 and a slope of ($1/K_0$). The value for ΔG_{ads} is then obtained by using K_0 from such an analysis in eq. (13.10).

Figures 13.18–13.22 display results of Langmuir analyses of adsorption isotherm data from steel–steel and steel–starch friction and hexadecane–water interfacial-tension measurements. As can be seen in figs. 13.18–13.22, in all cases plots of [veg oil]$^{-1}$ vs. θ^{-1} gave a straight line with an intercept close to 1. From linear regression analyses of the data, the exact values of the intercepts and slopes are obtained. From the slope, the equilibrium constant K_0 is determined and used to calculate the free energy of adsorption using eq. (13.10).

Table 13.2 compares the free energies of adsorption of some vegetable oils obtained from Langmuir analysis of the adsorption isotherms. Data from steel–steel and starch–steel boundary-friction measurements, as well as from hexadecane–water interfacial-tension measurements, are compared. The data in table 13.2 show a number of interesting features on the effect of oil and substrate properties on ΔG_{ads}.

The data derived from metal–metal boundary-friction measurements indicate that triglyceride vegetable oils adsorb more strongly on metal surfaces (larger negative ΔG_{ads} values) than simple methyl esters of fatty acids. The same data indicate that the monoester vegetable oil jojoba displayed ΔG_{ads} values between those of the triglycerides and the simple methyl esters. These differences have been attributed to the ability of the triglycerides to engage in multiple bonding to the metal surface, which will result in stronger adsorption of the triglycerides [22]. The slightly stronger adsorption of jojoba relative to the simple methyl ester of fatty acid was attributed to a stronger van der Waals attraction between the surface and the higher-molecular-weight jojoba relative to the low-molecular-weight methyl palmitate.

The data in table 13.2 show that the ΔG_{ads} values derived from hexadecane–water interfacial tension followed trends similar to those derived from the metal–metal boundary-friction measurements described above. This was attributed to the similarity in the energies of the surfaces of cleaned "technical" metals and water [21]. Technical metals are those commonly employed in the manufacture of metallic consumer and industrial products. The surface of cleaned technical metals generally comprises layers of oxides covered with a layer of adsorbed water, possibly by hydrogen bonding. As a result, the surface energy of a cleaned technical metal should resemble that of water. The data in table 13.2 support this similarity.

The data in table 13.2 show that the free energy of adsorption of the oils on the starch–oil composite is weaker (smaller negative number) than those on the steel or

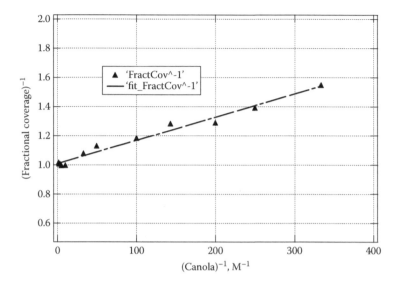

FIGURE 13.18 Langmuir analysis of adsorption isotherm derived from steel–steel friction due to canola oil in hexadecane.

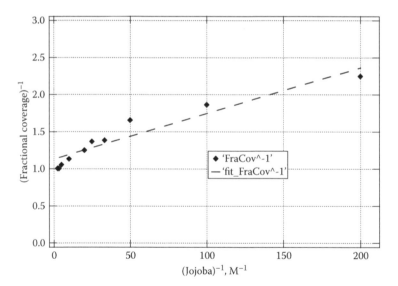

FIGURE 13.19 Langmuir analysis of adsorption isotherm derived from interfacial tension between water and hexadecane with solubilized jojoba oil.

water surface. Adsorption of the vegetable oils to the starch surface is due to H-bonding between the ester group of the oil and the hydroxyl group of the starch glucose units. Since starch is much more hydrophobic than steel or water, it is expected that adsorption of the vegetable oils on the starch composite will be weaker. The data in table 13.2 are consistent with this expectation.

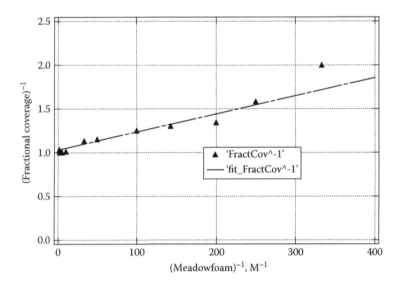

FIGURE 13.20 Langmuir analysis of adsorption isotherm derived from steel–steel friction due to meadowfoam oil in hexadecane.

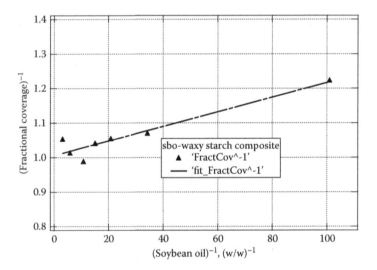

FIGURE 13.21 Langmuir analysis of adsorption isotherm derived from friction between steel and starch–soybean oil composite. The concentration of soybean oil in the composite is expressed in weight fraction (w/w).

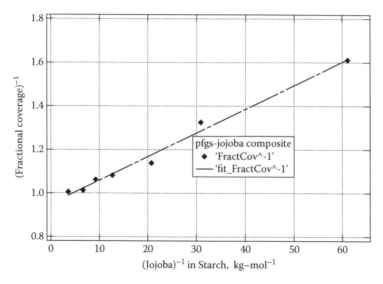

FIGURE 13.22 Langmuir analysis of adsorption isotherm derived from friction between steel and starch–jojoba oil composite.

TABLE 13.2
Comparison of ΔG_{ads} of Vegetable Oils on Solid and Liquid Surfaces from Langmuir Analysis of Friction and Interfacial-Tension Data

	ΔG_{ads}, kcal/mol		
	Friction (metal–metal)	Interfacial tension (hexadecane–water)	Friction (starch–metal)
Methyl palmitate	-2.70[a]	-3.06[d]	...
Safflower	-3.27[a]	-3.70[d]	...
Jojoba	-3.20[a]	-3.27[d]	-2.63[e]
Soybean	-3.60[b]	...	-2.96[e]
Meadowfoam	-3.57[c]	...	-2.41[e]

[a] Data from Biresaw, G., Adhvaryu, A., and Erhan, S. Z. 2003. *J. Am. Oil Chem. Soc.* 80: 697–704.
[b] Data from Biresaw, G., Adhvaryu, A., Erhan, S. Z., and Carriere, C. J. 2002. *J. Am. Oil Chem. Soc.* 79 (1): 53–58.
[c] Data from Adhvaryu, A., Biresaw, G., Sharma, B. K., and Erhan, S. Z. 2006. *Ind. Eng Chem. Res.* 45 (10): 3735–3740.
[d] Data from Biresaw, G. 2005. *J. Am. Oil Chem. Soc.* 82 (4): 285–92.
[e] Data from Biresaw, G., Kenar, J. A., Kurth, T. L., Felker, F. C., and Erhan, S. M. 2007. *Lubrication Sci.* 19 (1): 41–55.

13.5.5 Temkin Adsorption Model

One of the models that assumes lateral interaction is the Temkin model [40]. This model assumes a repulsive lateral interaction, i.e., $\alpha > 0$ in eq. (13.9). As a result, values of ΔG_{ads} obtained from the Temkin analysis of adsorption isotherms are expected to be larger (smaller negative number) than those from the Langmuir model. The relationship between adsorbate concentration in solution, c, and the fractional surface coverage, θ, for the Temkin model depends on the range of surface coverage being analyzed. For the range of $0.2 \le \theta \le 0.8$, the Temkin model predicts the following relationship [37, 38]:

$$\theta = \left(\frac{RT}{\alpha} \right) \ln \left(\frac{O_b}{K_0} \right) \tag{13.11}$$

Application of the Temkin model given in eq. (13.11) for analyzing adsorption isotherms similar to those given in figs. 13.15–13.17 requires replotting the data as $\ln(O_b)$ vs. θ. Linear regression analysis is then carried out on the data in the range $0.2 \le \theta \le 0.8$. The slope and the intercept from such analysis are then used to calculate α and K_0, respectively. The resulting K_0 is then used to calculate ΔG_0 using eq. (13.10), and the latter is used in eq. (13.9) to calculate ΔG_{ads}.

A typical Temkin analysis of friction-derived adsorption isotherm is illustrated in fig. 13.23. As can be seen in fig. 13.23, $\ln (O_b)$ vs. θ is linear in the range $0.2 \le \theta \le 0.8$, as was predicted by the model. Table 13.3 compares ΔG_{ads} data obtained from the Temkin model with those obtained using the Langmuir model. The data in table 13.3 are obtained from the analysis of adsorption isotherms derived from steel–steel and copper–copper boundary-friction measurements [19–23,37]. Three types of adsorbates are compared in table 13.3: triglycerides, esters of long-chain fatty alcohols and fatty acids, and methyl esters of fatty acids. As expected, the

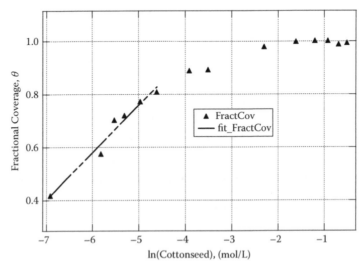

FIGURE 13.23 Temkin analysis of adsorption isotherm derived from steel–steel friction due to cottonseed oil in hexadecane.

TABLE 13.3

Comparison of ΔG_{ads} (kcal/mol) of Vegetable Oils Obtained from the Analysis of Metal–Metal Friction Data Using the Langmuir and Temkin Models

Oil	Metal	Geometry	ΔG_{ads} model	
			Langmuir	Temkin
Cottonseed	steel–steel	ball-on-disk	−3.68	−2.16
Canola	steel–steel	ball-on-disk	−3.81	−2.04
Meadowfom	steel–steel	ball-on-disk	−3.66	−2.28
Jojoba	steel–steel	ball-on-disk	−3.27	−1.31
Methyl oleate	steel–steel	ball-on-disk	−2.91	−1.02
Methyl oleate	Cu–Cu	ball-on-cylinder	—	−1.3
Methyl palmitate	steel–steel	ball-on-disk	−2.7	−1.63
Methyl laurate	steel–steel	ball-on-disk	−1.9	−0.6

Langmuir model predicts a stronger adsorption (larger negative number) than the Temkin model for all the adsorbates studied. This is consistent with the fact that the Temkin model assumes lateral repulsive interaction. Both models predict that triglycerides adsorb more strongly (larger negative values) than the monoesters. In addition, both models predict weaker adsorption by the short-chain monoesters compared with the long-chain monoester jojoba.

13.6 SUMMARY AND CONCLUSIONS

Vegetable oils are an important class of farm-based renewable raw materials that are actively being developed for applications in a variety of industrial and consumer products. They are obtained from a variety of seed crops and are considered highly promising substitutes for the resource-limited petroleum in a variety of applications. The chemical structure of vegetable oils comprises polar esters or triester groups attached to a long-chain hydrocarbon. Vegetable oils display a great deal of structural variability within the same crop (e.g., soybean), or between crops (soybean vs. canola), mainly due to variation in the structure of the fatty acid component of the ester. However, in spite of the complex structural variability, vegetable oils from specific crops have their own characteristic composition of fatty acid residues, thus making it possible to distinguish between vegetable oils from different crops.

Vegetable oils are considered functional fluids, since they have the ester functional group(s) and are also a liquid at room temperature. Such a property allows vegetable oils to be considered for a variety of applications where both functionality and fluidity are important. One such application area is lubrication, where vegetable oils, due to their functional fluid characteristics, have the potential to be applied in processes occurring in all lubrication regimes: boundary, hydrodynamic, and mixed.

The presence of both the polar ester group and the nonpolar hydrocarbon group in the same molecule imparts a vegetable oil with amphiphilic characteristics. As a result, vegetable oils can adsorb at interfaces where polar and nonpolar materials intersect. Examples of such interfaces include oil–water, oil–metal, and starch–metal.

The adsorption of vegetable oils at the interface changes the characteristics of the interface, which can be probed using a variety of methods. The degree of adsorption of vegetable oils at the interface is a function of the structure of the vegetable oil molecule, its concentration in the nonpolar material, and the properties of the polar and nonpolar materials intersecting at the interface.

In the work described here, adsorption of vegetable oils at metal–metal, water–hexadecane, and starch–metal interfaces was probed using boundary-friction and interfacial-tension methods. These properties were measured as a function of vegetable oil structure and concentration in hexadecane or starch. The resulting boundary-friction and interfacial-tension data were then analyzed using various adsorption models in order to quantify the free energy of adsorption of the vegetable oils on the various surfaces. The analysis showed that:

1. The adsorption of vegetable oils was highly dependent on the structure of the vegetable oils. Thus, (a) vegetable oils with triesters (triglycerides) had stronger adsorption than monoesters, which was attributed to multiple bonding by the triesters, and (b) low-molecular-weight simple monoesters displayed weaker adsorption than high-molecular-weight monoesters. This was attributed to stronger van der Waals attraction between the high-molecular-weight ester and the polar surface.
2. The adsorption of vegetable oils at metal surfaces was found to be similar to that on the surface of water. This was attributed to the similarities in the surface energies of water and cleaned "technical" metals. The surface of cleaned metals comprises oxides covered with at least one monolayer of adsorbed water, thus providing it with similar surface characteristics as the surface of a drop of water.
3. The adsorption of vegetable oils on starch is weaker than that on water or a metal surface. Vegetable oils adsorb on starch surface due to H-bonding between the ester group of the vegetable oil and the hydroxyl group of the starch glucose unit. However, because starch comprises hydrophobic segments, adsorption of vegetable oils on starch is expected to be weaker than that on water or a metal surface.
4. The free energy of adsorption of vegetable oils was found to be highly dependent on the adsorption model used to analyze the adsorption data. Thus, a stronger adsorption of vegetable oils on metal surfaces was predicted by the Langmuir than by the Temkin model. This was attributed to the fact that the Temkin model assumes a net repulsive lateral interaction, whereas the Langmuir model assumes no lateral interaction.

ACKNOWLEDGMENTS

The author expresses profound gratitude to Ms. Natalie LaFranzo for her help with the preparation of this chapter. The author is also grateful to AOCS Press for granting permission to use figs. 13.6, 13.7, and 13.12 in this chapter.

Product names are necessary to report factually on available data; however, the USDA neither guarantees nor warrants the standard of the product, and the use of a product name by the USDA implies no approval of the product to the exclusion of others that may also be suitable.

REFERENCES

1. Kaplan, D. L., ed. 1998. *Biopolymers from renewable resources.* Berlin: Springer-Verlag.
2. Lawton, J. W. 2000. In *Handbook of cereal science and technology.* 2nd ed. Ed. K. Kulp and J. G. Ponte. New York: Marcel Dekker.
3. Dwivedi, M. C., and Sapre, S. 2002. *J. Synthetic Lubrication* 19: 229–241.
4. Eskins, K., and Fanta, G. F. 1996. *Lipid Technol.* 8 (3): 53–55.
5. Erhan, S. Z., ed. 2005. *Industrial uses of vegetable oils.* Champaign, IL: AOCS Press.
6. Cermak, S., and Isbell, T. 2004. *Inform* 15 (8): 515–17.
7. Wool, R. P., Kusefoglu, S. K., Khot, S. N., Zhao, R., Palmese, G., Boyd, A., Fisher, C., Bandypadhyay, S., Paesano, A., Dhurjati, P., LaScala, J., Williams, G., Gibbons, K., Bryner, M., Rhinehart, J., Robinson, A., Wang, C., and Soultoukis, C. 1998. *Polym. Prepr., (Am. Chem. Soc., Div. Polym. Chem.)* 39: 90.
8. Mohanty, A. K., Misra, M., and Hinrichsen, G. 2000. *Macromol. Mater. Eng.* 276/277 (3/4): 1–24.
9. Guo, A., Cho, Y., and Petrovic, Z. S. 2000. *J. Polym. Sci. A: Polym. Chem.* 38: 3900–10.
10. Lodha, P., and Netravali, A. 2005. *Ind. Crops Prod.* 21: 49–64.
11. Becker, R., and Knorr, A. 1996. *Lubrication Sci.* 8: 95–117.
12. Fox, N. J., Simpson, A. K., and Stachowiak, G. W. 2001. *Lubrication Eng.* 57: 14–20.
13. Miles, P. 1998. *J. Synthetic Lubrication* 15: 43–52.
14. Adhvaryu, A., Erhan, S. Z., Liu, Z. S., and Perez, J. M. 2000. *Thermochim. Acta* 364: 87–97.
15. Schey, J. A. 1983. *Tribology in metalworking friction: Lubrication and wear.* Metals Park, OH: American Society of Metals.
16. Lawate, S. S., Lal, K., and Huang, C. 1997. In *Tribology data handbook.* Ed. E. R. Booser. Boca Raton, FL: CRC Press.
17. Biresaw, G. 2006. *J. Am. Oil Chem. Soc.* 83 (6): 559–66.
18. Biresaw, G., Adhvaryu, A., Erhan, S. Z., and Carriere, C. J. 2002. *J. Am. Oil Chem. Soc.* 79 (1): 53–58.
19. Kurth, T. L., Cermak, S. C., Byars, J. A., and Biresaw, G. 2005. *Proceedings of World Tribology Congress III*, WTC2005-64073, Washington, DC. CD-ROM.
20. Kurth, T. L., Biresaw, G., and Adhvaryu, A. 2005. *J. Am. Oil Chem. Soc.* 82 (4): 293–99.
21. Biresaw, G. 2005. *J. Am. Oil Chem. Soc.* 82 (4): 285–92.
22. Biresaw, G., Adhvaryu, A., and Erhan, S. Z. 2003. *J. Am. Oil Chem. Soc.* 80: 697–704.
23. Adhvaryu, A., Biresaw, G., Sharma, B. K., and Erhan, S. Z. 2006. *Ind. Eng Chem. Res.* 45 (10): 3735–3740.
24. Rotenberg, Y., Boruvka, L., and Neumann, A. W. 1983. *J. Colloid Interface Sci.* 93: 169–83.
25. Hiemenz, P. C. 1986. *Principles of colloid and surface chemistry.* 2nd ed. New York: Marcel Dekker.
26. Adamson, A. P., and Gast, A. W. 1997. *Physical chemistry of surfaces.* New York: Wiley.
27. van Oss, C. J., Chaudhury, M. K., and Good, R. J. 1987. *Adv. Colloid Interface Sci..* 28: 35–64.
28. Fanta, G. F., and Eskins, K. 1995. *Carbohydr. Polym.* 28: 171–75.

29. Eskins, E., Fanta, G. F., Felker, F. C., and Baker, F. 1996. *Carbohydr. Polym.* 29: 233–39.
30. Eskins, K., Fanta, G. F., and Felker, F. C. 2000. In *Designing crops for added value*. Ed. D. M. Peterson. Madison, WI: Am. Soc. of Agronomy.
31. Fanta, G. F., and Felker, F. C. 2004. Formation of hydrophilic polysaccharide coatings on hydrophobic substrates. U.S. Patent 6,709,763.
32. Fanta, G. F., Knutson, C. A., Eskins, K., and Felker, F. C. 2003. Starch microcapsules for delivery of active agents. U.S. Patent 6,669,962.
33. Cunningham, R. L., Gordon, S. H., Felker, F. C., and Eskins, K. 1998. *J. Appl. Polym. Sci.* 69: 351–53.
34. Erhan, S. M., Fanta, G. F., Eskins, K., Felker, F. C., and Muijs, H. 1999. *Program guide for Society of Tribologists and Lubrication Engineers 54th annual meeting*, Las Vegas.
35. Biresaw, G., and Erhan, S. M. 2002. *J. Am. Oil Chem. Soc.* 79 (3): 291–96.
36. Biresaw, G., Kenar, J. A., Kurth, T. L., Felker, F. C., and Erhan, S. M. 2007. *Lubrication Sci.* 19 (1): 41–55.
37. Jahanmir, S., and Beltzer, M. 1986. *ASLE Trans.* 29: 423–30.
38. Jahanmir, S., and Beltzer, M. 1986. *J. Tribol.* 108: 109–16.
39. Langmuir, I. 1918. *J. Am. Chem. Soc.* 40: 1361–402.
40. Temkin, M. I. 1941. *J. Phys. Chem. (USSR)* 15: 296–332.

14 Vegetable Oil Fatty Acid Composition and Carbonization Tendency on Steel Surfaces at Elevated Temperature
Morphology of Carbon Deposits

Ömer Gül and Leslie R. Rudnick

ABSTRACT

The objective of this study was to investigate the deposit-forming tendency of vegetable oils as a function of fatty acid profile. Vegetable oils were thermally and oxidatively stressed at high temperatures (250°C). The temperature regime of these studies was in the range of crown head and upper piston temperatures in a conventional and high performance spark ignition (SI) engine. Differences in deposit forming tendency of the oils studied were significant. These large differences in deposit-forming tendency under the same stressing conditions can be attributed to the differences in the chemical composition of the vegetable oils. The amount of carbonaceous deposit left on a stainless steel (SS) 304 strip was found to be directly proportional to the amount of palmitic and total saturated acids present in the vegetable oil. Carbon deposit was found to be independent of the amount of stearic acid present in the vegetable oil. Oxidative stability was found to be dependent on the total unsaturated fatty acid amount. Different deposit morphologies or surface coverages were observed from vegetable oils having different fatty acid compositions. In general, the SS 304 surface after thermo-oxidative stressing was covered more uniformly with higher oleic acid content vegetable oils, whereas a layered surface coverage was seen with vegetable oil containing higher palmitic acid content. A sponge-like deposit morphology was observed with coconut oil, a vegetable oil that is solid at room temperature having 90% total saturates.

14.1 INTRODUCTION

Most lubricants used in engine and industrial lubricants are mineral oils derived from petroleum sources. Synthetic lubricants offer significant advantages over mineral oils in several applications; however, their cost and slow biodegradability are offsetting disadvantages. Lately, interest has shifted to the replacement of mineral oils with vegetable oils in many applications [1], mainly because mineral oils have limited biodegradability and, therefore, have a greater tendency to persist in the environment [1]. Vegetable oils are both biodegradable and renewable, and they can replace mineral oils in many industrial applications where temperatures are moderate, for example, in the food industry, where leakage would have a great impact on the environment and where contact of the lubricant with food is an issue.

Vegetable oils have shown potential as biodegradable lubricants in applications that include engine oils, hydraulic fluids, and transmission oils [2–8]. Rudnick [8] and Erhan and Asadauskas [9] have reported on lubricant-base stocks based on vegetable oils. Environmentally acceptable hydraulic fluids, based on vegetable oil-base fluids, have been reported by several researchers [10–12].

Studies have reported that in normal equipment operation, including for bearings, gears, and piston rings, temperatures rarely go above 150°C [13, 14]. Automotive design aimed at fuel economy, however, has resulted in higher engine temperatures for passenger car engine oils and for high-performance diesel engine oils. This is partly due to smaller engines with smaller oil sumps aimed at reducing weight and aerodynamic drag. There are shifts to exhaust gas recirculation (EGR) and other modes of operation where engine temperatures are more severe, resulting in enhanced oil oxidation. Thus, lubricant fluid thermal and oxidative stabilities need to address these changes.

Normal sump temperatures are generally below 150°C. For example, engine sump temperatures under normal loads are on the order of 100°C–120°C for conventional automotive systems, and about 90°C for high-performance diesel engines. These temperatures are increasing with increases in performance and the constant desire to lower vehicle weight by designing smaller, hotter-running engines.

Based on reported temperatures in combustion engines, lubricants experience temperatures much higher than 150°C [13]. Lubricants, even if for only short residence times, experience temperatures through the autoxidation range (below 260°C) and into the intermediate range (260–480°C). Evaporation of lower boiling components begins at much lower temperatures [15].

These repetitive excursions into the combustion zone decrease the useful life of oil and result in carbonaceous deposits that cause wear and reduce performance. These may only represent transient temperatures for a small portion of the oil; however, it should be noted that, at the molecular level, these higher temperatures are sufficient to break covalent bonds, initiate free-radical reactions, and permit reactions to occur that would have very low rates at lower temperatures.

However, vegetable oils are known to have less oxidative stability than mineral oils and synthetic-base fluids. This is due to the fact that they contain unsaturated fatty acids having allylic hydrogens, which are oxidatively unstable compared with fully saturated hydrocarbons. Saturated vegetable oils, while more oxidatively stable, exhibit poor low-temperature properties relative to more-branched hydrocarbons. This results in some vegetable oils, notably palm and coconut oils, being cloudy or solid at room temperature, which makes them, on first inspection, poor candidates for lubricants.

From previous studies, it has been shown that the amount of residue deposited depends on the metal substrate and also on the chemical structure of the oil [15]. Thermal stability refers to the resistance of the oil to decomposition at elevated temperatures to form deleterious solid deposits. Oxidative stability refers to the resistance to degradation of the oil in the presence of air or oxygen.

This study examines the effects of thermal and oxidative stressing on the stability of several vegetable oil lubricants. By modifying a technique originally developed for thermal stressing of fuels [16], the oils in this study were thermally and oxidatively stressed, and the amounts of carbon residue deposited on a stainless steel metal coupon (SS 304) were measured.

14.2 EXPERIMENTAL

14.2.1 MATERIALS

Experiments were carried out using soybean oil, canola oil, refined corn oil, refined cottonseed oil, refined sunflower oil, refined palm oil, and refined coconut oil obtained from commercial sources. Biobased lubricant oils used in this study are given in table 14.1.

14.2.2 METHODS

14.2.2.1 Thermal Stressing

The vegetable oils were tested on stainless steel (SS) 304 strips. The metal strips, 15 cm × 3 mm × 0.6 mm, were cut and rinsed in acetone. Thermal stressing was carried out at 250°C oil temperature (approximately 260°C wall temperature) in the presence of SS 304 foil. The oil was subjected to stressing for 5 h at a 4-mL/min oil

---

TABLE 14.1
Biobased Lubricant Oils Used in This Study

Oil (refined)	Supplier
Soybean	Avatar Co., University Park, IL
Canola	Welch, Holmes and Clark Co., Newark, NJ
Corn	Welch, Holmes and Clark Co., Newark, NJ
Cottonseed	Welch, Holmes and Clark Co., Newark, NJ
Sunflower	Welch, Holmes and Clark Co., Newark, NJ
Palm	Fuji Vegetable Oil, Inc., Savannah, GA
Coconut	Fuji Vegetable Oil, Inc., Savannah, GA
Rapeseed	Welch, Holmes and Clark Co., Newark, NJ
Sesame	Welch, Holmes and Clark Co., Newark, NJ

Note: Palm and coconut oils were fully refined.

flow rate. The metal strip was placed in a 26.5-cm long by 0.635-cm o.d. glass-lined stainless steel tube reactor. Throughout the experiment, the reactor outlet temperature and oil flow rate were kept at 121°C and 4 mL/min, respectively, for 5 h of circulating oil. The stainless steel preheating section was 2 mm inside diameter (3.175 mm o.d.) and 61 cm in length. The oil residence time in this preheating zone was 22 s at a liquid oil flow rate of 4 mL/min. The oil residence time in the reactor (4 mm i.d., 6.35 mm o.d., and 31.75 cm length) was 59 s at the same oil flow rate. At the end of the reaction period (5 h), the foils were cooled under an argon flow in the reactor.

A detailed description of the flow reactor system used in these studies has been given earlier [16]. These earlier studies [16] on the thermal and oxidative stability of fresh jet fuel used a procedure whereby fuel was pumped once through the reactor during the experiment. In the present work, 100 mL of vegetable oil was used and recirculated through the reactor for the duration of the experiment. The reactor was allowed to run for 5 h, and then the metal strip was removed and analyzed. This represents a regime where the lubricant is repetitively stressed during the experiment.

14.2.2.2 Temperature-Programmed Oxidation (TPO)

In the above stressing experiments, deposits were collected on an SS 304 strip. After removing the strip from the reactor, the SS 304 strip was first carefully rinsed with acetone to remove residual vegetable oil without removing deposits.

The total amount of carbon deposits on the metal strips was determined using a LECO RC-412 Multiphase Carbon Analyzer (LECO Corp., St. Joseph, MO). Conventionally, the LECO RC-412 instrument has been used to measure the amount of deposition on metal surfaces [2]. In the carbon analyzer, carbon in the deposit is oxidized to carbon dioxide by reaction with ultrahigh purity O_2 in a furnace and over a CuO catalyst bed. The product, CO_2, is quantitatively measured by a calibrated infrared (IR) detector as a function of the temperature in the furnace. In this study, the deposits formed on the metal surface were characterized by temperature-programmed oxidation (TPO). The metal coupon with the deposit was heated at a rate of 30°C/min in flowing O_2 (750 mL/min) to a maximum temperature of 900°C with a hold period of 6 min at the final temperature. In this study, only the deposits collected on the metal strips were measured. The application of the technique has been previously reported [2]. The carbon deposits' morphologies were examined with an ISI-DS 130 dual-stage scanning electron microscope (SEM). To examine the morphology of the carbon deposits on the strips, 1-cm-long pieces were cut from the center of a 15-cm-long inconel strip.

14.2.2.3 Pressure Differential Scanning Calorimetry (PDSC)

The oxidative stability of the vegetable oils was determined using PDSC. The baseline and temperature calibration of the instrument were done using indium, which was used for the standard calibration temperature (156.6°C).

A small amount of oil (≈1 mg) sample was weighed into an aluminum flat-bottom sample pan. The uncovered pan containing the sample was placed on the left platform (sample platform) in the PDSC cell, and an empty pan was placed as a reference on the right platform. The cell was then capped, and all the valves were closed with the exception of the outlet valve. The cylinder regulator was then adjusted to deliver the desired pressure. The inlet valve was slowly opened; air was purged through the cell; and finally the pressure was allowed to reach a constant level (13.6 atm). In the constant-pressure operational mode, the outlet valve was opened and adjusted to give an oil flow of 20 mL/min.

For oxidation onset temperature measurements (temperature ramp) at 13.6 atmospheres pressure setting, a 10°C/min heating rate was used. The sample temperature was programmed to increase at a rate of 10°C/min from 100°C to 320°C. After the test was completed, the inlet valve on the cell was closed, and the pressure was released by opening the pressure-release valve.

14.3 RESULTS AND DISCUSSION

The thermal and oxidative stabilities of vegetable oils depend on the composition of the oil, the temperature, the nature of any metals, etc. Vegetable oils contain varying amounts of stearic, palmitic, oleic, linoleic, and linolenic acids. The chemical compositions of fatty acids of the vegetable oils tested are given in table 14.2, which is divided into two parts: vegetable oils that are liquid at room temperature (first seven samples) and vegetable oils that are solid at room temperature (the last two samples, palm and coconut oils). Carbon deposit amounts from thermal decomposition of vegetable oils on SS 304 are given in table 14.3.

Chemical structures of some of these fatty acids are shown in table 14.4. These fatty acids contain 16 (palmitic acid) or 18 carbon atoms with different amounts of unsaturation. Oleic acid (18:1), linoleic acid (18:2), and linolenic acid (18:3) have one, two, and three carbon-carbon double bonds, respectively. Stearic acid is a saturated fatty acid that has 18 carbon atoms and does not contain any carbon-carbon double bonds. The conventional designation of this is 18:0. Another fatty acid found in most vegetable oils is palmitic acid, which contains 16 carbon atoms. Palmitic acid also contains no carbon-carbon double bonds and is designated (16:0).

The vegetable oils tested in this study contained mostly palmitic, oleic, and linoleic, with lower concentrations of other fatty acids. Some vegetable oils, notably palm and coconut oils, are cloudy or solid at room temperature due to high amounts of saturated fatty acids (50.0% and 91.2% total saturated fatty acids in palm and coconut oil, respectively). This makes them poor candidates for low-temperature lubricant applications (table 14.2). On the other hand, the other vegetable oils in this study had 25% or lower total saturates, or conversely, these oils had 75% or higher total unsaturated fatty acids. High concentrations of unsaturated components impart the liquid character to these oils.

Vegetable oils are known to have less oxidative stability than mineral oils and synthetic-base fluids [16]. This is directly related to the fact that they contain unsaturated fatty acids. Chemically, it is the allylic hydrogens that are oxidatively unstable compared with fully saturated hydrocarbons. Saturated fatty acids, while more oxidatively stable, exhibit poor low-temperature properties relative to more-branched hydrocarbons.

TABLE 14.2
Fatty Acid Composition (%) of Vegetable Oils

Oil (refined)	Caprylic 8:0	Capric 10:0	Lauric 12:0	Myristic 14:0	Palmitic 16:0	Stearic 18:0	Total saturates	Oleic 18:1	Linoleic 18:2	Linolenic 18:3	Gadoleic 20:1	Erucic 22:1	Total unsaturates
Sunflower	...	...	...	...	9.5	3.4	**12.9**	28.6	58.6	...	...	...	**87.1**
Sesame	...	...	...	...	10.7	3.5	**14.2**	40.0	45.9	...	...	...	**85.8**
Rapeseed	...	...	...	...	5.5	1.4	**6.9**	23.6	14.6	5.6	7.5	41.4	**92.6**
Canola	...	...	...	...	10.0	2.9	**13.0**	58.4	20.4	8.3	...	...	**87.1**
Corn	...	...	...	...	12.1	1.7	**13.8**	29.2	57.0	...	...	...	**86.2**
Soybean	...	...	...	...	7.6	2.8	**10.3**	67.2	18.0	4.5	...	...	**89.7**
Cottonseed	...	...	...	...	21.9	2.2	**25.0**	19.3	55.7	...	...	...	**75.1**
Palm	...	...	...	2.4	44.4	3.4	**50.2**	36.8	10.6	...	...	...	**47.4**
Coconut	8.1	6.1	48.0	16.3	9.9	2.3	**90.6**	7.9	1.5	...	...	...	**9.4**

Note: Palm and coconut oils were fully refined.

TABLE 14.3
Thermal Oxidation Temperatures and Carbon Deposit Amounts of Biobased Lubricant Oils

Oil (refined)	Oxidation temperature (°C) as determined by pressure DSC	Carbon deposit ((g/cm2)
Soybean	148.3	38.5
Canola	173.6	5.3
Corn	180.4	31.2
Cottonseed	166.9	86.3
Sunflower	165.3	2.0
Palm	191.7	31.3
Coconut	210.3	43.2
Rapeseed	170.6	5.5
Sesame	180.9	3.1

Note: Palm and coconut oils were fully refined.

TABLE 14.4
Structures of Fatty Acids in Vegetable Oils

Fatty acid	Structure	Number of carbon atoms & unsaturation
Palmitic acid		16:0
Stearic acid		18:0
Oleic acid		18:1
Linoleic acid (omega-6)		18:2
Alpha linolenic acid (omega-3)		18:3

The quantities of carbon deposits obtained from the thermally stressed oils are given in fig. 14.1. A wide variation was observed in the amount of carbon deposits obtained from different vegetable oils. The amounts of carbon deposits, expressed as micrograms of carbon deposit per square centimeter of foil, are shown in table 14.3.

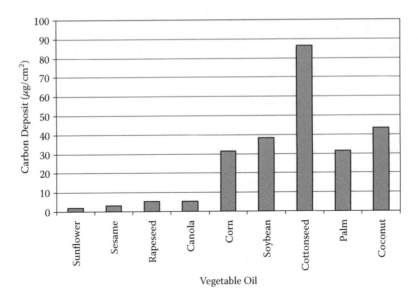

FIGURE 14.1   Carbon deposit amounts from vegetable oils used in this study.

Carbon deposit amounts varied from 2.0 to 86.3 µg/cm². Sunflower oil gave the least amount of carbon deposit, while cottonseed oil gave the highest carbon deposit.

This large difference in deposit-forming tendency under the same stressing conditions can be attributed to the differences in the chemical compositions of the vegetable oils. In this study, relationships between chemical composition and carbon deposit were sought. The chemical composition of the lubricant is an important parameter in the formation of deposits. Exposure of vegetable oils to high temperatures triggers free-radical reactions that lead to the deposition of carbonaceous solids on metal surfaces.

To assess the effect of chemical composition on carbon deposit, the relationship between chemical group concentration and the deposit amount was plotted. A plot of deposit versus palmitic acid for each oil shows that the carbon deposit amount increases as the palmitic acid concentration increases (fig. 14.2) ($R^2 = 0.87$). No relationship was found between carbon deposit and percent stearic acid (fig. 14.3).

When the amount of carbon deposit is plotted against the percent total saturates, a similar trend to that found for palmitic acid is observed, i.e., as the total percent of saturates increases, the carbon deposit amount increases (fig. 14.4) ($R^2 = 0.73$). On the other hand, an inverse correlation was seen between carbon deposit amount and total percent unsaturates (fig. 14.5) ($R^2 = 0.73$).

Oxidation temperatures of the vegetable oils used are given in fig. 14.6. The oxidation temperatures for vegetable oils, obtained using PDSC, were used to characterize the oxidative stability of the oils. The oxidation temperatures of the vegetable oils were found to be between 148.3°C and 210.3°C. Soybean oil had the lowest oxidation temperature, while coconut oil had the highest oxidation temperature among the oils tested.

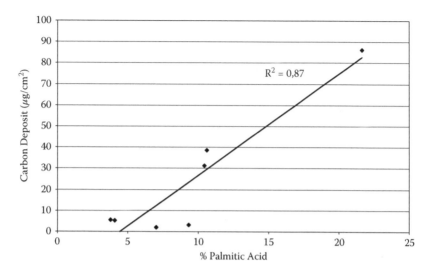

FIGURE 14.2 Percent palmitic acid vs. carbon deposit.

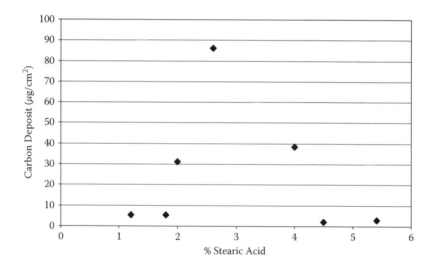

FIGURE 14.3 Percent stearic acid vs. carbon deposit.

When vegetable oil oxidation temperatures were plotted relative to the amounts of total unsaturates, an inverse relationship was observed, i.e., the total amount of unsaturation decreased as the oxidation temperature increased, as seen in fig. 14.7 ($R^2 = 0.69$). The mechanism of oxidation of vegetable oils is consistent with that of hydrocarbon oxidation. The allylic hydrogens of unsaturated fatty acids are the most susceptible sites for the initiation of oxidation.

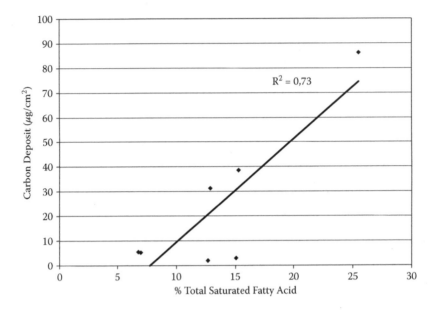

FIGURE 14.4   Percent total saturated fatty acid vs. carbon deposit.

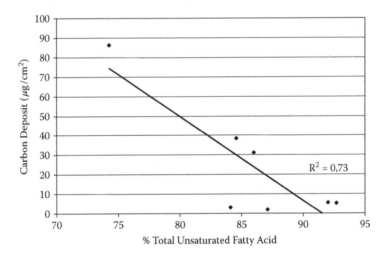

FIGURE 14.5   Percent total unsaturated fatty acid vs. carbon deposit.

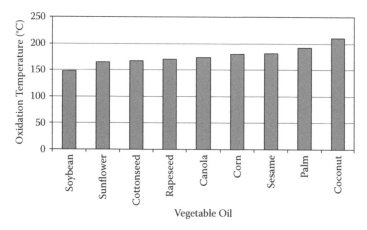

FIGURE 14.6   Vegetable oil oxidation temperature as determined by PDSC.

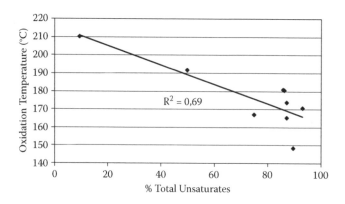

FIGURE 14.7.   Percent total unsaturates vs. vegetable oil oxidation temperature.

14.3 TEMPERATURE-PROGRAMMED OXIDATION (TPO) AND SCANNING ELECTRON MICROSCOPY (SEM) EVALUATION OF CARBON DEPOSITS

In this study, the solid carbon deposit morphology observed was different for different vegetable oil lubricants, depending on their chemical composition. The stressing

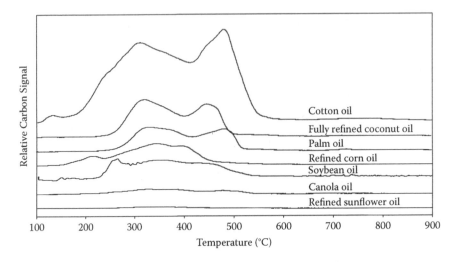

FIGURE 14.8 TPO profiles of carbon deposits from thermal stressing of vegetable oils on stainless steel 304 at 250°C for a flow rate of 4 mL/min for 5 h.

temperature is also a significant parameter having a major influence not only on the rate, but also on the type of carbonaceous deposit [17–19].

TPO profiles show peaks as a function of temperature (fig. 14.8). The peaks observed in these plots are related to the nature of the deposits. Peak intensities are directly proportional to the carbon deposit amount. Observations from the TPO and SEM images of the deposits are discussed below.

14.3.1 TPO PROFILES EVALUATION

TPO profiles show lower temperature peaks, which are between 100°C and 200°C. These low-temperature peaks correspond to a less-ordered (more reactive and possibly hydrogen-rich chemisorbed carbonaceous substance) deposit that may result from secondary deposition processes promoted by the presence of the incipient carbon [16]. The secondary deposition refers to thermally driven deposit growth on incipient carbon deposit that leads to the thickening and coating of the deposits by pyrolytic carbon formation [20]. Reactive deposits burn off at the lower temperatures, while less-reactive deposits burn off at higher temperatures.

High-temperature peaks (600°C) in the TPO profiles indicate the presence of relatively low-reactive carbon deposits. Highly ordered structures in the deposits, having pregraphitic or graphitic order, would reduce oxidation reactivity compared with amorphous carbon with no apparent structural order [21]. Of the vegetable oils examined in this study, no deposits of this type were observed based on the TPO profiles.

Each vegetable oil gave a different TPO profile with different relative carbon signals. Figure 14.8 shows the TPO profiles of the carbon deposits from thermal stressing of the different vegetable oils on SS 304 at 250°C and 4 mL/min flow rate for 5 h. Most of the vegetable oils gave two major peaks: one between 300°C and 350°C (amorphous carbon), and one between 450°C and 500°C (more-ordered amorphous carbon).

Cottonseed oil, which has the highest amount of palmitic acid and the lowest amount of oleic acid, gave the highest carbon signal among the oils tested. Sunflower oil, which has a lower amount of palmitic acid and the highest amount of linoleic acid, gave the lowest carbon signal. Cottonseed oil gave the dominant peaks at 300°C and 500°C and gave a relatively high carbon signal at higher TPO temperature (500°C).

Lower temperature peak (≈300°C) intensity to higher temperature peak (≈500°C) intensity ratio decreases from the bottom (TPO profile of sunflower) to the top (TPO profile of cottonseed oil) in fig. 14.8. Higher burn-off peak intensity implies that a more-ordered carbon deposit was formed.

14.3.2 SEM Images Evaluation

Higher oleic acid-containing vegetable oils (e.g., soybean and canola oils) showed more uniform coverage on the SS 304 strip surface (figs. 14.9 and 14.10, respectively). Corn

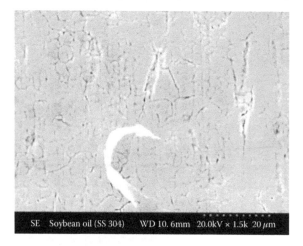

FIGURE 14.9 SEM image of carbon deposit from soybean oil.

FIGURE 14.10 SEM image of carbon deposit from canola oil.

oil, a vegetable oil with a higher linoleic acid content, gave a uniformly covered surface and some agglomeration on the covered surface (fig. 14.11).

The SEM image from the thermo-oxidative stressing of the cottonseed oil sample shows layered surface coverage. Of the oils studied, this vegetable oil has the highest palmitic acid content. A defect structure of layered coverage from the thermo-oxidative stressing of cottonseed oil can be seen in fig. 14.12.

The two vegetable oils that are solid at room temperature produced deposits with different surface coverage and different microstructure. Palm oil covered the surface more uniformly (fig. 14.13), while coconut oil produced a spongelike surface coverage (fig. 14.14). Fatty acid compositions of these two vegetable oils are quite different

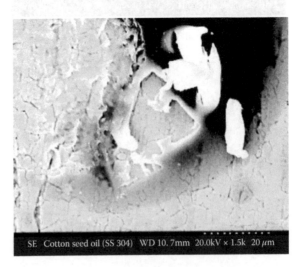

FIGURE 14.11 SEM image of carbon deposit from corn oil.

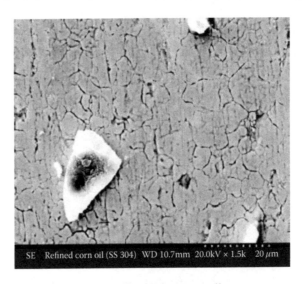

FIGURE 14.12 SEM image of carbon deposit from cottonseed oil.

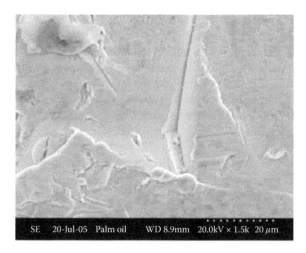

FIGURE 14.13   SEM image of carbon deposit from palm oil.

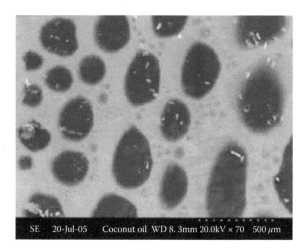

FIGURE 14.14   SEM image of carbon deposit from coconut oil.

(table 14.2). Palm oil has higher palmitic, oleic, and linoleic acid contents, but coconut oil has higher total saturates (sum of all saturated fatty acid components).

Hydrocarbon composition and stressing conditions both affect the type and the amount of deposit. Different optical textures were observed in samples of deposits formed with different vegetable oils; for example, the deposits formed from cottonseed and coconut oils are obviously quite different (figs. 14.12 and 14.14).

14.4   CONCLUSIONS

In this work, the relationships between the amount of carbon deposit or deposit morphology and the chemical composition of the vegetable oil were sought. The differences in carbon-forming tendencies for the oils studied were significant. This large difference in deposit-forming tendency under the same stressing conditions can be attributed to the differences in the chemical compositions of the vegetable oils. In this study, it was found that a vegetable oil containing higher amounts of palmitic acid or total saturates gave a higher amount of carbon deposits at 250°C. Oxidation temperatures of the vegetable oils used were found to be dependent on the amount of total unsaturated fatty acids. Vegetable oils with different fatty acid contents showed different deposit morphologies or surface coverages. The SS 304 coupon surface was covered more uniformly with vegetable oils having a higher oleic acid content. Layered surface coverage was seen with the vegetable oil containing higher palmitic acid content. A spongelike deposit morphology was observed with coconut oil, which is solid at room temperature and contains 90% total saturate.

ACKNOWLEDGMENTS

Special thanks to Welch, Holmes and Clarke Co., Avatar Co., and Fuji Vegetable Oil, Inc., for providing the materials used in this work. We also thank Quinta Nwanosike and Motunrayo Kemiki for their efforts in these studies.

REFERENCES

1. Rudnick, L. R., and Erhan, S. Z. 2006. In *Synthetics, mineral oils and bio-based lubricants: Chemistry and applications.* Ed. L. R. Rudnick. Boca Raton, FL: CRC Press.
2. Permsuwan, A., Picken, D. J., Seare, K. D. R., and Fox, M. F. 1996. *Int. J. Ambient Energy* 17: 157–61.
3. Honary, L. A. T. 1996. *Bioresource Technol.* 56: 41–47.
4. Adamczewska, J. Z., and Wilson, D. 1997. *J. Synthetic Lubrication* 14: 129–42.
5. Arnsek, A., and Vizintin, J. 1999. *J. Synthetic Lubrication* 16: 281–96.
6. Arnsek, A., and Vizintin, J. 1999. *Lubrication Eng.* 55 (8): 11–18.
7. Arnsek, A., and Vizintin, J. 2001. *Lubrication Eng.* 57: 17–21.
8. Rudnick, L. R. 2002. In *Bio-based industrial fluids and lubricants.* Ed. S. Z. Erhan and J. M. Perez. Champaign, IL: AOCS Press.
9. Erhan, S. Z., and Asadauskas, S. 2000. *Ind. Crops Products.* 11: 277–82.
10. Rhee, I. 1996. *NLGI Spokesman* 60 (5): 28–35.
11. Bisht, R. P. S., Sivasankaran, G. A., and Bhatia, V. K. 1989. *J. Sci. Ind. Res.* 48: 174–80.
12. Lavate, S. 2002. In *Bio-based industrial fluids and lubricants.* Ed. S. Z. Erhan and J. M. Perez. Champaign, IL: AOCS Press.
13. Rudnick, L. R., Buchanan, R. P., and Medina, F. 2006. *J. Synthetic Lubrication* 23: 11–26.
14. Sukirno, M. M. 1998. *Proceedings of ASME/STLE Tribology Conference.* Toronto, ON.
15. Kopsch, H. 1995. *Thermal methods of petroleum analysis.* Weinheim, Germany: VCH.
16. Gül, Ö., Rudnick, L. R., and Schobert, H. H. 2006. *Energy Fuels* 20 (6): 2478–85.
17. Trimm, D. L. 1983. In *Pyrolysis: Theory and industrial practice.* Ed. L. F. Albright, B. L. Crynes, and W. H. Corcoran. New York: Academic Press.
18. Albright, L. F., and Marek, J. C. 1988. *Ind. Eng. Chem. Res.* 27: 755–59.

19. Baker, R. T. K., Yates, D. J. C., and Dumesic, J. A. 1982. *ACS Symposium Series 202.* Ed. L.F. Albright and R.T. Baker. Washington, DC: American Chemical Society.
20. Li, J., and Eser, S. 1995. In *Carbon '95, Extended Abstracts, 22nd Biennial Conf. on Carbon.* San Diego: American Carbon Society.
21. Altin, O., and Eser, S. 2001. *Ind. Eng. Chem. Res.* 40: 596–603.

15 Biobased Greases
Soap Structure and Composition Effects on Tribological Properties

Brajendra K. Sharma, Kenneth M. Doll, and Sevim Z. Erhan

ABSTRACT

The review on bio-based greases contains 58 references. Biobased grease is part of the lubricant market, and will likely grow considerably due to economic, environmental, and legislative factors. There are several factors to consider when formulating a biobased grease. The lubricating base fluid, the thickener, and property enhancing additives all need to be tailored for an individual application. For example, the choice of the metal cation in the thickener is especially important in extreme pressure and temperature conditions. If a biobased soap stock is used, the fatty acid chain length and cation choice are the main factors to consider. Greases can be tested by a variety of methods, and should be evaluated for oxidative stability, rheology, low temperature properties, friction and wear properties, and hardness. Using these tests, it is possible to formulate acceptable biobased grease as a petroleum grease replacement with the possibility of improved lubrication properties.

15.1 INTRODUCTION

A lubricant is a substance designed to separate moving parts, thereby minimizing friction and reducing wear. Greases are lubricating fluids that have been thickened with a gelling agent [1]. A functioning grease should remain in contact with, and lubricate, the moving surfaces. It should not leak out under gravity or centrifugal action, or be squeezed out under applied pressure. Greases are the most versatile lubricant and are usually the first choice in lubricant selection due to their ease of application, lower maintenance requirements, fewer sealing requirements, greater effectiveness in stop and start conditions, and overall lower cost. A properly selected grease will self-regulate lubricant feeding. When the lubricant film between wearing surfaces thins, the resulting heat softens the adjacent grease, causing expansion and a release of oil to restore film thickness. Grease structure and composition undergo significant changes during its operational life, mostly from shearing and oxidation. Therefore, the usefulness of grease in a particular application is controlled, to a large extent, by the ability of the grease to sustain changes in temperature, high pressure, chemically diverse operating environments, and large shearing forces.

Greases contain three basic active ingredients: a base oil, a thickener, and additives. Different biobased and petroleum-based lubricating oils, thickeners, and additives are shown in table 15.1. The base oil can be a mineral oil, a synthetic oil, or a vegetable oil. Thickeners can be inorganic (silica and bentonite clay) or organic (polyurea) materials or metal soaps. The metal soap can be a reaction product of a metal-based material (oxide, hydroxide, carbonate, or bicarbonate) and carboxylic acid or its ester, or it can be added separately. Commonly used soap-type greases are based on calcium, lithium, aluminum, and sodium. The additives for greases are similar to those used in lubricating oils.

TABLE 15.1
Composition of Grease

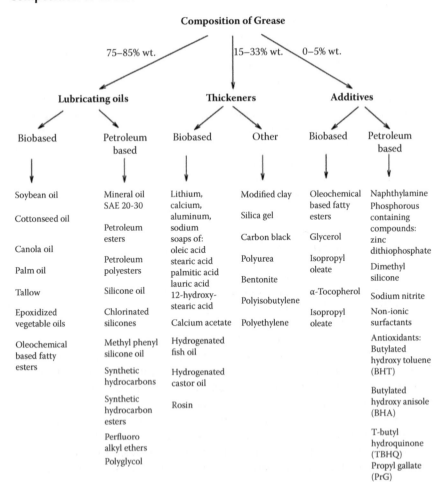

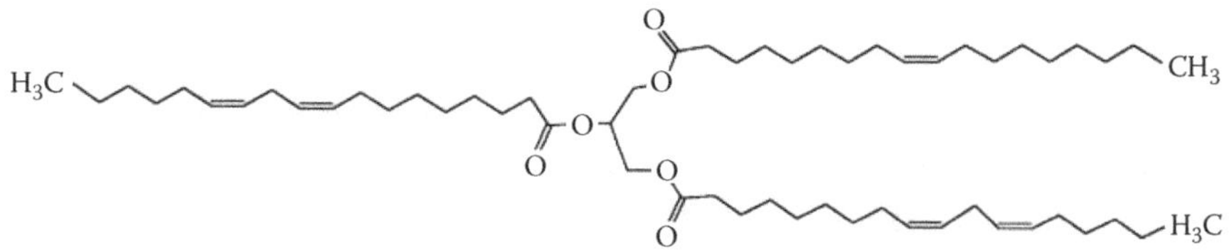

FIGURE 15.1 Natural triacylglyceride structure of vegetable oils. The double bonds of each chain vary, with an average of 4.2 double bonds contained in soybean oil.

Greases have a recorded history that dates back to 1400 B.C. with the use of tallow to lubricate chariot wheels. In 2002, there were 47 million gallons of grease sold in the United States alone. Of that, over 68% was considered either a conventional lithium grease or a lithium-complex grease [2]. The first greases were biobased, with no significant manufacture of petroleum grease until 1859. However, petroleum greases now dominate the marketplace, comprising an estimated 98% in 2004 [3].

Recently, due in part to increasing petroleum prices, the use of biobased oleochemicals as lubricant fluids [4–15], metalworking fluids, and greases [16–18] has increased dramatically. Vegetable oil (see fig. 15.1) contains a triglyceride structure with a mixture of saturated and unsaturated fatty chains. Grives [19] has discussed commercial methods for biodegradable grease preparation using different biobased and other thickeners.

To formulate a successful biobased grease, there are two major factors to consider: the friction-reducing ability of the compound and its stability to thermal oxidation [20]. The lubricity of oleochemicals [4, 5, 21–23] is well known in the fuel [9, 24, 25], textile [26], and other industries [22, 27, 28]. However, the significant amount of unsaturation in linoleic and linolenic fatty acid chains makes them particularly susceptible to oxidation. This is especially problematic in greases, where oxidation leads to carbonization and crust formation on the substrate surface. Doubly allylic positions—sites directly between two adjacent double bonds—are particularly easy to oxidize, as shown in kinetic studies [29]. Two methods of overcoming this problem have been attempted. The addition of antioxidants such as butylated hydroxy toluene (BHT), t-butylhydroquinone (TBHQ), or dialkyldithiocarbamates has been studied both experimentally [30–32] and theoretically [31]. Although a significant effect was observed, a potentially more advantageous approach is through chemical modification of the vegetable oil itself [17, 32–35].

Vegetable oil modification has been accomplished through hydrogenation [36], transesterification, epoxidation [37–40], metathesis [41–43], and alkylation, or a combination of chemistries, all yielding a more oxidatively stable fluid. Several variables in lubricant formulation have been studied recently [16, 17, 44]. Different fatty acid soaps were used in combination with a soybean oil with a low linoleic acid content. It was found that the soap structure was very important to the performance of the grease, and variations in soap structure were ultimately displayed in the final properties of the grease. These results point to a significant surfactant effect on grease performance, and a suitable composition of grease must be carefully formulated to give good performance properties capable of use in multifunctional products.

Notwithstanding the tremendous significance of biodegradable greases, only scanty information is available on the relationship between their composition and performance properties.

15.2 PHYSICOCHEMICAL AND PERFORMANCE PROPERTIES

Grease performance properties are measured by a variety of methods.

15.2.1 HARDNESS/CONSISTENCY

The physical and chemical behaviors of grease are largely controlled by the consistency or hardness. The hardness of grease is its resistance to deformation by an applied force and is generally measured by the ASTM D 217 and D 1403 methods. In order to standardize grease hardness measurements, the National Lubricating Grease Institute (NLGI) has separated grease into nine classifications, ranging from the softest, NLGI 000, to the hardest, NLGI 6. The grease is tested by penetrating its surface with a standard metal cone. If the cone sinks in 47.5 mm or more, the very soft grease is classified as NLGI 000, the softest classification. If it only sinks 8.5 mm or less, the hard grease is classified as NLGI 6, the hardest classification. Results discussed herein were obtained using a microprocessor-based digital penetrometer from Koehler Instrument Co. (Bohemia, NY). The ASTM D 217 method was followed, in which a stainless steel-tipped brass cone initially touching the grease surface was allowed to drop and penetrate freely for 5 s through the grease medium. The penetration value was digitally displayed after each test. The test was repeated three times to obtain an average value, and the corresponding NLGI grade was then used to classify the greases.

15.2.2 RHEOLOGY

Using a controlled-stress rheometer, the sample is introduced between an appropriate cone and plate. When appropriate strain is applied, the shear stress, yield stress, and viscosity can be measured, and the storage modulus and loss modulus can then be calculated from various mathematical models. The rheology is more difficult to measure in greases, which slip at the rheometer cone. If this happens, the measurement does not give properties representative of the bulk material. Sample fracture is also a problem if the grease is too thick. In this case, the grease cracks along the edges of the rheometer cone, and the resulting air gaps convolute the observed data.

15.2.3 OXIDATION PROPERTIES

Oxidation of grease results in insoluble gum, sludge, and deposits and ultimately leads to decreased metal wettability, sluggish operation, reduced wear protection, and increased corrosion. Excessive temperatures result in accelerated oxidation or even carbonization, where grease hardens or forms a crust as a result of evaporation of base fluid and increased thickener concentration. Therefore, higher evaporation rates require more frequent relubrication. A number of studies have been reported on various aspects of thermo-oxidative stability [45, 46]. The oxidative stability of grease and the oxidation inhibition capacity of antioxidant additives can be evaluated using various methods, such as the thin-film micro-oxidation (TFMO) test, pressurized differential scanning calorimetry (PDSC), and the rotary bomb oxidation test (RBOT).

15.2.3.1 TFMO

The deposit-forming tendencies depend on the oxidation stability of the grease. One of the methods to measure deposit-forming tendencies and volatility of lubricants is the TFMO method [47]. Generally, a small amount of the grease sample is accurately weighed and spread on an activated low-carbon steel catalyst surface to make a thin film on the metal. The sample is usually oxidized under a steady stream of dry air in a bottomless glass reactor on a hot plate. A constant temperature, ranging between 100°C and 225°C, is used for a period of 2 h. To study the time effect, the greases can also be oxidized for periods of 2, 4, 6, 8, and 24 h. After the stipulated time, the coupon is cooled and weighed to determine percent weight loss, also termed as volatile loss. In one experiment, a standard deviation of 0.08 was obtained for a TFMO test on grease samples at 150°C performed in quadruplicate for 2 h. Coefficients of variation (CoV) of 4.5%–5.5% and percent weight loss values that vary within ±0.1 of the average are reported in the literature [48] for NLGI grease samples on steel pans at 150°C for 24 h.

15.2.3.2 PDSC

PDSC is one of several benchtop tests (rotary bomb oxidation test, turbine oil oxidation test, peroxide estimation, active oxygen method, Rancimat method) for measuring the oxidative stability of greases [4, 5, 16]. The high pressure in PDSC inhibits the volatilization loss of lubricants and saturates the sample with oxygen. This results in an acceleration of oxidation as well as a sharpening of the exotherm compared with normal DSC, allowing the use of lower test temperatures or shorter test times. The PDSC experimental conditions can be varied by changes in DSC pans (hermetic, nonhermetic, pinhole hermetic, pans for solid fat index), gases (air or oxygen), pressures (ranging from 1379 to 3449 kPa), flow rates of gas, and/or isothermal temperatures (depending on the oxidation stability of the grease). Typical experiments were conducted as follows: The equipment used was a PDSC 2910 thermal analyzer from TA Instruments (New Castle, DE). The module was temperature calibrated using the melting point of indium metal (156.6°C) at a 10°C/min heating rate. A small amount of sample (<2.0 mg) was oxidized in an open aluminum pan. PDSC experiments were run either in an isothermal mode to measure oxidation induction time (OIT) or in a programmed temperature mode to measure the onset temperature (OT) of grease oxidation. For isothermal experiments, the PDSC cell was pressurized to 2068 kPa with air, and the temperature was ramped at 40°C/min to 90°C. The experiment was continued at this isothermal temperature until the appearance of a thermal peak corresponding to the oxidation of the sample. The OIT of grease samples containing additives were calculated by extrapolating the tangent drawn on the steepest slope of the corresponding exotherm to the baseline. The OIT is defined as the time when additives no longer prevent oxidation and a rapid increase in the rate of oxidation is observed in the system. For the temperature-ramping experiment, a constant scanning rate of 10°C/min was used until the exotherm peak appears. The OT of oxidation is then calculated from this peak. OT is defined as the temperature when a rapid increase in the rate of oxidation is observed in the system and is obtained by extrapolating the tangent drawn on the steepest slope of the reaction exotherm. A high OT or OIT would suggest a high oxidation stability of the grease. All grease samples were run in triplicate to obtain average OT and OIT values with standard deviation <1.0°C.

15.2.3.3 RBOT

The RBOT method determines the resistance of lubricating greases to oxidation when stored statically in an oxygen atmosphere in a sealed system at an elevated temperature under the conditions of the test. RBOT is usually conducted on grease samples containing antioxidant additives. The test conditions were similar to those described in the ASTM D 942 method. Typically, a small amount of sample was weighed into a Pyrex® glass vessel and then enclosed in a bomb. The bombs were pressurized to 758 kPa with extra-dry oxygen and oxidized in a constant-temperature oil bath set at 99°C. The pressure in the bomb was recorded over 6000 min. The degree of oxidation after 6000 min was determined by the corresponding pressure drop. High pressure drops correspond to higher amounts of oxygen consumption and increased rates of oxidation. This also allows determination of the efficacy of antioxidant additives. Grease samples were usually run in duplicate, and the average pressure drop was reported.

15.2.4 Low-Temperature Properties

At low temperatures, grease hardens, leading to poor pumpability and rheological properties. Typically the pour point of base oil is considered the low-temperature limit of grease. Below this temperature, the base oil will not flow properly and therefore will not provide sufficient lubrication.

15.2.5 Friction and Wear Properties

Important performance properties such as adhesion, rheology, and lubrication are largely dependent on the grease hardness and its ability to maintain a stable lubricating film at the metal contact zone. There are a variety of bench test methods for measuring friction and wear properties of a grease. The method of choice depends on the end-use application. Two commonly used bench test methods are described here.

15.2.5.1 Four-Ball Wear Test

This test is designed to study the antiwear properties of greases under sliding point contact using a four-ball test geometry (Model Multi-Specimen, Falex Corp., Sugar Grove, IL). The test zone is a rotating top ball (52100 steel, 12.7-mm diameter, 64-66 Rc [Rockwell C hardness; typically a cutting blade will have Rc 62–67], and extreme polish) in contact with three identical balls clamped in a cup containing enough grease sample to cover the stationary balls. Appropriate load is applied from below, and the top ball is rotated at a set speed for a preset length of time. Prior to the test, the balls are thoroughly cleaned with methylene chloride followed by hexane before each experiment. The instrument allows automatic acquisition and display of data such as torque, normal load, sliding speed, and chamber temperature of test grease in real time. The test conditions are similar to the ASTM D 2266 method, which allows the system to attain a set speed of 1200 rpm before a load of 392 N is applied at 75°C for 60 min. Duplicate tests are always done with a new set of balls. Scar diameters on balls are measured with a digital optical microscope. Two measurements, perpendicular to each other, are recorded for each scar on a ball, and the average of six measurements for three balls is taken in each case. The scar diameter is reported in millimeters, and the standard deviation of six measurements is typically less than 0.04 mm.

15.2.5.2 Ball-on-Disk Friction Test

This is a good test method to study boundary lubrication properties of grease samples using a multispecimen friction test apparatus from FALEX (Sugar Grove, IL). Ball-on-disk experiments are often carried out at room temperature under low speed, 6.22 mm·s^{-1} (5 rpm), and high load, 1778 N. The lower specimen was a 1018 steel disk with Rc 15–25 hardness, while the upper specimen ball has similar specifications as in the four-ball test method. The coefficient of friction (CoF) values were measured and reported as averages of three independent experiments. Typical CoF values varied by ±0.02 from the average.

15.3 DISCUSSION

The surfactant in grease is a truly multipurpose component. It has to function as a viscosity modifier, an emulsifier/dispersant, and a lubricant adjuvant, all in an interconnected manner. First, the surfactant, or soap, has to thicken the lubricating fluid in order to form a grease with the proper viscosity and hardness characteristics. However, a grease differs from viscous oil by having a true colloidal nature.

The grease emulsion has two different mechanisms of lubrication. The first, and minor component, is that it simply supplies the lubricating fluid to the substrate surface in a timely manner. However, the major lubrication factor involves the flow of the entire grease compound into the lubricated area. The lubrication effectiveness is controlled by the joint design, the amount of applied grease, and the rheology of the grease. If it is too hard, it will not flow properly to give sufficient surface coverage. If it is too soft, it will flow too rapidly, reducing effectiveness. An effective grease must have high resistance to initial flow, but a very low resistance to flow after the yield stress is reached. A grease emulsion can be formed with either spherical droplets, more common when inorganic thickeners are used, or longer fibrous droplets, more common when oleochemical thickeners are used. It is these droplets and their resistance to deformation that cause the rheology behavior of the grease. It has also been suggested that a twisting ribbon structure of these droplets improves grease properties [47].

The surfactant or soap used in a grease is critical in the temperature dependence of its rheology and hardness behavior. For example, calcium soap greases become liquid in nature (often called the dropping point, or less commonly called the dripping point) at less than 100°C, whereas lithium soap greases can still be effective at up to 200°C. Some specially formulated calcium sulfonate greases have even higher dropping points.

The surfactant must also operate against the coalescence of drops to form a separate oil layer. When this happens, the grease loses its ability to perform its sealing function, leading to a loss of the lubricant. The Davies equation [49] for drop coalescence, eq. (15.1), shows a relation of drop coalescence to the energy barrier of coalescence. A surfactant structure that will increase this barrier will result in better grease performance. This barrier can result from both physical and electrostatic means.

$$\frac{d\bar{V}}{dt} = Ae^{\frac{-E}{kt}} \tag{15.1}$$

where

$\dfrac{d\bar{V}}{dt}$ = rate of increase in mean drop volume

A = collision factor, dependent on the oil

$-E$ = energy barrier to coalescence

kt = Boltzmann's constant × time

The ζ-potential, the potential of a charged droplet or particle measured by its kinetic motion in an electric field, is one common way of looking at the electrostatic stabilization given by a surfactant. In aqueous solutions, a ζ-potential of 25 mV in magnitude is considered a strong stabilizing factor in stopping the agglomeration of particles [50]. It is conveniently measured in aqueous solutions, and less commonly measured in organic solutions [51]. However, a convenient method to measure the ζ-potential in industrial pastes and slurries [52] has recently been developed. Even the new method, which utilizes electroacoustic spectroscopy, is limited by its requirement of monodispersed particle size. Only the general conclusions that a longer fatty acid chain will help increase droplet stability can be made. The counterion is also important. In a system using either oleate or dodecanoate as the anions, various metal cations were studied. Al^{3+} displayed superior stabilization properties, and Ca^{2+} was also good, but Mg^{2+} performed poorly [53].

A good surfactant can also serve as an adjuvant to the lubrication process itself. The adsorption of stearate ion on metal oxide films has been studied [54, 55]. It was shown that the equilibrium for stearate adsorption on a metal surface was high, and the kinetics reasonably fast. Stearate, along with the appropriate cation, has even been shown to form multilayered structures on metal oxides. This added adsorption gives additional protection to the substrate. Adsorbed surfactants alone were able to reduce the friction of one system by ≈80% [56]. This was rationalized by the extended chains and their ability to easily orient along the direction of shear force.

If a fatty acid soap/surfactant is chosen for the biobased grease thickener, there are three components to consider, as shown in fig. 15.2: the hydrophobe, the hydrophile, and the metal cation. The hydrophobe has to make the surfactant compatible with the lubrication fluid. Low-temperature properties are also controlled to a large extent by

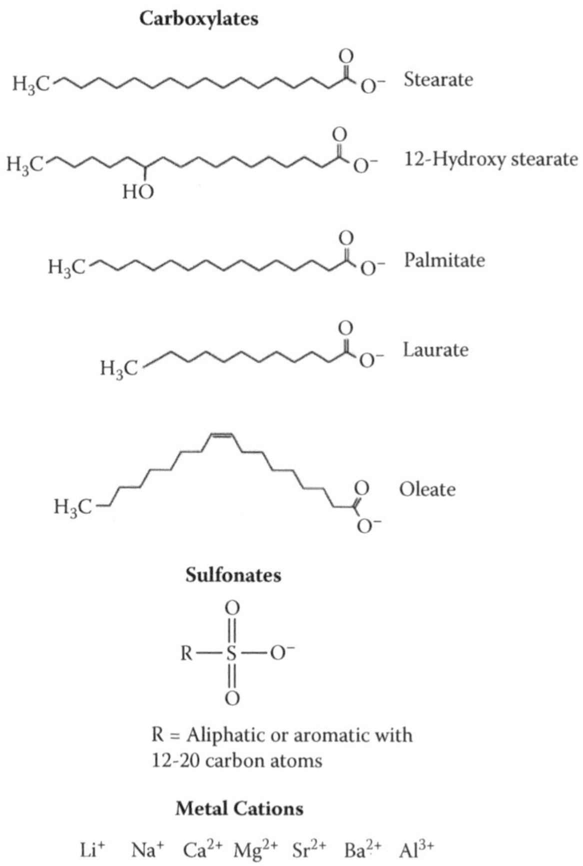
Carboxylates
H_3C
O
O^-
Stearate
12-Hydroxy stearate
HO
Palmitate
Laurate
Oleate
Sulfonates
R—S—O^-
R = Aliphatic or aromatic with
12-20 carbon atoms
Metal Cations
Li^+ Na^+ Ca^{2+} Mg^{2+} Sr^{2+} Ba^{2+} Al^{3+}

FIGURE 15.2 Structures of common ionic thickeners used in greases.

the hydrophobe. More branches in the fatty acid chain will generally result in a less crystalline grease that will display superior lubrication under low-temperature conditions. The chain length is also important. Chain lengths of C_{12}–C_{18} are commonly used. It was found that volatile loss, obtained using TFMO and a high-temperature oven test, and coefficient of friction, obtained using a ball-on-disk test, both increase with increasing fatty acid chain length of the thickener. As the chain length of fatty acid increases beyond C_{12}, there is a significant increase in grease hardness, with an optimum reached at C_{18} chain length, yielding grease with an NLGI reading of 2–3. The hypothesis for this behavior is that the soap fibers derived from short chain fatty acids are not well developed and sufficiently structured to hold and stabilize the base oil within its mesh structure. Longer fatty acid chain lengths in the metal soap form stronger interlocking fibers, resulting in a harder grease matrix. The trade-off here is that although longer chains initially [16] give a harder grease, they have a somewhat higher rate of breakdown under extreme oxidative conditions. Unsaturation in the hydrophobe is generally undesirable, resulting in polymerization and grease instability. Lithium oleate-based grease thus has a higher coefficient of friction and a lower onset temperature compared with lithium stearate grease [16]. A hydroxyl moiety on the grease [47] has been shown to be effective in creating a grease that has greater interaction with oleochemical lubricating oils and increased stability.

The most common hydrophile is the carboxylate group. This is easily obtained by simple saponification of animal or vegetable fats, which is sometimes performed in situ during the manufacture of the grease [3] (see fig. 15.3). This helps create an intimate mixture of grease and lubricating oil. Another commonly used soap in grease thickeners is calcium sulfonate. It has been used successfully in extreme pressure and temperature conditions since the late 1980s. It displays good lubricity, oxidation resistance, and rust inhibition.

The metal counter cation is probably the biggest factor in the thermal properties of grease. The dropping points of aluminum and calcium greases are the lowest at $\approx 90°C$–$100°C$. Barium, sodium, lithium, strontium, and mixed cation greases have much higher dropping points, often near or above 200°C [1]. The cation also affects the lubricity. In general [57], group II metals, especially Ca^{2+} and Ba^{2+}, reduce friction further than greases of group I metals, Na^+ and Li^+. Calcium-based thickeners provide medium to good shear stability, good water resistance, and medium to poor high-temperature stability, while sodium-based thickeners provide medium to good shear stability, poor water resistance, and good high-temperature stability. The greases prepared with lithium-based thickeners have good shear stability and high-temperature stability, but medium water resistance. By using calcium-complex thickeners, the high-temperature stability can be improved, and similarly, the water resistance of lithium-based greases can be improved by using lithium-complex thickeners. The mechanical stability of a lithium grease is much better than that of a calcium grease, while the antirust properties of a calcium grease are far better than

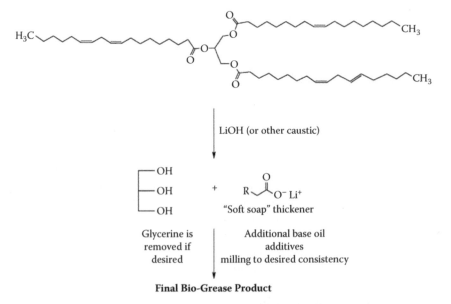

FIGURE 15.3 Transesterification of vegetable oil, forming the biogrease in situ.

those of a lithium grease. From this, it is easy to see the importance of matching the desired grease application with the cation. While preparing the thickener, the ratio of metal to fatty acid in the soap structure also affects the thermo-oxidative stability and lubricity of the grease. It was found that an optimum ratio of 1:0.75 for lithium hydroxide to stearic acid in grease thickener provided higher PDSC onset temperature and lower CoF compared with the grease with 1:1 ratio of metal to fatty acid.

Once the thickener is selected, the next variable is the thickener concentration. The final rheology of the product grease is dramatically affected by this concentration, as well as by the initial lubricating oil. In one study [58], the concentration of the surfactant, lithium 12-hydroxy stearate, was varied from less than 2 wt% up to a final concentration of 16 wt%. The product's physical properties varied considerably. The observed viscosity at a shear rate of 0.01 s^{-1} increased from ≈10 Pa·s up to more than 10,000 Pa·s. The viscoelastic properties were also tremendously affected. At 4 wt% or less thickener, the samples displayed a loss modulus greater than the storage modulus at all shear strain levels. When the concentration of thickener was increased beyond 4 wt%, an as-expected behavior was obtained and a critical strain could be calculated. A general conclusion was reached that <5 wt% thickener was not enough to thicken the grease sufficiently, but >25 wt% thickener caused the formation of cracks in the lubricant. The optimal concentration also varies, depending on oleochemical selection. It was shown [16] that a grease formulated from low linoleic soybean oil and lithium oleate was more effective in reducing friction as its thickener concentration was varied. The concentration of thickener was changed from 33 wt% down to 20 wt%, resulting in coefficient of friction measurements of 0.363 and 0.215, respectively. The corresponding hardness also changed from a rating of NLGI 4 down to an NLGI rating of 2. Thus soap with a low thickener content results in a soft grease that is capable of wetting the metal surface better during the tribochemical process. Under the given geometry of the wear test, harder greases with higher thickener content were pushed aside by the rotating ball, leading to a "starved lubrication" situation. The oxidation stability of the grease with a lower thickener concentration is higher. This is due to better interaction between soap fiber structure and the confined base oil with 20 wt% thickener concentration. The fiber–base-oil interaction profile of the grease matrix is not fully developed in the grease with 33 wt% thickener and thus results in "bleeding" of oil at test temperatures. This free oil is more amenable to oxidation and thermal evaporation loss compared with the oil confined in the grease matrix.

Most naturally occurring vegetable oils contain a triacylglyceride structure, as shown in fig. 15.1. This structure [5, 59] has been described as a tuning fork and is a good lubricant, even without any chemical modification [22, 28]. Additionally, the fatty acid chains on the oil itself can become an oleochemical thickener. During manufacture, the appropriate metal cation and a base are added to the mixture, and glycerol is removed, as shown in fig. 15.3.

The oxidation stability of soybean-oil–based greases can be further improved by using a more oxidatively stable oil, such as epoxy soybean oil (ESO). The ESO has low iodine value and high viscosity compared with regular soybean oil (SBO) due to the presence of epoxy rings in place of double bonds. The reduction in double bonds of ESO provides more oxidative stability to its structure. The PDSC onset

temperatures of most ESO-based greases are in the range of 256°C–276°C, while
those of soybean-oil-based greases are in the range of 175°C–208°C (18). Similarly,
the RBOT pressure drop in the case of ESO-based grease is 110–137 kPa in 6000
min compared with 172 kPa in 242 min for SBO-based grease. One experiment
on an ESO-based grease displayed a four-ball wear scar as low as 0.541 mm compared with 0.64 mm for an SBO-based grease under the same conditions. These
highly oxidatively stable greases can be used in high-temperature applications, as
they deliver excellent oxidative stability, effective lubrication, high wear protection,
and good friction reduction performance on par or better than most commercially
available mineral-oil– and vegetable-oil–based greases. High-oleic vegetable oils
can also be used if even higher oxidation stability of grease is required for a particular application.

The use of suitable additives also improves the oxidation stability and friction
wear properties of greases. For example, an SBO-based grease with no additive has
a PDSC OT of 120°C, oxidation induction time of 3 min, and RBOT pressure drop
of 228 kPa. If a soybean-oil–based grease with 4 wt% antioxidant is used, an OT of
160°C, an OIT of 366 min, and an RBOT pressure drop of 172 kPa are observed [16].
The coefficient of friction, obtained using the ball-on-disk experiment, and wear-
scar diameter, obtained using the four-ball test, are also improved with increased
additive concentrations. Overall, high concentrations of additives have been shown
to form a better grease with a longer life compared with samples without additives or
with low additive levels. This may be because the polar vegetable oils and additives
compete for the metal surfaces, and a higher proportion of additive will increase its
rate of diffusion to the metal surface.

Biodegradation of a biobased grease is also significantly higher than that of a
petrochemically based grease. The biodegradation also depends on the components
of the grease. In biobased greases, the vegetable oil used is not toxic to aquatic
organisms and biodegrades relatively fast and completely. As shown in table 15.1, the
biobased thickeners such as metal soaps are commonly natural fatty acids based, and
are thus biodegradable to a large extent, with only the inorganic metal remaining.
The additives constitute a very diverse range of chemicals that possess a wide range
of biodegradability and aquatic toxicity. They are usually present in small quantity in
the grease, and their biodegradation depends on their chemistry. The biodegradable
greases are especially important in areas where significant amounts of grease are
lost to the environment, such as in forestry.

15.4 CONCLUSION

Biobased grease is a limited but currently successful part of the lubricant market that
has a large growth potential due to economic, environmental, and legislative factors.
There is not a single formulation of grease or grease thickener that will be successful for all applications. The initial choice of lubricating fluid, as well as tailoring the
thickener to the application and lubricating oil, is critical. The choice of use of an
inorganic or soap thickener is especially important in extreme pressure and temperature conditions. If a biobased soap stock is used, the fatty acid chain length and cation
choice are the main factors to consider. Greases can be tested by a variety of methods

and should be evaluated for oxidative stability, rheology, low-temperature properties, friction and wear properties, and hardness. Using these tests, it is possible to formulate acceptable petroleum grease replacements and sometimes even improve lubrication properties. This is one more step toward the renewable economy of the future.

ACKNOWLEDGMENT

The use of trade, firm, or corporation names in this publication is for the information and convenience of the reader. Such use does not constitute an official endorsement or approval by the United States Department of Agriculture or the Agricultural Research Service of any product or service to the exclusion of others that may be suitable.

REFERENCES

1. Booser, E. R. 1995. In *Lasers to mass spectrometry: Kirk–Othmer encyclopedia of chemical technology.* Ed. J. I. Kroschwitz and M. Howe-Grant. New York: John Wiley and Sons.
2. Persuad, M. 2004. *2004–2005 Lubricant industry sourcebook: Lubes and greases.* Falls Church, VA: LNG Publishing.
3. Lansdown, A. R. 2004. *Lubrication and lubrication science.* 3rd ed. New York: ASME Press.
4. Adhvaryu, A., Biresaw, G., Sharma, B. K., and Erhan, S. Z. 2006. *Ind. Eng. Chem. Res.* 45: 3735–40.
5. Adhvaryu, A., Sharma, B. K., Hwang, H. S., Erhan, S. Z., and Perez, J. M. 2006. *Ind. Eng. Chem. Res.* 45: 928–33.
6. Hwang, H.-S., and Erhan, S. Z. 2002. In *Biobased industrial fluids and lubricants.* Ed. S. Z. Erhan and J. M. Perez. Champaign, IL: AOCS Press.
7. Hwang, H.-S., Adhvaryu, A., and Erhan, S. Z. 2003. *J. Am. Oil Chem. Soc.* 80: 811–15.
8. Kurth, T. L., Sharma, B. K., Doll, K. M., and Erhan, S. Z. 2007. *Chem. Eng. Commun.* 194: 1065–77.
9. Bhatnagar, A. K., Kaul, S., Chhibber, V. K., and Gupta, A. K. 2006. *Energy Fuels* 20: 1341–44.
10. Erhan, S. Z., Adhvaryu, A., and Sharma, B. K. 2006. U.S. Patent Application No. 11/165857, filed June 24, 2005 and published May 4, 2006.
11. Durak, E., and Karaosmanoglu, F. 2004. *Energy Sources* 26: 611–25.
12. Basu, H. N., Robley, E. M., and Norris, M. E. 1994. *J. Am. Oil Chem. Soc.* 71: 1227–30.
13. Basu, H. N., Robley, E. M., and Norris, M. E. 1995. *J. Am. Oil Chem. Soc.* 72: 257.
14. Sun, Z., and Li, C. 2005. *Polym. Prepr.* 46: 585.
15. Rowland, R. G., and Migdal, C. A. 2006. 2006/0090393 A1, May 4, 2006. U.S. Patent Application No. 11/165857.
16. Sharma, B. K., Adhvaryu, A., Perez, J. M., and Erhan, S. Z. 2005. *J. Agric. Food Chem.* 53: 2961–68.
17. Sharma, B. K., Adhvaryu, A., Liu, Z., and Erhan, S. Z. 2006. *J. Am. Oil Chem. Soc.* 83: 129–36.
18. Sharma, B. K., Adhvaryu, A., Perez, J. M., and Erhan, S. Z. 2006. *J. Agric. Food Chem.* 54: 7594–99.
19. Grives, P. R. 2000. *NLGI Spokesman* 63: 25–29.
20. Wang, C., and Erhan, S. 1999. *J. Am. Oil Chem. Soc.* 76: 1211–16.

21. Svejgaard, T. 2005. *INFORM* 16: 675–76.
22. Gawrilow, I. 2004. *INFORM* 15: 702–5.
23. Foglia, T. A. 2006. *INFORM* 17: 14–15.
24. Knothe, G. 2005. In *The biodiesel handbook*. Ed. G. Knothe. Champaign, IL: AOCS Press.
25. Wan Nik, W. B., Ani, F. N., Masjuki, H. H., and Eng Giap, S. G. 2005. *Ind. Crops Prod.* 22: 249–55.
26. Proffitt, Jr., T. J., and Patterson, H. T. 1988. *J. Am. Oil Chem. Soc* 65: 1682–94.
27. Hillion, G., and Proriol, D. 2003. *OCL—Oleagineux Corps Gras Lipides* 10: 370–72.
28. Pratap, A. P., Kadam, A. S., and Bhowmick, D. N. 2005. *INFORM* 16: 282–85.
29. Cosgrove, J. P., Church, D. F., and Pryor, W. A. 1987. *Lipids* 22: 299–304.
30. Ruger, C. W., Klinker, E. J., and Hammond, E. G. 2002. *J. Am. Oil Chem. Soc.* 79: 733–36.
31. Zhang, Y. Y., Ren, T. H., Wang, H. D., and Yi, M. R. 2004. *Lubrication Sci.* 16: 385–92.
32. Erhan, S. Z., Sharma, B. K., and Perez, J. M. 2006. *Ind. Crops Prod.* 24: 292–99.
33. Erhan, S. Z., Adhvaryu, A., and Sharma, B. K. 2005. In *Synthetics, mineral oils, and bio-based lubricants chemistry and technology*. Ed. L. R. Rudnick. Boca Raton, FL: CRC Press.
34. Biswas, A., Adhvaryu, A., Stevenson, D. G., Sharma, B. K., Willet, J. L., and Erhan, S. Z. 2007. *Ind. Crops Prod.* 25: 1–7.
35. Hwang, H.-S., and Erhan, S. Z. 2001. *J. Am. Oil Chem. Soc.* 78: 1179–84.
36. King, J. W., Holliday, R. L., List, G. R., and Snyder, J. M. 2001. *J. Am. Oil Chem. Soc.* 78: 107–13.
37. Findley, T. W., Swern, D., and Scanlan, J. T. 1945. *J. Am. Chem. Soc.* 67: 412–14.
38. Crocco, G. L., Shum, W. F., Zajacek, J. G., and Kesling, H. S. J. 1992. 5166372, Nov. 24, 1992. U.S. Patent Application No. 5166372.
39. Carlson, K. D., Kleiman, R., and Bagby, M. O. 1994. *J. Am. Oil Chem. Soc.* 71: 175–82.
40. Campanella, A., Baltanás, M. A., Capel-Sánchez, M. C., Campos-Martín, J. M., and Fierro, J. L. G. 2004. *Green Chem.* 6: 330–34.
41. Erhan, S. Z., Bagby, M. O., and Nelsen, T. C. 1997. *J. Am. Oil Chem. Soc.* 74: 703–6.
42. Holser, R. A., Doll, K. M., and Erhan, S. Z. 2006. *Fuel* 85: 393–95 (2006).
43. Verkuijlen, E., Kapteijn, F., Mol, J. C., and Boelhouwer, C. 1977. *Chem. Commun.* 198–99.
44. Adhvaryu, A., Sharma, B. K., and Erhan, S. Z. 2002. In *Industrial uses of vegetable oils*. Ed. S. Z. Erhan. Champaign, IL: AOCS Press.
45. Araki, C., Kanzaki, H., and Taguchi, T. 1995. *NLGI Spokesman* 59: 15–23.
46. Honary, L. A. T. 2001. *NLGI Spokesman* 65: 18–27.
47. Hagemann, J. W., and Rothfus, J. A. 1991. *J. Am. Oil Chem. Soc.* 68: 139–43.
48. Harris, J. W. 2002. *NLGI Spokesman* 65: 18.
49. Rosen, M. J. 2004. *Surfactants and interfacial phenomena*. 3rd. ed. Hoboken, NJ: John Wiley and Sons.
50. Roland, I., Piel, G., Delattre, L., and Evrard, B. 2003. *Int. J. Pharm.* 263: 85–94.
51. Parra-Barraza, H., Hernandez-Montiel, D., Lizardi, J., Hernandez, J., Herrera Urbina, R., and Valdez, M. A. 2003. *Fuel* 82: 869–74.
52. Hunter, R. J. 2001. *Colloids Surfaces A* 195: 205–14.
53. Oliveira, A. P., and Torem, M. L. 1996. *Colloids Surfaces A* 110: 75–85.
54. Capelle, H. A., Britcher, L. G., and Morris, G. E. 2003. *J. Colloid Interface Sci.* 268: 293–300.
55. Polat, M. 2006. *J. Colloid Interface Sci.* 298: 593–601.

56. Qian, L., Charlot, M., Perez, E., Luengo, G., Potter, A., and Cazeneuve, C. 2004. *J. Phys. Chem. B* 108: 18608–14.
57. Nakonechnaya, M. B., Lyubinin, I. A., Khalyavaka, E. P., and Mnishchenko, G. G. 1983. *Chem. Technol. Fuels Oils* 19: 198–200.
58. Yeong, S. K., Luckham, P. F., and Tadros, T. F. 2004. *J. Colloid Interface Sci.* 274: 285–93.
59. Adhvaryu, A., Erhan, S. Z., and Perez, J. M. 2003. *Thermochim. Acta* 395: 191–200.

Part V

Surfactant Adsorption and Aggregation: Relevance to Tribological Phenomena

16 How to Design a Surfactant System for Lubrication

Bengt Kronberg

ABSTRACT

It has been found that surfactants, when aggregating in their lamellar liquid crystalline form, provide good lubrication of surfaces. This contribution discusses the conditions for obtaining lamellar liquid crystalline phases at surfaces. Beside the surfactant molecular structure itself, other parameter of importance are co-surfactants, co- and counter-ions and the presence of solubilizates.

16.1 INTRODUCTION

Many of us have experienced the difficulty of catching a wet soap bar at the bottom of a wet bath tub. The cause of the slippery soap bar is the formation of lamellar liquid crystalline phases between the soap bar and the surface of the bathtub. The structure of a lamellar liquid crystalline phase is similar to that of graphite, in the sense that the binding between the lamellae is weak compared with the lateral binding in the lamellae. The sliding movement is, therefore, expected to occur between the layers [1, 2].

Our first question is, what shape of the surfactant self-assembly would be the most beneficial for good lubrication, i.e., a low friction coefficient? A variety of different shapes can occur, and some of these are depicted in fig. 16.1. One might argue

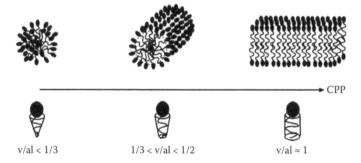

FIGURE 16.1 Surfactant molecules self-assemble into various aggregate shapes, depending on the surfactant molecular structure as described by the critical packing parameter (CPP), which is the ratio of the molecular volume (v) divided by its length (l) times the cross-sectional area of the head group (a): v/al.

327

that the formation of micelles between two sliding surfaces would be akin to a ball bearing and hence would facilitate lubrication. On the other hand, when we have the presence of a lamellar liquid crystalline phase between the two sliding surfaces, there will be a weak boundary in the hydrocarbon tail part. Hence a lamellar liquid crystalline phase should give similar lubrication properties as ordinary lubrication oils. However, it is not only the self-assembly structure, but also the load-carrying capacity that are determinant for a low friction between two hydrophilic surfaces. This leads us to suggest that important parameters are the extent of cohesion and packing of the hydrocarbon chains [3–5]. Closely related are the solvency [6] and melting [7] of the surfactant hydrocarbon chains. Finally, the interaction between the surfactant head group and the surface is also an important parameter [8–10]. This chapter will focus on how to obtain the right surfactant aggregate structure for lubrication. We will not discuss the other important parameters.

16.2 ADSORPTION OF SURFACTANTS AT HYDROPHILIC SURFACES

Before the advent of the atomic-force microscope (AFM) and other sophisticated techniques, such as neutron scattering, one believed that the adsorption of ionic surfactants on oppositely charged surfaces adhered to the following pattern with increasing surfactant concentration. At very low surfactant concentrations, far below one one-hundredth of the critical micelle concentration (CMC), the surfactants are present in the solution as unimers, and they adsorb at the surface by an ion-exchange mechanism. This mechanism provides only a very small driving force, and hence the surfactant concentration at the surface only increases very slowly with an increase in surfactant solution concentration. At a critical concentration, the surfactants at the surface start to interact, and it is here that the old and new interpretations differ. In older schools, one pictured the surfactants forming a monolayer at the surface, with their charged head groups directed toward the oppositely charged surface and the hydrocarbon chains protruding into the aqueous solution. This picture would explain the fact that it is possible to hydrophobize a charged surface by the addition of oppositely charged surfactants. Recent research has shown, however, that this molecular picture is not correct. It has been shown that surfactants form micelles at the surface at very low surfactant concentrations, and there is no monolayer formation. This is shown in fig. 16.2, which presents AFM pictures of cetyl trimethyl ammonium bromide (CTAB) surfactant self-assemblies on a mica surface at three decades of surfactant concentration [11]. Note that the critical concentration, where surfactant aggregates form at the surface, is on the order of one one-hundredth of the critical micelle concentration (CMC) in the bulk of the surfactant. Therefore, it is not necessary that the surfactant concentration be as high as the CMC, provided that the molecules adsorb at the surface in question. We will later show that mesophases, such as lamellar liquid crystalline phases, occur at surfaces long before (i.e., at far lower concentrations) they appear in solution. Hence, the presence of a surface can be sufficient to induce a "lubrication entity" in the form of surfactant aggregates at the surface.

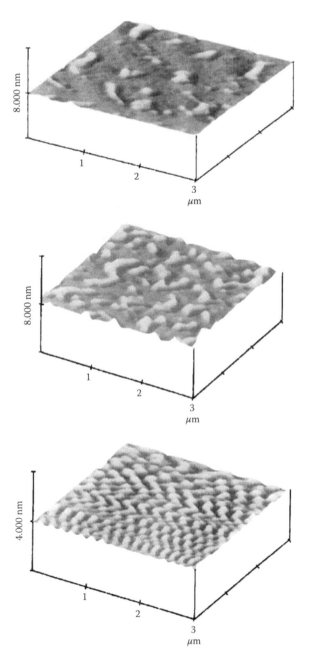

FIGURE 16.2 Surfactant molecules at a polar surface self-assemble into aggregates, even at very low surfactant concentration. A monolayer of surfactants, where hydrophobic layers with the surfactant head group facing the surface and the hydrophobic tail facing the aqueous phase, is not formed at the surface, as was earlier believed. The system is cetyl trimethyl ammonium bromide (CTAB), adsorbed on mica, in equilibrium with the bulk concentrations of 10^{-5} M, 10^{-4} M, and 10^{-3} M. (From Sharma, B. G., Basu, S., and Sharma, M. M. 1996. *Langmuir* 12: 6506. With permission.)

16.3 THE IMPORTANCE OF SURFACTANT AGGREGATE STRUCTURE IN LUBRICATION PERFORMANCE

We will first give an account of the importance of the structure of the assemblies formed at the surface for lubrication performance. After that we present a review of the various parameters that are at play in altering these structures and, hence, their lubrication properties.

Figure 16.3 presents the measured friction force versus load for four different surfactant systems, where the structure of the surfactant has been varied systematically in order to alter the critical packing parameter (CPP) [12]. The four surfactants are gemini surfactants with hydrocarbon chain lengths of 12 carbon atoms and where the length of the spacer, between the chains, has been varied from 3 to 12 carbon atoms (see fig. 16.3b). The surfactant with a spacer length of 12 carbon atoms was found to be the most hydrophilic, forming micelles, while the surfactant with a spacer length of 3 carbon atoms was the most hydrophobic, forming lamellar

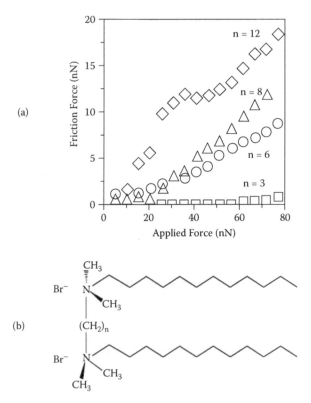

FIGURE 16.3 (a) Friction force versus applied load for four different surfactants. The lowest friction coefficient (slope of the curve) is obtained from the surfactant with the highest CPP, most likely forming lamellar liquid crystalline phases at the surface. (b) The molecular structure of the investigated surfactants. One of the surfaces was tungsten, and the other was gold. (From Boschkova, K., Feiler, A., Kronberg, B., and Stålgren, J. J. R. 2002. *Langmuir* 18: 7930. With permission.)

liquid crystalline phases. A likely reason for this apparent counterintuitive result is that, for short spacers, the effective charge is less than two, hence rendering a less hydrophilic surfactant.

Figure 16.3a reveals that it is the latter surfactant, with 3 carbon atoms as a spacer, that gives by far the lowest friction. We hence conclude that the concept of micelles forming "ball-bearing spheres" between the surfaces is not correct. A likely reason for the failure of surfactant micelles to lubricate is their lack of load-carrying capacity.

We attribute the low friction of the lamellar liquid crystalline phases to the repulsion between the surfactant layers in the structure, culminating in a load-carrying capacity. This repulsion will naturally cause a low friction if the planes are sheared. We do, however, expect an increase in the friction if the surfaces are sheared at a faster rate, due to a disruption of the lamellar order, resulting in a higher friction. This slipping between the layers is the reason for the low friction. Hence, a surfactant system should be designed such that a lamellar liquid crystalline layer is formed at the interface between the sliding layers in order to obtain a low friction.

Figure 16.4a shows the normal force of didodecyl dimethyl ammonium bromide (DDAB) and dodecyl trimethyl ammonium bromide (DTAB) as a function of distance between friction surfaces [13]. DDAB forms lamellar liquid crystalline phases at the surface, while DTAB forms oblate ellipsoid structures [14]. We note in passing that neither system exhibits a double-layer repulsion at longer distances, indicating a nearly neutralized charge at the anionic mica surface. Figure 16.4b shows the lubrication results of these two surfactant systems as obtained by elastohydrodynamic (EHD) film thickness measurements [15]. It is observed that, in both systems, the film thickness increases with rolling speed. We also note that both systems form much thicker films under dynamic conditions compared with static conditions. The DTAB surfactant forms a 6.8–8.0-nm-thick film at low rolling speeds, but at higher speeds the film thickness is the same as would be obtained for pure water, as indicated by the straight solid line. The DDAB system, on the other hand, forms a 60–80-nm-thick film that increases continuously with rolling speed, indicating that this system is capable of spontaneously forming induced surfactant aggregate structures during shear. It is also noted that this system has a good load-carrying capability, and it can, hence, be used for technical purposes.

We thus conclude that, in order to achieve good lubricating properties of a surfactant system, it should be able to form lamellar liquid crystalline aggregates at the surfaces that are sheared. Since the aggregate shape is determined by the surfactant's molecular structure, it is concluded that this structure plays a detrimental role in the lubrication properties of surfactant systems. We suggest viewing the results in terms of load-carrying capability. The best load-carrying capability is obtained when there is an intact lamellar liquid crystalline layer between the surfaces. However, if the surfactants assemble in micellar aggregates or the like, the load-carrying capability is drastically reduced. In fact, it has been shown, in the system above, that the addition of a small amount of DTAB to a DDAB system completely ruins the lubrication properties. Therefore, it is reasonable to assume that the formation of "water bridges" between the two surfaces will reduce the load-carrying capability such that the lubrication improvement by the surfactant ceases.

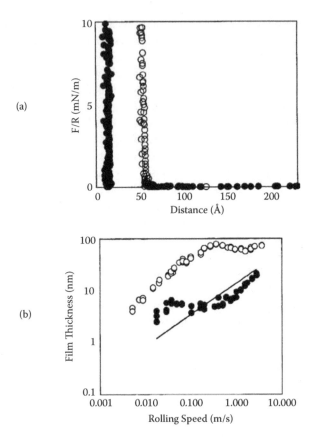

FIGURE 16.4 (a) Normal force versus distance between two surfaces for the systems didodecyl ammonium bromide (DDAB) (open symbols) and dodecyl trimethyl ammonium bromide (DTAB) (filled symbols). (b) Lubrication results from a fully flooded elastohydrodynamic (EHD) rig showing film thickness versus rolling speed (symbols are the same as in fig. a). The solid line is the film thickness of pure water. (From Boschkova, K., Kronberg, B., Stålgren, J. J. R., Persson, K., and Ratoi Salagean, M. 2002. *Langmuir* 18: 1680. With permission.)

The ability to form a lamellar liquid crystalline film depends on the spontaneous curvature of the surfactant aggregates, or the CPP, which is a convenient and intuitive description of the surfactant molecular structure. Kabalnov and Wennerström [16] have shown that, for the formation of a "water bridge" between two water droplets, a large free energy is required for a surfactant with a high CPP, while the free energy required for a surfactant with a low CPP is lower. Hence, the stability of a surfactant double layer increases with an increase of the CPP of the surfactant.

16.4 FACTORS INFLUENCING THE SELF-ASSEMBLY STRUCTURE AT SURFACES

We now have come to the conclusion that lamellar liquid crystalline phases, present at the surface, are the most suited for obtaining lubrication. We will now investigate the various parameters that will have an effect on the self-assembly structure at surfaces so that we can design good lubrication systems. First, it is concluded that it is the relative packing geometry of the surfactant that determines the curvature of the self-assemblies. This has been worked out in detail in numerous publications [17]. As depicted in fig. 16.1, the general pattern when one gradually increases the CPP of the surfactant system is that spherical micelles form at low (ca. 0.3) CPP values. The aggregate structure then shifts to hexagonal and, in turn, to lamellar phases as the CPP increases. Note that if the surfactant hydrocarbon chains are very flexible, the lamellar phase will be replaced by a bicontinuous microemulsion phase. In solution, the following alterations will increase the CPP of an ionic surfactant system:

- Change the hydrocarbon chain to a longer one
- Change the hydrocarbon chain to two chains
- Change from a linear to a branched hydrocarbon chain
- Add a solubilizate such as a long-chain alcohol
- Add a nonionic surfactant
- Add salt
- Add a small amount of ionic surfactant with opposite charge

For nonionic surfactant systems with ethylene oxide (EO) as the polar part, we have the same dependence regarding the hydrocarbon chain. We also have an increase of the CPP with the following alterations:

- Change the EO chains to shorter ones
- Increase the temperature
- Add salt
- Add a solubilizate such as a long-chain alcohol
- Add a shorter EO chain nonionic surfactant

We note that for nonionic surfactants, the addition of salt is about 100 times less efficient compared with ionic surfactants. In the latter case, there is a remarkable effect even if the salt concentration is on the order of 10 mM, while for nonionic surfactants there is an effect for salt concentrations even on the order of 1 M. In the latter case, the mechanism is that salt dehydrates the EO chain, while in the former case the mechanism is that the salt increases the counterion binding and hence decreases the ionic repulsion between the surfactant head groups. Thus, by knowing the mechanisms, we conclude that, for the ionic systems, adding a divalent counterion would be much more efficient compared with a monovalent counterion. For the nonionic surfactant systems, the anion is the deciding partner, while changing the cation hardly affects the CPP of the system. Here di- and trivalent anions are much more efficient compared with monovalent anions.

These general rules are also likely to be valid for the surfactant self-assembly structures at surfaces. There are, however, some complicating factors due to the presence of the surface. First, the surface might induce a preferred structure, and second, the surface might partake in the formation of self-assembled structures, and hence there are counterions from the surface to take into account. An example of the first case, where the surface induces a preferred self-assembly structure at the interface, is shown in fig. 16.5. Here it is seen that for the system of cationic surfactants on graphite, the surface prefers a hexagonal structure, while the preferred aggregate structure would be micellar (image A) or lamellar (image C) [18]. A second example is given in fig. 16.6, which shows the adsorption of octadecyl trimethyl ammonium bromide at mica and graphite surfaces. We note, again, that the graphite surface induces a hexagonal surfactant self-assembly structure, while a featureless lamellar structure is formed on the mica surface [19]. An example of the second case is shown in figs. 16.7a and 16.7b, where the counterion is salicylate [18]. Salicylate counterion is known to bind to the micellar surface, and hence it induces a larger CPP, which is seen in the figure.

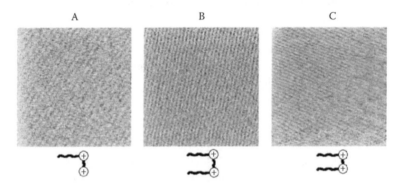

FIGURE 16.5 AFM images of three different surfactants on graphite, showing a hexagonal aggregate structure of all three systems despite the fact that they should display micellar (image A) and lamellar (image C), as they do on mica [18]. The surfactants are gemini surfactants with linear hydrocarbons and quaternary ammonium head groups with the following chain lengths of tail-spacer-tail; 18-3-1 (image A), 12-6-12 (image B), and 12-2-12 (image C). (From Manne, S., Schaffer, T. E., Huo, Q., Hansma, P. K., Stucky, D. E., and Aksay, L. A. 1997. *Langmuir* 13: 6382. With permission.)

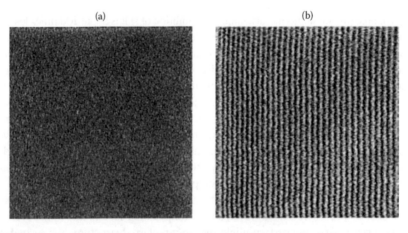

FIGURE 16.6 AFM images of octadecyl trimethyl ammonium bromide on mica (a) and graphite (b) surfaces, showing that graphite induces a hexagonal surfactant aggregate structure, while the mica surface renders a featureless lamellar structure. (From Liu, J.-F., and Ducker, W. A. 1999. *J. Phys Chem. B* 103: 8558. With permission.)

Another example is shown in fig. 16.8, where cesium (Cs) ions are used as co-ions for cetyl trimethyl ammonium chloride (CTAC) adsorbed on a mica surface [20]. Here we note that the CPP decreases with increasing Cs ion concentration. The interpretation is that the Cs cation competes with the ammonium for the surface

(a)   (b)

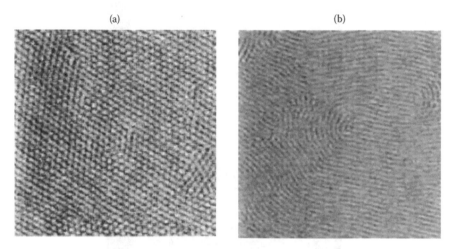

FIGURE 16.7 AFM images of gemini surfactant, with chain length of tail-space-tail of 18-3-1, on mica surface without (a) and with (b) salicylate counterions, showing that salicylate increases the CPP of the surfactant and, thus, that the aggregates go from micellar to hexagonal. (From Manne, S., Schaffer, T. E., Huo, Q., Hansma, P. K., Stucky, D. E., and Aksay, L. A. 1997. *Langmuir* 13: 6382. With permission.)

(a)

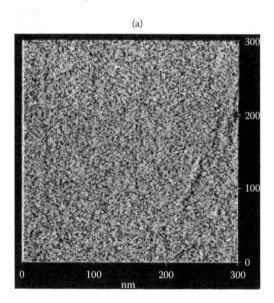

FIGURE 16.8 AFM images of cetyl trimethyl ammonium chloride (CTAC) on mica. The figures show the effect of increasing Cs salt concentration: (a) 0 mM, (b) 34 mM, and (c) 100 mM. (From Lamont, R. E., and Ducker, W. A. 1998. *J. Am. Chem. Soc.* 120: 7602. With permission.)

(b)

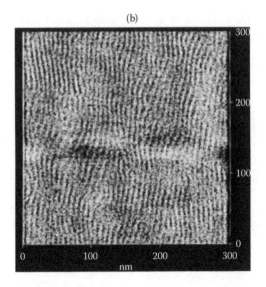

(c)

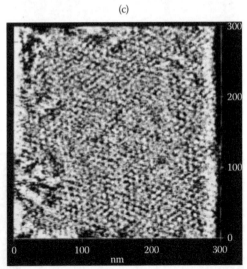

FIGURE 16.8 (CONTINUED)   AFM images of cetyl trimethyl ammonium chloride (CTAC) on mica. The figures show the effect of increasing Cs salt concentration: (a) 0 mM, (b) 34 mM, and (c) 100 mM. (From Lamont, R. E., and Ducker, W. A. 1998. *J. Am. Chem. Soc.* 120: 7602. With permission.)

sites, rendering a lower effective surface charge, which, in turn, leads to a lower CPP of the surfactants forming aggregates at the surface.

Finally, we conclude that the surface charge is also an important factor. In the AFM image shown in fig. 16.9, it is seen that the packing of tetradecyl trimethyl ammonium

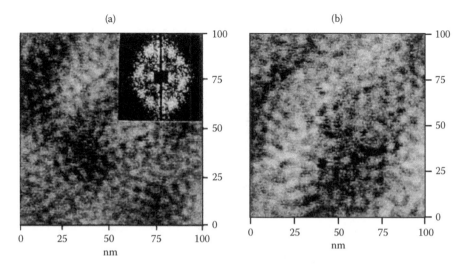

FIGURE 16.9 AFM images of tetradecyl trimethyl ammonium bromide (TTAB) on silica. (a) pH = 6.3 and (b) pH = 2.9, illustrating the effect of surface charge on the surfactant aggregates. (From Manne, S., and Gaub, H. E. 1995. *Science* 270: 1480. With permission.)

bromide (TTAB) micelles is more dense on silica at a pH of 6.3 compared with silica in equilibrium with a solution at pH = 2.9, where the surface charge density is lower [21].

An interesting way to increase the CPP of a system is to add a solubilizate to the surfactant aggregates on the surface. This is shown in fig. 16.10, where naphthalene is added to the hexagonal layer of TTAB surfactant molecules, eventually forming a bilayer at the surface [22]. We note that different solubilizates affect the surfactant's self-assembly structure differently. Polar solubilizates will partition toward the interface of the self-assembled aggregates, thus increasing the CPP, as shown in fig. 16.10. Nonpolar solubilizates, however, partition into the center of the aggregates, which, in turn, will decrease the CPP of the system and hence decrease the lubrication. Hence, the lubrication properties of a system forming micelles at the surface can be improved by adding a polar solubilizate that will partition at the water/aggregate interface.

16.5 CONCLUSION

Lamellar liquid crystalline phases should be formed at the surface in order to attain good lubrication of a surfactant system. Surfactant molecular structure is one of the critical variables affecting the shape of the surfactant aggregates and, hence, the lubrication properties of surfactant systems. The aggregate shape is also determined by other factors, such as hydrophilic/hydrophobic additives, salt, solubilizates, and temperature. In addition, the aggregates present at the surface are also sensitive to surface properties as well as to co-ions.

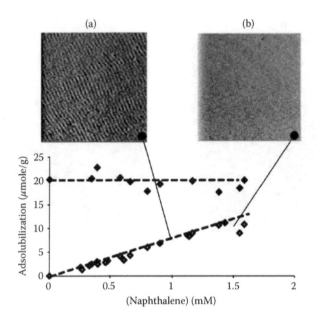

FIGURE 16.10 Illustration showing that solubilization alters the morphology at the surface. At low naphthalene concentrations, the system shows a hexagonal structure (image a), while at higher concentrations the system displays a featureless lamellar structure (image b). The system is tetradecyl trimethyl ammonium bromide (TTAB) on mica with added naphthalene. (From Kovacs, L., and Warr, G. G. 2002. *Langmuir* 18: 4790. With permission.)

ACKNOWLEDGMENT

Stimulating discussions with Katrin Boschkova are gratefully acknowledged.

REFERENCES

1. O'Shea, S. J., Welland, M. E., and Rayment, T. 1992. *Appl. Phys. Lett.* 61: 2240.
2. Fuller, S., Li, Y., Tiddy, G. J., and Wyn-Jones, E. 1995. *Langmuir* 11: 1980.
3. Vakarelski, I. U., Brown, S. C., Rabinovich, Y. I., and Moudgil, B. M. 2004. *Langmuir* 20: 1724.
4. Xiao, X., Hu, J., Charych, D. H., and Salmeron, M. 1996. *Langmuir* 12: 235.
5. Lee, S., Scon, Y.-S., Colorado, Jr., R., Guenard, R. L., Lee, T. R., and Perry, S. S. 2000. *Langmuir* 16: 2220.
6. Clear, S. C., and Nealy, P. F. 2001. *Langmuir* 17: 720.
7. Israelachvili, J. N., Chen, Y.-L., and Yoshizawa, H. 1994. *J. Adhesion Sci. Technol.* 8: 1231.
8. Liu, Y., Evans, F. D., Song, Q., and Grainger, D. W. 1996. *Langmuir* 12: 1235.
9. Frisbe, C. D., Roznyai, L. F., Noy, A., Wrighton, M. S., and Lieber, C. M. 1994. *Science* 265: 2071.
10. Overney, R., and Meyer, E. 1993. *MRS Bulletin* 18: 26.
11. Sharma, B. G., Basu, S., and Sharma, M. M. 1996. *Langmuir* 12: 6506.
12. Boschkova, K., Feiler, A., Kronberg, B., and Stålgren, J. J. R. 2002. *Langmuir* 18: 7930.

13 Boschkova, K., Kronberg, B., Stålgren, J. J. R., Persson, K., and Ratoi Salagean, M. 2002. *Langmuir* 18: 1680.

14. Bergström, M., and Pedersen, J. S. 1999. *Phys. Chem. Chem. Phys.* 1: 4437.

15. Johnston, G. J., Wayte, R., and Spikes, H. A. 1991. *Tribol. Trans.* 34: 187.

16. Kabalnov, A., and Wennerström, H. 1996. *Langmuir* 12: 276.

17. Mitchell, D. J., and Ninham, B. W. 1981. *J. Chem. Soc., Faraday Trans.* 77 (2): 601.

18. Manne, S., Schaffer, T. E., Huo, Q., Hansma, P. K., Stucky, D. E., and Aksay, L. A. 1997. *Langmuir* 13: 6382.

19. Liu, J.-F., and Ducker, W. A. 1999. *J. Phys Chem. B* 103: 8558.

20. Lamont, R. E., and Ducker, W. A. 1998. *J. Am. Chem. Soc.* 120: 7602.

21. Manne, S., and Gaub, H. E. 1995. *Science* 270: 1480.

22. Kovacs, L., and Warr, G. G. 2002. *Langmuir* 18: 4790.

17 Aqueous Solutions of Oxyethylated Fatty Alcohols as Model Lubricating Substances

Marian Wlodzimierz Sulek

ABSTRACT

The aim of this investigation is to present application possibilities of aqueous solutions of surface active agents as lubricating substances. The intended application targets are working and hydraulic fluids.

Water can be used as a lubricant base due to its high thermal capacity and conduction, non-flammability, being environment-friendly and availability. Water also has a number of negative properties, such as corrosive properties and insufficient lubricating properties which, however, can be alleviated by means of appropriate additives. At the current stage of investigation, simple binary solutions with oxyethylated fatty alcohols as additives are being tested. They are soluble in water in which, at low concentrations, they form micellar solutions while at high concentrations they form lyotropic liquid crystals. The physicochemical properties of those solutions as well as their behavior at the solid-solution interface have been presented in literature.

Twelve oxyethylated fatty alcohols with various lengths of alkyl chain and ethylene oxide were used in this investigation. In view of numerous literature data available, physicochemical tests were limited to measurements of surface tension, wettability and viscosity. Microscope photographs were taken in polarized light in order to confirm the appearance of liquid crystalline structures. As expected, formation of micelles was observed at low concentrations, whereas mesophases (hexagonal and lamellar) were identified at concentrations of about 50% to 70%.

Tribological investigations were carried out by means of a four-ball tester and a T-11 tester with a ball-on-disk friction pair. Antiseizure properties (scuffing load—P_t, seizure load—P_{oz}, and limiting pressure of seizure—p_{oz}) as well as motion resistance and wear at a constant load were determined using a four-ball tester. It has been found that ethoxylates used as additives significantly modify tribological properties. The measured P_t, P_{oz}, p_{oz} values increase by as much as several times compared with water. The coefficient of friction and wear measured at a constant load decrease considerably to about half relative to water.

The effects of additive concentration, compound structure and friction pair material on tribological properties were discussed on the basis of the results obtained. A comparison of the simple binary systems with commercial lubricant compositions based on mineral oils was made. The result of the comparison indicates that aqueous solutions may be applied in real tribological systems.

17.1 INTRODUCTION

Aqueous solutions of surfactants may be used as lubricants in working and hydraulic fluids [1–13]. They are much more stable than the emulsions and microemulsions used so far. The problem with water used as a lubricant base consists in its low lubricity. Common additives employed in mineral bases are usually insoluble in water and, quite often, do not meet ecological safety criteria.

The effectiveness of additives in the friction zone is connected with their surface activity. Therefore, additives are being sought that, under friction conditions, form a lubricant film that leads to a decrease in the coefficient of friction and wear and has a high load-carrying capacity. In addition, the additives should be efficient at low concentrations, such as 1%.

Aqueous solutions of oxyethylated alcohols were selected as model lubricating substances. They are produced from vegetable raw materials on an industrial scale and are relatively inexpensive. Alcohol ethoxylates exhibit high surface activity [14–22]. They form micelles and liquid crystal structures in aqueous solutions [16, 23–26]. Aqueous solutions of oxyethylated alcohols can be used as inflammable, ecological lubricating substances, particularly in the friction pairs that are in direct contact with the environment, people, foodstuffs, pharmaceuticals, and cosmetics.

17.2 OXYETHYLATED FATTY ALCOHOLS

Alcohols are chemical compounds containing a hydroxyl group that determines a number of their properties. In the hydroxyl group, the hydrogen atom is attached to a strongly electronegative oxygen atom, and that may lead to the formation of hydrogen bonds with atoms having free electron pairs. The solubility of aliphatic alcohols depends on the length of the alkyl chain. If it contains only up to three carbon atoms, the alcohols are readily soluble in water. If it contains three to four carbon atoms, it displays limited solubility, while with more than five carbon atoms it is insoluble. The alcohols with an alkyl chain containing 8 to 22 carbon atoms are termed fatty alcohols. They are insoluble in water, while their oxyethylated derivatives are water soluble. Ethoxylates of fatty alcohols comprise a large group of nonionic surfactant compounds with great practical importance. They result from a reaction of ethylene oxide fatty alcohols acquired from renewable sources and of petrochemical origin. The reaction takes place in the presence of a catalyst according to the mechanism shown in fig. 17.1. It consists in breaking of the oxygen–hydrogen bond and introducing a certain number of ethylene oxide monomer units in its place.

Oxyethylated alcohols are prepared at 130°C–180°C in the presence of a catalyst whose concentration is on the order of a few tenths of a percent. Various kinds of catalysts are used. Earlier, basic catalysts (sodium hydroxide) were commonly used, which yielded a wide distribution of ethoxymers. Currently, acidic catalysts with a considerably narrower distribution of ethoxymers and a lower content of unreacted alcohol are used. It follows from this comparison that products obtained by means of different syntheses may have different compositions due to the presence of both different ethoxymers and residual free alcohol, which is often hard to vaporize. Additionally, alcohol as a reactant may be a mixture of compounds with various alkyl chain lengths. Therefore, particularly in the case of ethoxylates prepared from natural sources on an industrial scale, one may have to deal with a mixture of compounds with various ethylene oxides (m) and alkyl (n) chain lengths.

In the REO_n notation used here, R is the alkyl chain with n carbon atoms in the chain, while EO_m is the number of moles (m) of the combined ethylene oxide (EO). As a result of oxyethylation, fatty alcohols become amphiphilic, where the ethylene oxide chain is the hydrophilic part and the alkyl chain is the hydrophobic part. The general formula and a spherical model of lauryl alcohol oxyethylated with 3 moles and 5 moles of ethylene oxide are presented, respectively, in figs. 17.2 and 17.3. Using fig. 17.3, it is possible to geometrically estimate a relative proportion of the hydrophilic part (ethylene oxide chain) and of the hydrophobic part (alkyl chain) in the oxyethylated alcohol molecule.

$$R - OH \ + \ mCH_2CH_2O \longrightarrow R - O(CH_2CH_2O)_mH$$

FIGURE 17.1 Reaction scheme for preparation of oxyethylated alcohols.

FIGURE 17.2 General formula of oxyethylated alcohols.

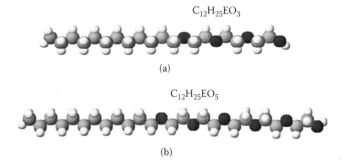

FIGURE 17.3 Spherical model of lauryl alcohol oxyethylated with (a) 3 moles and (b) 5 moles of ethylene oxide.

The solubility of oxyethylated fatty alcohols results from the formation of hydrogen bonds between water and a free electron pair on the oxygen atom of the ether group. The dehydration process connected with an increase in the temperature of the solution is important from the point of view of application, e.g., in tribology. Above a certain temperature, called the cloud point, hydrogen bonds are broken, and the compound loses its amphiphilic properties. As a result of desolvation, alcohol relatively increases its hydrophobic properties. These observations can be confirmed by the decrease in the critical micelle concentration (CMC) values with increasing temperature [27–31]. The dehydration process is reversible and, with a decrease in temperature, repeated hydration of molecules may occur at a certain temperature, called the clear point. Both the cloud point and the clear point depend on the structure of the compound, particularly on a relative proportion of the hydrophilic and hydrophobic parts [32–38].

Oxyethylated alcohols are marketed under trade names: Brij (produced by Sigma-Aldrich, Germany), Genapol (produced by Hoechst, Germany), Emulsogen (produced by Henkel, Germany), and Rokanol (produced by Rokita, Poland). Brij 35 ($C_{12}H_{25}EO_{23}$), Brij 56 ($C_{16}H_{33}EO_{10}$), Brij 58 ($C_{16}H_{33}EO_{20}$), Brij 97 ($C_{18}H_{35}EO_{10}$), Brij 98 ($C_{18}H_{35}EO_{20}$), Brij 76 ($C_{18}H_{37}EO_{10}$), Brij 78 ($C_{18}H_{37}EO_{20}$), and Brij 700 ($C_{18}H_{37}EO_{100}$) were used in this investigation.

17.3 SURFACE ACTIVITY OF OXYETHYLATED FATTY ALCOHOLS

The contact of a lubricating substance with a solid is particularly significant from a tribological point of view. Oxyethylated alcohols are nonionic surfactants, and their interactions with the surface are basically quite specific (hydrogen bonds). The contribution of universal (electrostatic) interactions is considerably smaller, as these are very weak dispersion interactions. In the solution in contact with a solid, one can distinguish the surface phase and the bulk phase. Due to adsorption from solutions, the surface phase is enriched with the component that has a stronger affinity for the surface. It is a characteristic of adsorption from solutions on a solid surface that individual components compete for "free sites" on the surface. At this point, one should not confuse adsorption with absorption, the latter of which may lead to penetration of the components into the solid.

The adsorption of surfactants at the solid–solution interface results in changes in such surface properties as charge and wettability, as well as in hydrophilic and hydrophobic properties. Positively or negatively charged centers may appear on the surface, or the surface may be charge free. The adsorption of the surfactants that form micellar solutions is of particular interest. It turns out that micelles can form both in the bulk phase as well as in the surface phase. The former are termed micelles, whereas the latter are called hemimicelles. At low concentrations, monomers adsorb on the surface, and they may react through the hydrophilic or the hydrophobic part, depending on the kind of surface. Hemimicelles are formed at a clearly defined concentration called the CSAC, the critical surface aggregation concentration.

Studies have shown that surface micelles are formed at lower concentrations than the ones in the bulk phase. The formation of micelles in the bulk phase is determined by the critical micelle concentration (CMC) [35–37]. As the value of CSAC < CMC, it can be assumed that the formation of aggregates on the surface does not result from the adsorption of micelles from the bulk phase [38]. It is accepted, however, that the process of micellization on the surface follows a mechanism similar to micelle formation in the bulk phase after taking into account interactions with the surface. In both cases, the driving force is the entropy connected with the hydrophobic effect [39–42]. A similar mechanism of the formation of bulk and surface micelles has been confirmed by calorimetric investigations. At concentrations below the CSAC, one can observe exothermic effects whose values correspond to universal or hydrogen bond interactions. After exceeding the CSAC, endothermic effects appear, and these are comparable with the heats of micelle formation in the bulk phase. Due to the complexity of the processes, hemimicelles have different spatial forms and different numbers of monomers, and their concentration in the surface phase may also be different. However, geometrical similarity to the micelles in the bulk phase cannot be excluded. The process of formation of hemimicelles terminates when the CMC value in the bulk is reached. Above this concentration, the chemical potential in the solution does not change, so there is no thermodynamic driving force that would stimulate further adsorption. At this stage, the process of micelle formation in the bulk phase begins.

Adsorption properties and the ability to form surface structures have been the subject of a number of studies employing various measurement techniques. The most important ones include: adsorption isotherm analysis [43, 44], x-ray and neutron scattering [45], nuclear magnetic resonance (NMR) spectroscopy [46], ellipsometric measurements [14, 15, 47–49], atomic force microscopy (AFM) [16, 17, 50–55], calorimetric methods [14, 39, 56–58], and measurements of electrokinetic potential and wetting angle [37].

In dilute solutions, oxyethylated alcohols are present as monomers whose characteristic structural components are a hydrophilic part (ethylene oxide chain) and a hydrophobic part (aliphatic chain). With an increase in concentration, the monomers undergo spontaneous self-aggregation, forming micelles in the surface phase (hemimicelles) at lower concentrations, while at higher concentrations forming micelles in the bulk phase. Figure 17.4 shows a schematic example of a spherical micelle.

The term "micelle" is very general and may be associated with any soluble, spontaneous, and reversible aggregate made up of amphiphilic molecules or their ions. The aggregates may have a closed structure of a size that is on the order of twice the length of the molecules. Micelles are thermodynamically stable entities characterized by a relatively simple structure.

Studies on micellar solutions began in the first half of the 20th century, and a summary of the results obtained was presented in Hartley's book [59] published in 1936, which describes many currently used terms. Further research focused primarily on micelle structures and properties, monomer–micelle equilibrium, and solubilization by micelles [60]. The research turned out to be extremely interesting, and the results were used in a number of applications.

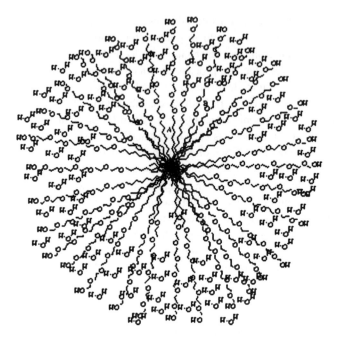

FIGURE 17.4 Model of a spherical micelle solvated with water molecules.

The formation of micelles is by itself interesting scientifically. In the solution, micelles constitute a peculiar donor of monomers. If the number of monomers is reduced, their loss may be compensated by decomposition of the micelles. The main cause of monomer self-aggregation is the so-called hydrophobic effect, whose driving force is the unfavorable interactions between the hydrophobic chains and water. As a result, the amphiphilic molecules "migrate" to the interface or form micelles. The micelles do not constitute stationary systems, but undergo constant changes. The changes are connected with the size, shape, and number of aggregates. A state of dynamic equilibrium between micelles and monomers occurs, and its position can be shifted as a result of changes in external factors, e.g., temperature or pressure.

The formation of a significant number of micelles is determined by the concentration range, which is called the critical micelle concentration (CMC). Above the CMC, there appears an equilibrium between individual monomers and micelles, and this system is thermodynamically stable. The CMC is the most important parameter defining micellar solutions, and its value depends on a number of factors, the most important being monomer structure, solvent properties, temperature, and concentration. The value of the CMC for micelles depends mainly on the alkyl chain length and, to a lesser extent, on the structure of the hydrophilic part.

The formation of micelles in solutions may be accompanied by changes in a number of macroscopic properties of solutions as a function of concentration, e.g., surface and interfacial tension, density, and equivalent conductivity. The character of these functional dependences changes above the CMC. After the CMC value is exceeded, new micelles form, whereas the concentration of monomers and their thermodynamic potential do not change. Thus, no changes should take place at the interface.

Oxyethylated alcohols are amphiphilic compounds, and they display a tendency for self-aggregation, i.e., the formation of micelles. The geometry of micelles and the number of aggregates depend on the structure of the ethoxylates [16, 23, 61]. Spherical micelles are formed by alcohols with short alkyl chains and at lower temperatures. Longer-chain surfactants at higher temperatures form cylindrical micelles. In the case of oxyethylates, an increase in the length of the alkyl chain increases the number of aggregates, while their number is reduced with an increase in the length of the ethylene oxide chain [62, 63].

For a given ethoxymer, the CMC value increases with an increase in the length of the ethylene oxide chain [62, 63]. This is due to an increase in the solubility of the compounds, which is connected with the formation of hydrogen bonds of water molecules with the oxygen from the ether group. Hydrogen bonding is an exchange interaction with a low energy on the order of 20 to 30 kJ/mol on average. This energy is comparable to thermal energy and, therefore, a temperature increase results in breaking of the bonds, and dehydration occurs. From a thermodynamic point of view, solvation of micelles rather than individual monomers is more advantageous. Figure 17.4 shows a diagram of solvation of micelles by water molecules.

Ethylene oxide mers close to the alkyl chain are tightly packed, and that makes it difficult for water to access them. It is assumed that the first two ether groups are not solvated. Therefore, if the number of attached ethylene oxide molecules is sufficiently large, the micelles are stable and do not decompose with a temperature rise. For the homologous series the CMC value decreases [62, 63], which can be interpreted as a result of the increased hydrophobic property of the compounds and increased tendency for micelle formation.

The effect of the alkyl chain length on the CMC value is considerably stronger than the effect of ethylene oxide [62, 63]. This comparison gives an idea to what extent the presence of homologues in the postreaction mixture may affect the surface activity and concentration at which surface and bulk micelles form. The range of experimentally determined CMC values depends not only on the methodology used, but also on the share of chemical entities in the mixture of compounds used in the investigation. A possible comparison should aim at pointing out change gradations and not at determination of absolute values of CMC for individual compounds. Activity and other physicochemical properties are characteristics of commercial products.

As CMC > CSAC, above the CMC in the solution there exists an equilibrium between surface micelles, bulk micelles, and monomers. The formation of bulk micelles by molecules is a sign of surface activity of the compounds, as they form after reaching saturation conditions in the surface phase. The measured values of surface tension and wettability have been accepted as a measure of surface activity of compounds. Surface tension of the aqueous solutions was measured by the "ring-detachment" method using a TD1 Lauda tensiometer. The quantity measured is the force needed to detach a platinum ring from the surface of the solution. Wettability was determined by measuring the angle of wetting on a steel surface with solutions of amphiphilic compounds. The kind of steel (LH15) and its surface characteristics were the same as those used in tribological experiments. A G10/DSA 10 device produced by Krüss with the software for mathematical analysis of drop shape was used in the investigation to determine the wetting angle.

17.3.1 Surface Tension

The measurements of both surface tension and wetting angle were carried out at a constant temperature. The arithmetic mean of at least three measurements was taken for each experiment. The measure of error was taken as the standard deviation of the arithmetic mean, taking into account the Student's t distribution for confidence level 0.90.

The measurements of surface tension and wettability angle were carried out for aqueous solutions of oxyethylated lauryl, cetyl, oleyl, and stearyl alcohols. The alcohols differed in the lengths of their alkyl and ethylene oxide chains. It should be noted that oleyl and stearyl alcohols have the same number of carbon atoms; however, oleyl alcohol has a double bond. Alcohols with a given chain length differed in the degree of oxyethylation. The measurements were conducted for aqueous solutions of ethoxylates at the concentrations of 0.0001, 0.001, 0.01, 0.1, and 1 wt%. The surface tension of the water was relatively high (72 mN/m), and the addition of a surfactant even at a low concentration resulted in a reduction of its value. For chemical compounds used as solution components, changes in surface tension as a function of their concentration may be used to determine the CMC values. Above the CMC, changes in σ (surface tension) should be small, as the formation of the adsorption layer is terminated, and micelles form in the bulk phase. The dependencies $\sigma(c)$ are more complex for solutions of commercial products, due to the fact that the solutions contain a large number of compounds with various lengths of ethylene oxide and alkyl chains. Figure 17.5 shows examples of measured σ values for solutions of oxyethylated lauryl alcohols as a function of their concentration. The largest changes were observed for the concentration of about 0.01 wt%, above which no changes in σ values can be noticed.

Analysis of the results obtained shows that above the concentration of 0.01 wt%, there appears an equilibrium state between the surface phase and the bulk phase, whereas the effect of the concentration of a compound on surface tension is small.

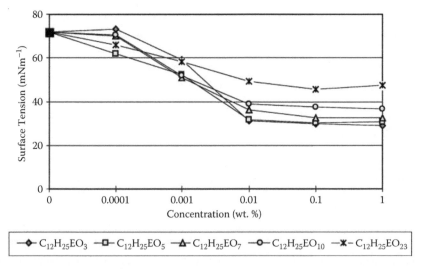

FIGURE 17.5 Surface tension changes for solutions of lauryl alcohols with various degrees of oxyethylation as a function of their concentration.

This principle applies to all the alcohols tested. They display high surface activity, and surface tension of their solutions decreases as much as 2.5-fold relative to water (72 mN/m). Due to the dispersion of homologues and ethoxymers, it is hard to determine the CMC values on the basis of surface tension. One may say, however, that after exceeding the concentration of 0.01 wt%, micelles appear in the solution, and their concentration is significant. The differences in surface tension originate from changes in the structure of the molecule, namely, the length of the ethylene oxide and alkyl chains and the presence of an unsaturated bond (oleyl alcohol).

The comparison of surface activity of individual alcohols was performed for their 1 wt% solutions for two reasons. First, at this concentration micelles formed both in the surface phase and in the bulk phase, and they appeared in relatively high concentrations. No changes in surface tension should be observed. Second, this is a predicted concentration of alcohols as additives in model lubricating substances. Changes in the surface tension of 1 wt% alcohol solutions as a function of the degree of oxyethylation and the nature of the alkyl chain are presented in fig. 17.6.

The analysis of the results (fig. 17.6a) indicates that an increase in the degree of oxy-ethylation results in an increase in surface tension, which is a direct result of increased hydration as well as increased solubility of the compounds. At the same time, the value of the CMC increases [62, 63]. The increase in the alkyl chain length causes an increase in the hydrophobic properties of the molecule and, consequently, an increase in its surface activity and a reduction in surface tension. These changes are accompanied by a decrease in the CMC value [62, 63]. The effect of the length of the alkyl chain of alcohols on the value of surface tension of their solutions can be analyzed for oxyethylation degrees of 10 and 20 (fig. 17.6b). The difference in the σ value between the longest-chain and the shortest-chain alcohols is about 4 mN/m. For $m = 10$, the observed increase in σ (fig. 17.6b) is about 4 mN/m, and for $m = 20$ there is a drop of about 4 mN/m. Relatively slightly lower values are displayed by 1 wt% solutions of oleyl alcohol, which, unlike the others, has a double bond. It follows unequivocally from figs. 17.6a and 17.6b that the oxyethylation degree is primarily responsible for the ability of the compounds to reduce surface tension. The adsorbed film formed on solid surfaces by nonionic surfactants is very sensitive to even relatively small changes in concentration, temperature, and the structure of the adsorbate molecules. A significant role is also played by interactions of the adsorbate with the solvent and the adsorbent. An increase in interactions of alcohols with the solvent, due to the increased oxyethyla-tion degree, may lead not only to increased solubility, which causes the CMC increase [62, 63], but also to a change in orientation and arrangement of surfactant molecules on solid surfaces [64]. The more hydrophobic the adsorbate (increased alkyl chain length), the more the hydrophilic group is dislodged from the surface. This favors the formation of hemimicelles whose ethylene oxide chains are oriented towards the solution. Due to such a configuration, the amount of adsorbate adsorbed per unit area decreases with an increase in the length of the ethylene oxide chain in an alcohol molecule [43, 65].

17.3.2 WETTABILITY OF STEEL

Surface-active agents reduce surface tension. This quantity (σ) should also have an influence on the wettability of solids. However, this is not a simple correlation and,

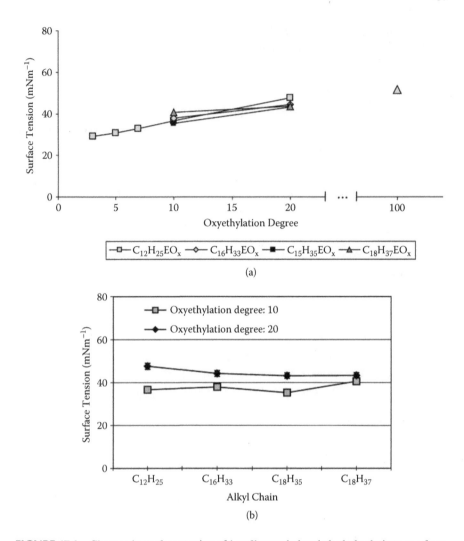

FIGURE 17.6 Changes in surface tension of 1 wt% oxyethylated alcohol solutions as a function of (a) oxyethylation degree and (b) length and nature of the alkyl chain.

therefore, the measurement of wettability of LH15 steel, which is also a friction-pair material, was carried out. The sessile drop method was used in the measurements of wetting angle (θ). For water, the angle was 89°. The wettability of aqueous solutions of oxyethylated alcohols is higher than that of water. Using solutions of oxyethylated lauryl alcohols as an example, fig. 17.7 shows the effect of concentration of these compounds on the values of wetting angle for steel.

The wetting angle on steel surfaces by solutions of oxyethylated lauryl, oleyl, and stearyl alcohols displays a significant reduction at concentrations as low as 0.0001 wt%, and in the case of solutions of oxyethylated cetyl alcohols at about 0.001 wt%. Wetting-angle changes were determined within the concentration range of 0.0001 to 1 wt%. The concentration value of 1 wt% was chosen for practical reasons, as this

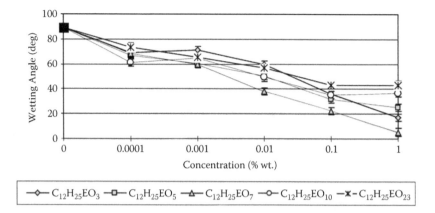

FIGURE 17.7 Changes in wetting angles on steel as a function of concentration for different oxyethylated lauryl alcohols in aqueous solutions.

is a common concentration of additives in a lubricant composition. Addition of an oxyethylated alcohol to water leads to a reduction in the wetting angle on steel at the lowest concentrations, and at 1 wt% it decreases from about 1.5-fold to about 18-fold relative to water (89°). The effect of oxyethylation degree and the nature of the alkyl chain on the wettability of steel by 1 wt% solutions of oxyethylated alcohols is shown in fig. 17.8.

A drop in wettability can be observed with an increase of the ethylene oxide chain (fig. 17.8b). The $C_{12}EO_7$ ethoxylate whose solution displays an unexpectedly low wetting angle is an exception. An increase in the alkyl chain length causes a reduction in wettability both for $m = 10$ and for $m = 20$. The differences between individual ethoxymers are not really large, except for cetyl alcohols (fig. 17.8a), whose 1 wt% solutions show an unexpectedly high reduction in wettability when the oxyethylation degree is increased from 10 to 20. The uncharacteristic behavior of oxyethylated lauryl alcohols ($C_{15}H_{25}EO_7$) and cetyl alcohol ($C_{16}H_{33}EO_{20}$) should not be treated as an experimental error, but as a result of the optimal sharing of the hydrophobic and hydrophilic parts (constant HLB or hydrophobic-lipophilic balance), or as a result of the dispersion of homologues and ethoxymers. The general tendency for changes in steel wettability is consistent with surface tension changes and may be interpreted in the same way, i.e., in terms of adsorption of ethoxylates on the solid surface.

With a few exceptions, an increase in the degree of oxyethylation causes an increase in both surface tension and wetting angle (figs. 17.6 and 17.8). This may be explained in terms of the increased hydrophilicity of the compound and, hence, its solubility. A detailed analysis of the results indicates that the problem is more complicated. To solve it, a comparison of the measured σ and θ values was made for a group of homologues with increasing alkyl chain lengths and at two oxyethylation degrees: 10 and 20 (figs. 17.6 and 17.8). An increase in the proportion of the hydrophobic part should result in a reduction of both σ and θ values. The behavior of compounds investigated is different. The values increase with increasing alkyl chain length. These changes may be interpreted on the basis of the mechanism of adsorption of ethoxylates from their aqueous solutions. The increased hydrophobicity of a

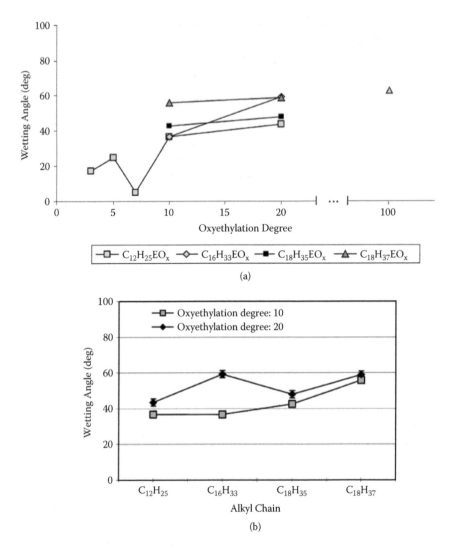

FIGURE 17.8 Dependence of wetting angle on steel by 1 wt% solutions of oxyethylated alcohols as a function of (a) oxyethylation degree and (b) length and nature of the alkyl chain.

compound may lead to a displacement of the hydrophilic group away from the surface and toward the solution. The surface micelles produced will be oriented with their ethylene oxide chains toward the solution, and they will be stabilized by hydration with water molecules. As a result, with increasing ethylene oxide chain length, the affinity of oxyethylated alcohols for the surface decreases, and the surface phase loses this component. The results obtained by other authors who studied the influence of the ethylene oxide chain length on adsorption lend credence to this hypothesis. It turns out that the amount of an adsorbed ethoxylate is reduced with an increase in oxyethylation degree [43, 65]. This reasoning is slightly weakened by the exceptions to the presented general rules observed for certain solutions. The simplest explanation seems to be that the commercial products used are mixtures of compounds of various alkyl and ethylene oxide chain lengths. The precise proportions of individual entities in the compound mixture are not known. The results obtained thus characterize the commercial raw material tested. Exceptions from the general rules may also be explained by a specific "matching" of the hydrophilic and hydrophobic parts of the ethoxylate (constant HLB). In order to confirm the hypotheses, the investigations will be extended to cover other oxyethylates.

17.4 TRIBOLOGICAL PROPERTIES OF AQUEOUS SOLUTIONS OF OXYETHYLATED FATTY ALCOHOLS

For a specific mating pair and friction conditions, tribological properties depend on the lubricant type. Lubricant compositions are a complex mixture of a base and additives that perform various functions. An important role of the adsorption of components in lubrication, particularly the boundary and mixed ones, has been emphasized in a number of publications. Adsorption results in the formation of a thin lubricant film that prevents adhesive welding, which, in turn, leads to an excessive increase in motion resistance and wear and may cause seizure [47, 66–78].

Aqueous solutions of amphiphilic compounds that display high affinity for the surface were used as model lubricating substances. Both in the surface phase and in the bulk phase, they form micelles that form lyotropic liquid crystals (LLC). The formation of surface structures under static conditions results in a lubricant film under friction conditions [79]. Under dynamic conditions, the structures of surface aggregates and their arrangement may change; however, direct experimental observations of the changes are difficult or practically impossible. They are determined indirectly and for friction-pair materials not used in practice. The thickness of a lubricant film may change from 2 to 20 nm, and water-filled defects may form, just as in the bulk phase [79–84]. Liquid crystal structures have high viscosity, as much as 10^6 times higher than that of water, and their rheology characteristics may be similar to those of lubricating greases. Hence, there is a danger of a too slow reappearance of lubricant film.

The study of aqueous solutions of surface-active agents is interesting from both research and application points of view [1–13]. This chapter attempts to correlate the tribological properties with molecular structure, surface activity, and micelle and LLC formation ability both in the surface phase and in the bulk phase. The physicochemical properties are determined under static conditions, and relating them to dynamic conditions is difficult. Pressures, relative movement, and dissipation of energy in the form of heat all occur under friction conditions.

The result of this study should form the basis for applications of water-based lubricating substances with surfactants as additives. Aqueous solutions are used because they are nonflammable and safe when applied. The former property makes them useful in machines operating in fire danger zones, e.g., in mining and metallurgy. The use of traditional mineral-oil-based lubricants may result in soil and water contamination. Decontamination and utilization of lubricating substances are technologically complicated and costly. The safety criterion is also crucial for machine operators, who are exposed to the toxic action of lubricants via their digestive and respiratory systems and through their skin. In manufacturing operations, the final product should also be protected against penetration by toxic lubricants during the manufacturing process. This is particularly important for the production of medicines, cosmetics, and foodstuffs. For the needs of these branches of industry, it is possible to develop aqueous solutions of selected amphiphilic compounds that are physiologically neutral and thus can become a component of the product.

However, water as a lubricating substance has a number of drawbacks. Its lubricating properties are insufficient, and it may cause corrosion and cavitation. Corrosion can be limited by adding corrosion inhibitors to solutions or by using such lubricants in mating pairs made of polymers, ceramics, or stainless steel. Cavitation can be prevented by suitable construction designs in which the flow will be adjusted to water properties. The water environment is also susceptible to bacterial growth. Such growth can be considerably limited by adding bactericidal compounds to the lubricant compositions. Such an activity is manifested by cationic surfactants, which also show the ability to improve the lubricity of water. In the case of water-based lubricants, the improvement in the tribological properties of water is of particular importance. This can be accomplished by using specially matched additives or additive packages. It appears that a number of surfactants improve the lubricity of water [1–13]. However, a search for an optimal water-based lubricant composition is connected not only with an improvement in tribological properties. The surfactants tested are also those that can perform other functions, such as modifications of rheological properties, inhibition of corrosion, and biocidal action. Therefore, the most likely approach would involve the presence of several surfactants acting synergistically in the composition.

Aqueous solutions have so far been used mainly as hydraulic and working fluids. They are found as emulsions and microemulsions. They are unstable, but this can be improved by the addition of emulsifiers; however, these have a harmful environmental effect. There are also problems with the preparation and utilization of emulsions. It is suggested that, in these applications, emulsions should be replaced with solutions of amphiphilic compounds that do not have such disadvantages.

A group of compounds whose aqueous solutions will act as model lubricating substances are presented on the basis of literature data. A number of the author's own investigation results are also presented. The aim of the investigation was to study the surface activity of compounds (section 17.3). The complexity of the behavior of compounds at the interface increases under friction conditions under which high pressures and temperatures occur, as well as the relative motion of the friction-pair elements.

For the purposes of tribological investigations, the T02 tester produced at the Institute of Sustainable Technologies in Radom (Poland) was used. The experimental methodology was also developed at this Institute [85]. The friction pair in the four-ball machine (tester T02) is made up of four balls, one of which is fixed in a spindle rotating at a preset angular velocity (upper part of the friction pair) and the three remaining balls are stationary. The stationary balls press against the rotating ball. The balls are 0.5 in. in diameter and are made of bearing steel (LH15). The surface roughness of the balls was determined to be $R_a = 0.032$ μm. Before the experiment, the balls were washed in various solvents and dried. Special attention was paid to the proper washing of the handle and the spindle, as even trace amounts of compounds can dramatically affect the measurement results. Two independent experimental procedures were employed to determine the antiseizure property and ability to modify friction and wear coefficients at a constant load. The individual tests are called seizure tests and constant-load tests.

Antiseizure properties were determined in a load range of 0 to 7.2 kN. The rate of load increase was 409 N/s, whereas the rotational speed was 500 rpm. The measured quantity was the friction force moment (M_T), which was plotted as a function of increasing load (fig. 17.9). Scuffing load (P_t) was determined by analyzing the curve. At this load, a sudden increase in the friction force moment occurs. The seizure load (P_{oz}) is the highest load reached during the test. It does not represent seizure in the physical sense or welding of the balls; rather, it represents the point at which the friction force moment exceeds 10 Nm. After each test, the wear-scar diameter was measured in two directions perpendicular to each other. The measured quantity was an arithmetic mean. The measurements were carried out at least three times, and the quantity d in eq. (17.1) is an arithmetic mean of the measurements. It was used to determine the limiting pressure of seizure (p_{oz}) based on the equation

$$p_{oz} = 0.52 \cdot \frac{P_{oz}}{d^2} \tag{17.1}$$

in which the constant coefficient 0.52 results from the distribution of forces among the four balls in contact. The experiment was repeated at least three times at variable loads. The measure of error of the quantities was a standard deviation of the arithmetic mean, taking into account the Student's t distribution at the confidence level of 0.90. The error in the complex quantity (p_{oz}) was determined by the total differential method, whereas the errors in individual quantities p_{oz} and d are represented by standard deviation. A set of three quantities—scuffing load (P_t), seizure load (P_{oz}), and the limiting pressure of seizure (p_{oz})—characterizes the antiseizure properties of lubricating substances.

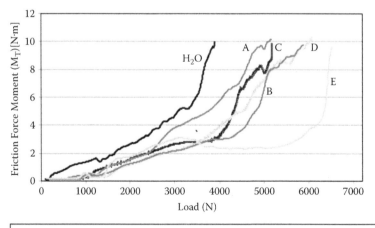

$A = C_{12}H_{25}EO_3 \quad B = C_{12}H_{25}EO_5 \quad C = C_{12}H_{25}EO_7 \quad D = C_{12}H_{25}EO_{10} \quad E = C_{12}H_{25}EO_{23}$

FIGURE 17.9 Changes in friction force moment (M_T) as a function of load (N) for 0.5 wt% solutions of lauryl alcohols.

The test at a constant load makes it possible to measure the changes in the friction force moment (M_T) as a function of friction time. Hence, the coefficient of friction (μ) was calculated from the equation

$$\mu = 222.47 \cdot \frac{M_T}{P} \tag{17.2}$$

The constant coefficient 222.47 results from the distribution of load P in the friction pair. The friction test was followed by measurements of wear-scar diameters of the three lower balls in two directions perpendicular to each other, and an arithmetic mean was calculated. The measurement error was determined on the basis of the Student's t distribution. (The results of the tests carried out at a constant load are discussed later, in section 17.4.2, and are shown in fig. 17.16 as coefficients of friction vs. time and in figs. 17.17 and 17.19, respectively, as coefficients of friction and wear-scar diameter vs. the chemistry of compound and its concentration in solution.)

The same group of compounds was used in tribological investigations using an analogous notation. The group included the following oxyethylated alcohols: lauryl ($C_{12}H_{25}EO_3$, $C_{12}H_{25}EO_5$, $C_{12}H_{25}EO_7$, $C_{12}H_{25}EO_{10}$, $C_{12}H_{25}EO_{23}$), cetyl ($C_{16}H_{33}EO_{10}$, $C_{16}H_{33}EO_{20}$), oleyl ($C_{18}H_{35}EO_{10}$, $C_{18}H_{35}EO_{20}$), and stearyl ($C_{18}H_{37}EO_{10}$, $C_{18}H_{37}EO_{20}$). Aqueous solutions of amphiphilic compounds prepared by weight were used as lubricating substances.

17.4.1 ANTISEIZURE PROPERTIES

Antiseizure properties were determined according to the methodology presented in the previous section. Changes in the friction force moment (M_T) as a function of increasing load (P) form the basis for the determination of individual quantities. As an example, this dependence for 0.5 wt% solutions of lauryl alcohols is shown in fig. 17.9.

The individual curves represent 0.5 wt% solutions of alcohols of various oxyethylation degrees. Studies were also carried out for water, which represents a reference system. The course of changes observed is relatively complicated, but it is possible to notice three intervals that differ in the rate of increase in the friction force moment. A slight increase can be observed at low loads, a moderate one at intermediate loads, and a rapid one ending with seizure at the friction force moment 10 N·m. Three quantities will be used to assess antiseizure properties: scuffing load (P_t), maximum seizure load (P_{oz}), and the limiting pressure of seizure (p_{oz}). Seizure tests were carried out in the presence of oxyethylated lauryl alcohol solutions at concentrations of 0.1, 0.5, 1, 4, and 10 wt%, and for cetyl, oleyl, and stearyl alcohols at concentrations of 0.1, 1, and 10 wt%.

The scuffing load is the lowest load at which there occurs a pronounced increase in the friction force moment, which indicates breaking of a lubricant film. The value of scuffing load for water (200 N) is several times lower than that for oxyethylated alcohol solutions, which exceeds 1000 N. The highest value was observed for a 1 wt% solution of $C_{12}H_{25}EO_{23}$ (ca. 1500 N). High antiseizure efficiency of compounds can be observed even at the lowest concentrations (0.1 wt%). The dependence of

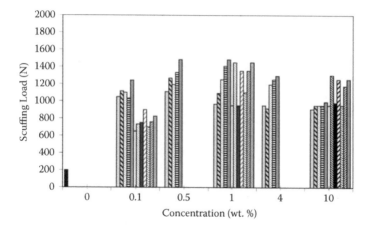

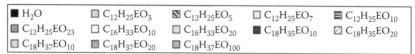

FIGURE 17.10 Changes in scuffing load (P_t) as a function of the nature and concentration of oxyethylated alcohols as a component of model water-based lubricating substances. (Data obtained using tester T02.)

scuffing load on the type of component of an aqueous solution and its concentration is presented in fig. 17.10.

The effect of concentration on the P_t value can be best studied for solutions of oxyethylated lauryl alcohols due to the number of test results presented. For these solutions, the scuffing load reaches maximum values at concentrations of 0.5 wt% ($C_{12}H_{25}EO_3$, $C_{12}H_{25}EO_5$) and 1 wt% ($C_{12}H_{25}EO_7$, $C_{12}H_{25}EO_{10}$, $C_{12}H_{25}EO_{23}$). The shift of the maximum values of P_t toward higher concentrations may be associated with an increase in the solubility of oxyethylated alcohols and on the CMC values with an increase in the degree of oxyethylation [62, 63].

Based on the results presented in figs. 17.10 and 17.11a, it can be concluded that P_t values increase practically for all concentrations and all oxyethylated alcohols with an increase in the degree of oxyethylation. Based on this, it can be stated that oxyethylated alcohols with a higher number of ethylene oxide units form a more stable lubricant film. One should look for a correspondence with the surface activity of solutions of these compounds. It turns out that the tendency for changes in activity whose measures are surface tension and wettability is opposite (figs. 17.6 and 17.8), and ethoxylates with higher m values form, under static conditions, an adsorbed film with a small concentration of additives [43, 65]. This apparent discrepancy may be explained as being due to the breaking of hydrogen bonds and dehydration of oxyethylated alcohols with an increase in temperature [62, 63]. The compounds lose their amphiphilic properties and are precipitated from solutions above the cloud point. Both the degree of oxyethylation and the alkyl chain length [62, 63] affect the cloud-point value. Under friction conditions, particularly in seizure tests at high loads, a large part of mechanical energy changes into thermal energy. This results

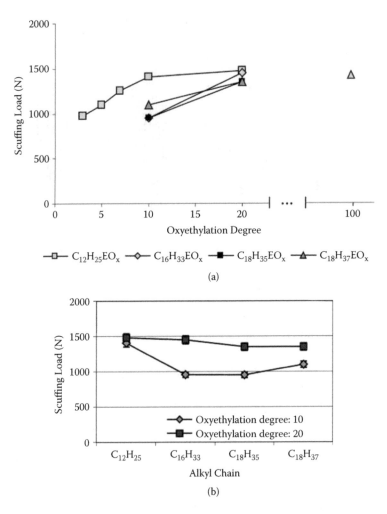

FIGURE 17.11 Changes in scuffing load (P_t) of 1 wt% oxyethylated alcohol solutions as a function of (a) oxyethylation degree and (b) length and nature of the alkyl chain. (Data obtained using tester T02.)

in a temperature increase, which creates conditions conducive to dehydration. Alcohols deprived of amphiphilic properties do not form structures in surfaces and bulk phases and lose their surface activity. As oxyethylated alcohols with a higher oxyethylation degree have a higher cloud point [62, 63], they will be capable of forming a stable lubricant film at higher temperatures.

The other factor that causes reduction in the cloud point is the increase of the alkyl chain length in oxyethylated alcohols. Following the assumed interpretation, one can expect a correlation between the seizure load and the increase in the hydrophobic properties of oxyethylated alcohols. The dependences can be conveniently analyzed on the basis of the results shown in fig. 17.11b, which clearly point to a significant effect of the increase in hydrophilicity of the compound. The differences between the P_t values for solutions of compounds with oxyethylation degrees of 20 and 10 are as much as 500 N. Increase in chain length results in a drop in the cloud point [62, 63] and, following the reasoning above, it should lead to a reduction in the P_t value. Scuffing load does indeed decrease both for $m = 10$ and for $m = 20$, except for the $C_{18}H_{37}EO_{10}$ solution. The changes for individual ethoxymers are not too large and are well correlated with the weaker effect of the alkyl chain length on the cloud point [62, 63]. Thus, an increase in the degree of alcohol oxyethylation mainly determines the stability of the lubricant film.

Seizure load (P_{oz}) is the highest value of the pressure at which the friction force moment exceeds 10 N·m. In the four-ball machine used in the experiments, the loads range from zero to 7200 N. Several of the solutions tested reached the maximum load value without undergoing seizure. The 7200-N value is nearly two times higher than the seizure load for water (3700 N). The dependence of seizure load on concentration and type of compound is presented in fig. 17.12.

Oxyethylated lauryl alcohol solutions are the most common group of lubricating substances tested. The changes in seizure load observed as a function of oxyethylated alcohol concentration are not monotonic, and the maximum reached for individual ethoxylates shifts toward higher concentrations, reaching the maximum values for 0.5 wt% ($C_{12}H_{25}EO_3$), 1 wt% ($C_{12}H_{25}EO_5$), and 4 wt% ($C_{12}H_{25}EO_7, C_{12}H_{25}EO_{10}, C_{12}H_{25}EO_{23}$) solutions. The absolute maximum value obtained increases with increasing oxyethylation degree not only for oxyethylated lauryl alcohols, but also for the other longer alkyl chain oxyethylated alcohols. The influence of ethylene oxide and alkyl chain lengths can be analyzed based on the dependence shown in fig. 17.13.

For most of the solutions examined, an increase in the degree of oxyethylation results in an increase in seizure load. The P_{oz} value for $C_{12}H_{25}EO_7$ solutions, which

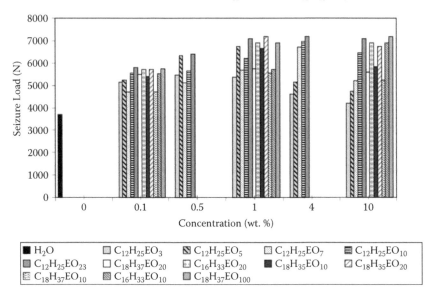

FIGURE 17.12 Changes in seizure load (P_{oz}) as a function of concentration and nature of oxyethylated alcohol as a component of aqueous solutions. (Data obtained using tester T02.)

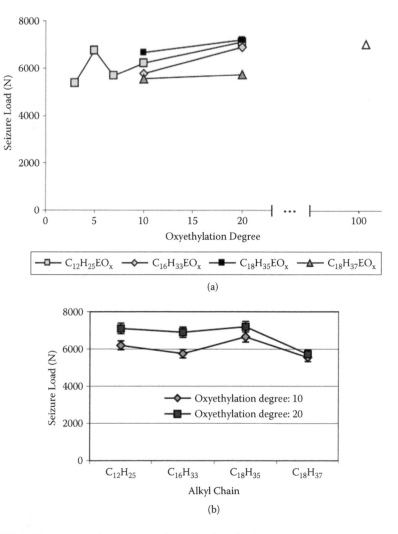

FIGURE 17.13 Changes in seizure load (P_{oz}) for 1 wt% oxyethylated alcohol solutions as a function of (a) oxyethylation degree and (b) alkyl chain length and type of compound. (Data obtained using tester T02.)

is inconsistent, is much higher than for the solutions of the other oxyethylated alcohols. The change in the degree of oxyethylation of alcohols from 10 to 20 leads to an increase in P_{oz}, and the differences between the ethoxylates reach a value of 1150 N. Seizure load decreases with an increase in the length of the alkyl chain both for $m = 10$ and for $m = 20$. Maximum differences for individual ethoxymers reach 650 N ($m = 10$) and 1400 N ($m = 20$) with the exception of oxyethylated oleyl alcohols, whose P_t value is higher than one might expect. At all concentrations, the solutions of oxyethylated oleyl alcohols have higher seizure load values than those for stearyl alcohols (at the same oxyethylation degree), which shows that the double bond has a beneficial effect. They are comparable even with an alcohol with a degree of oxyethylation of 100 (figs. 17.12 and 17.13a). The character of seizure load changes with an increase in the degree of oxyethylation, and the effect of the increase in the number of carbon atoms in the alkyl chain was analogous to the one for scuffing load. Hence, the interpretation of the results obtained can be analogous. Additives with a higher degree of oxyethylation, which undergo dehydration at higher temperatures, are the most effective ones in preventing seizure. The effect of the increase in the hydrophilic portion on the P_{oz} value may be stronger than on the P_t, as the seizure load is determined at higher pressures, at which more heat is generated, and the system reaches higher temperatures.

Oxyethylated fatty alcohols used as additives are found to be effective in increasing seizure load. Several of their solutions reach the seizure load of 7200 N, which is the maximum measurement range for the tester used. The degree of oxyethylation has a decisive influence on reaching these extreme loads at a given number of carbon atoms in the chain, whereas an increase in the length of the chain in the molecule, at a given oxyethylation degree, results in a reduced value of P_{oz}, but this is not as effective as the increase in the hydrophilic part of an alcohol.

The limiting pressure of seizure (p_{oz}) represents the pressure present within a friction pair at the maximum seizure load (P_{oz}). It is a function of two variables: seizure load (P_{oz}) and wear-scar diameter (d) measured in a "state of seizure" on the stationary balls. It is less accurate than the other two quantities, and its interpretation can only be done with less confidence. Oxyethylated alcohol solutions display relatively high values of limiting pressure of seizure that, relative to water (200 N), increase from twofold to as much as fivefold. The dependence of the limiting pressure of seizure on the concentration and kind of ethoxylate is shown in fig. 17.14.

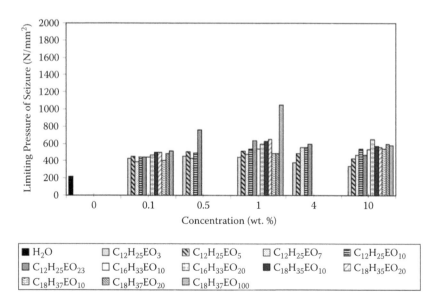

FIGURE 17.14 Changes in the limiting pressure of seizure (p_{oz}) as a function of concentration and the nature of oxyethylated alcohol as a component of aqueous solutions. (Data obtained using tester T02.)

An increase in the p_{oz} value as a function of concentration can be observed for solutions of the compounds. An exception is the stearyl alcohol oxyethylated with 100 moles of ethylene oxide, which reaches a value of about 1000 N·mm^{-2}. On the basis of the results obtained, one can discuss the effect of ethylene oxide and alkyl chain length on the limiting pressure of seizure (fig. 17.15).

An increase in the degree of oxyethylation results in an increase in p_{oz}. The p_{oz} values decrease as a function of chain length, and the maximum difference equals 50 N·mm^{-2} for $m = 10$ and 150 N·mm^{-2} for $m = 20$. An exception is the oxyethylated oleyl alcohol, which has a single unsaturated bond in the alkyl chain. A comparison

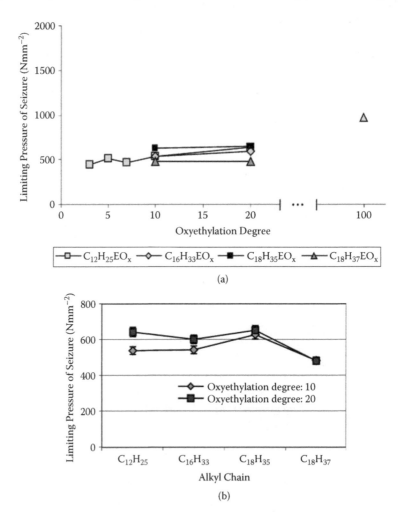

(a)

(b)

FIGURE 17.15 Changes in the limiting pressure of seizure (p_{oz}) for 1 wt% solutions as a function of (a) oxyethylation degree and (b) length and nature of the alkyl chain. (Data obtained using tester T02.)

with stearyl alcohol, which differs from oleyl alcohol only in the presence of a double bond, is particularly interesting. The character of changes in the p_{oz} value is similar to the changes in P_{oz} presented in figs. 17.12 and 17.13 and may be interpreted similarly.

Antiseizure properties of oxyethylated alcohol solutions were described in terms of the values of P_t (figs. 17.10 and 17.11), P_{oz} (figs. 17.12 and 17.13), and p_{oz} (figs. 17.14 and 17.15). Surface-active agents were selected in a way that would make it possible to examine the effects of their structures on the tribological characteristics. It turns out that an increase in the hydrophilicity of the ethylene oxide chain (increase in m) causes an increase in antiseizure properties, whereas an increase in the hydrophobicity of the alkyl chain (increase in n) causes a reduction in these properties. The effect of the latter factor is weaker. The relations indicate that hydration of oxyethylates is the dominant process [62, 63]. The necessary condition for favorable antiseizure properties is the ability of the compound to undergo hydration and, thus, to maintain amphiphilic properties at higher temperatures.

Summing up, all oxyethylated alcohols meet the criterion as additives that can significantly improve the antiseizure properties of water. Based on the results obtained, these properties can be associated with the structure of the compounds and depend on the changes in hydration together with temperature.

17.4.2 TESTS AT A CONSTANT LOAD

The tests were carried out at a constant load of 2 kN, which is relatively small and corresponds to a moderate increase in the friction force moment (fig. 17.9). Examples of changes in the coefficient of friction as a function of time are given in fig. 17.16.

The coefficient of friction as a function of time practically does not change for individual solutions. Therefore, a value obtained after 900 s was taken as a measure of μ. The values shown in fig. 17.17 are the results of two averagings, initially in 20-s

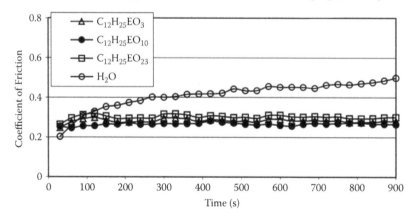

FIGURE 17.16 Changes in the value of the coefficient of friction for water and 1 wt% solutions of oxyethylated lauryl alcohols as a function of time at a load of 2 kN. (Data obtained using tester T02.)

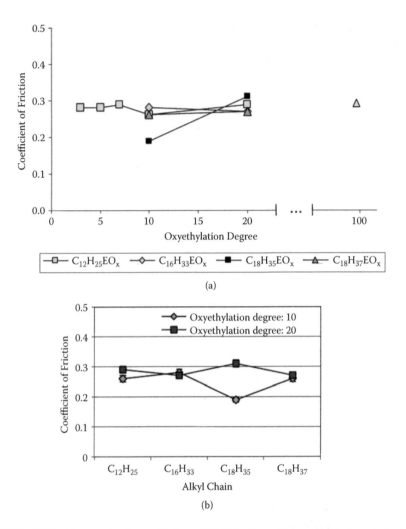

FIGURE 17.17 Changes in the coefficient of friction of 1 wt% oxyethylated alcohol solutions as a function of (a) oxyethylation degree and (b) length and nature of the alkyl chain at a constant load of 2 kN. (Data obtained using tester T02.)

intervals, and averaging was done after at least three independent tests. The changes in $\mu(t)$ proceed differently for water as a lubricating substance (fig. 17.16). In the initial time interval (100 s), the change rate is relatively high, but in the following intervals the rate decreases. From 300 s, a steady rise can be observed, and after 900 s the mean value equals 0.47. A graphic presentation of friction coefficients as a function of concentration and kind of compound is given in fig. 17.18.

The measured coefficient of friction can decrease by more than 2.5-fold relative to water ($C_{12}H_{25}EO_7$ and $C_{12}H_{25}EO_{23}$ solutions). The changes are not monotonic as a function of concentration, with 0.1 wt% solutions having, on average, the highest values of μ, and with 1 wt% solutions having the lowest values. For 1 wt% solutions, the friction coefficient μ does not practically depend on the degree of oxyethylation

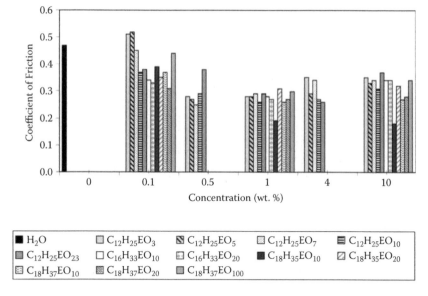

Legend:
- ■ H_2O
- □ $C_{12}H_{25}EO_3$
- ⊠ $C_{12}H_{25}EO_5$
- ▨ $C_{12}H_{25}EO_7$
- ▤ $C_{12}H_{25}EO_{10}$
- ▨ $C_{12}H_{25}EO_{23}$
- □ $C_{16}H_{33}EO_{10}$
- ▦ $C_{16}H_{33}EO_{20}$
- ■ $C_{18}H_{35}EO_{10}$
- ▨ $C_{18}H_{35}EO_{20}$
- ▨ $C_{18}H_{37}EO_{10}$
- ▨ $C_{18}H_{37}EO_{20}$
- □ $C_{18}H_{37}EO_{100}$

FIGURE 17.18 Changes in the coefficient of friction of aqueous solutions of oxyethylated alcohols as a function of concentration at a load of 2 kN. (Data obtained using tester T02.)

and alkyl chain length, and μ ranges from 0.26 to 0.30. An exception was observed for the solutions of oleyl alcohol oxyethylated with 10 moles of ethylene oxide, for which μ assumes considerably lower values, and 4 and 10 wt% concentrations, where μ was also considerably lower. However, solutions of oleyl alcohol oxyethylated with 20 moles of ethylene oxide show a value within the predicted range. The effects of the lengths of ethylene oxide and alkyl chains can be discussed on the basis of fig. 17.17.

Except for the previously noted solution of oxyethylated oleyl alcohol ($C_{18}H_{35}EO_{10}$), no pronounced effect of the increase in alkyl chain length or oxyethylation degree on the resistance to motion can be observed. The measured values are comparable within the margin of error.

The dependence of wear-scar diameter as a function of concentration and the type of additive is presented in fig. 17.19. The wear-scar diameter for the majority of solutions of oxyethylated alcohols and their concentrations decreases up to two-fold relative to water (1.8 mm). Only for 0.1 wt% solutions are the wear-scar values greater than 1, with a maximum value of 1.6 (0.1 wt% solution of $C_{12}H_{25}EO_{23}$). For concentrations equal to and above 0.5 wt%, the measured d values range from 0.9 to 1.1. Even oxyethylated oleyl alcohols do not exhibit any deviations from this. The effect of oxyethylation degree and the type of alkyl chain is analyzed on the basis of the results in fig. 17.20. There is no pronounced effect of the structure of the compound on wear under constant loading conditions.

The results of constant-load tests do not display regularities similar to the ones observed in seizure tests. The former were carried out under "milder conditions" than the seizure tests. There was no rapid increase in loading (409 N/s), and the constant load used (2 kN) was comparable to the seizure load. Under such conditions one may expect a stable lubricant film. The temperature in the friction node should

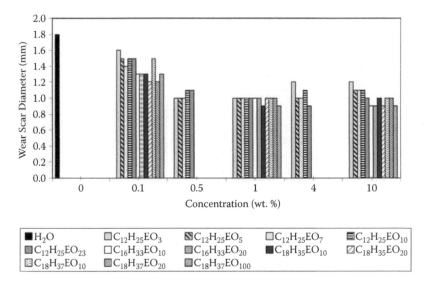

FIGURE 17.19 Changes in wear-scar diameter in the presence of aqueous solutions of oxyethylated alcohols as a function of concentration at a load of 2 kN. (Data obtained using tester T02.)

not exceed the cloud point. The reduction in resistance to motion and wear, relative to water, results primarily from the surface activity of oxyethylated alcohols.

17.5 SUMMARY

Water-based lubricating substances have been used mainly as hydraulic and working fluids. Their range of application will likely be widened in the future, with additives playing an increasingly important role in water-based lubricants. The tested additives were oxyethylated alcohols which, at relatively low concentrations (a few thousandths of a percent), show significant surface activity that produces micellar and liquid crystalline structures. Laboratory results suggest that the required properties of molecules can be "designed" by varying the lengths of ethylene oxide and alkyl chains.

In the case of oxyethylated alcohol solutions, there is a pronounced increase in surface tension and wettability resulting from an increase in the length of the ethylene oxide chain and, to a lesser extent, from an increase in the length of the alkyl chain. The results obtained and the literature data present a relatively complex picture of the interactions in the surface phase. The most probable hypothesis is that a change in the arrangement of molecules takes place due to an increase in the length of the alkyl chain. The structures favored are those in which ethylene oxide mers do not stick to the surface. As a result, both the interaction of oxyethylated alcohol molecules with the surface and the amount of adsorbed compound decrease.

The formation of an adsorbed film under static conditions may result in the formation of a lubricant film under dynamic friction conditions. It turns out that aqueous solutions of ethoxylates exhibit very good lubricating properties, which was

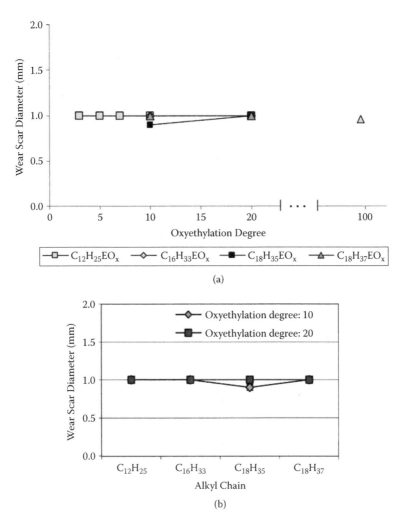

FIGURE 17.20   Changes in wear-scar diameter in the presence of 1 wt% solutions of alcohols as a function of (a) oxyethylation degree and (b) length and nature of alkyl chain at a constant load of 2 kN. (Data obtained using tester T02.)

confirmed in both seizure tests and constant-load tests. The aim of this investigation was to try to correlate the structure of a compound, its surface activity, and, particularly, the proportion of the hydrophilic and hydrophobic parts in an ethoxylate molecule with tribological properties.

Antiseizure properties were characterized by scuffing load (P_t), seizure load (P_{oz}), and the limiting pressure of seizure (p_{oz}). The tendency of changes in these three quantities as a function of alkyl and ethylene oxide chain lengths was analogous for most compounds. The measured values show a good correlation with the activity of the compounds that results from the alkyl chain length. Its increase caused a reduction in the amount of adsorbed additive and unfavorably affected the stability of the lubricant film. Contrary to all expectations, an increase in hydrophilicity of a compound, which caused a decrease in its surface activity, produced a considerable improvement in tribological properties. This probably results from the dehydration of the oxyethylated alcohol molecules associated with the increase in temperature. At higher temperatures, oxyethylated alcohols with a higher oxyethylation degree retain their amphiphilic properties and are more effective in preventing seizure.

Tests at a constant load of 2 kN do not exhibit changes characteristic of seizure tests. Although both friction coefficient and wear-scar diameter decrease more than twofold relative to water, there is no pronounced effect of alkyl and ethylene oxide chains on the measured values. This is due to the relatively low temperature of the lubricant during these kinds of tests, in which surfactants do not undergo significant dehydration.

Amphiphilic compounds used as additives considerably improve the tribological properties of water [1–13]. The hypothesis of the action of amphiphilic compounds formulated on the basis of experimental results assumes that ordered micellar structures form in the surface phase, and these fill the microirregularities and produce a lubricant film with a liquid crystalline structure. This results in an increase in the real area of contact, reduces pressure, and increases the load-carrying capacity. In a liquid crystalline structure, slip planes are present that considerably reduce the resistance to motion, similar to lubricants containing graphite and molybdenum disulfide. The results presented here confirm this hypothesis and provide information on the influence of the structure of the compounds on the surface activity and tribological properties of their aqueous solutions.

REFERENCES

1. Sulek, M. W., and Wasilewski, T. 2005. *Tribol. Lett.* 18: 197–205.
2. Sulek, M. W., and Wasilewski, T. 2006. *Wear* 260: 193–204.
3. Sulek, M. W., and Wasilewski, T. 2004. *Tribologie Schmierungstechnik* 1: 9–13 (in German).
4. Sulek, M. W., and Wasilewski, T. 2005. *Tribologie Schmierungstechnik* 3: 24–33 (in German).
5. Sulek, M. W., and Wasilewski, T. 2002. *Int. J. Appl. Mech. Eng.* 7: 189–95.
6. Sulek, M. W., and Bocho-Janiszewska, A. 2006. *Tribol. Lett.* 24 (3): 187–94.
7. Sulek, M. W., and Bocho-Janiszewska, A. 2006. *Tribologie Schmierungstechnik* 1: 51–56 (in German).
8. Sulek, M. W., and Bocho-Janiszewska, A. 2003. *Tribol. Lett.* 15: 301–7.
9. Biresaw, G., ed., 1998. *Tribology and the liquid-crystalline state.* Washington, DC: American Chemical Society.
10. Plaza, S., Margielewski, L., Celichowski, G., Wesolowski, R. W., and Stanecka, R. 2001. *Wear* 249: 1077–89.
11. Wei, J., and Xue, Q. 1996. *Wear* 199: 157–59.
12. Wei, J., and Xue, Q. 1993. *Wear* 162–64: 229–33.
13. Ratoi, M., and Spikes, H. A. 1999. *Tribol. Trans.* 42: 479–86.
14. Denoyel, R. 2002. *Colloids Surfaces* 205: 61–71.
15. Tiberg, F. 1996. *J. Chem. Soc. Faraday Trans.* 92: 531–38.
16. Patrick, H. N., and Warr, G. G. 2000. *Colloids Surfaces* 162: 149–57.
17. Grant, L. M., Ederth, T., and Tiberg, F. 2000. *Langmuir* 16: 2285–91.
18. Somasundaran, P., and Krishnakumar, S. 1997. *Colloids Surfaces* 123–24: 491–513.
19. Somasundaran, P., and Huang, L. 2000. *Adv. Colloid Interface Sci.* 88: 179–208.

20. Warr, G. G. 2000. *Curr. Opinion Colloid Interface Sci.* 5: 88–94.
21. Levitz, P. E. 2002. *Geoscience* 334: 665–73.
22. Luciani, L., Denoyel, R., and Rouquerol, J. 2001. *Colloids Surfaces* 178: 297–312.
23. Kunieda, H., Kaneko, M., López-Quintela, M. A., and Tsukahara, M. 2004. *Langmuir* 20: 2164–71.
24. Liljekvist, P., and Kronberg, B. 2000. *J. Colloid Interface Sci.* 222: 159–64.
25. Berni, M. G., Lawrence, C. J., and Machin, D. 2002. *Adv. Colloid Interface Sci.* 98: 217–43.
26. Müller, S., Börschig, C., Gronowski, W., Schmidt, C., and Roux, D. 1999. *Langmuir* 15: 7558–64.
27. Karlström, G. 1985. *J. Phys. Chem.* 89: 4962–64.
28. Kjellander, R. 1982. *J. Chem. Soc. Faraday Trans.* 78 (2): 2025–42.
29. Goldstein, R. E. 1984. *J. Chem. Phys.* 80: 5340–41.
30. Matsuyama, A., and Tanaka, F. 1990. *Phys. Rev. Lett.* 65: 341–44.
31. Bekiranov, S., Bruinsma, R., and Pincus, P. 1993. *Europhys. Lett.* 24: 183–88.
32. Huibers, P. D. T., Shah, D. O., and Katritzky, A. R. 1997. *J. Colloid Interface Sci.* 193: 132–36.
33. Katritzky, A. R., van Os, N. M., Haak, J. R., and Rupert, L. A. M. 1993. *Physicochemical properties of selected anionic cationic and nonionic surfactants.* Amsterdam: Elsevier.
34. Meguro, K., Ueno, M., and Esumi, K. 1987. *Nonionic surfactants.* Vol. 23. Ed. M. Schick. New York: Marcel Dekker.
35. Levitz, P., van Damme, H., and Keravis, D. 1984. *J. Phys. Chem.* 88: 2228–35.
36. Levitz, P., and van Damme, H. 1986. *J. Phys. Chem.* 90: 1302–10.
37. Levitz, P. E. 2002. *Colloids Surfaces* 205: 31–38.
38. Griffith, J. C., and Alexander, A. E. 1967. *J. Colloid Interface Sci.* 25: 311–16.
39. Kiraly, Z., Börner, R. H., and Findenegg, G. H. 1997. *Langmuir* 13: 3308–15.
40. Nevskaia, D. M., Guerrero-Ruiz, A., and López-Gonzales, J. 1996. *J. Colloid Interface Sci.* 181: 571–80.
41. Cases, J. M., Villieras, F., Michot, L. J., and Bersilon, J. L. 2002. *Colloids Surfaces* 205: 85–99.
42. Denoyel, R., and Rouquerol, J. 1991. *J. Colloid Interface Sci.* 143: 555–72.
43. Partyka, S., Zaini, S., Lindheimer, M., and Brun, B. 1984. *Colloids Surfaces* 12: 255–70.
44. Gellan, A., and Rochester, C. H. 1985. *J. Chem. Soc., Faraday Trans.* 81 (1): 2235–46.
45. Howse, J., Steitz, R., Pannek, M., Simon, P., Schubert, D. W., and Findenegg, G. H. 2001. *Phys. Chem. Chem. Phys.* 3: 4044–51.
46. Söderlind, E., and Stilbs, P. 1993. *Langmuir* 9: 1678–83.
47. Brinck, J., Jönsson, B., and Tiberg, F. 1998. *Langmuir* 14: 1058–71.
48. Brinck, J., and Tiberg, F. 1996. *Langmuir* 12: 5042–47.
49. Tiberg, F., Jönsson, B., Tang, J., and Lindman, B. 1994. *Langmuir* 10: 2294–2300.
50. Wanless, E. J., Davey, T. W., and Ducker, W. A. 1997. *Langmuir* 13: 4223–28.
51. Wall, J. F., and Zukoski, Ch. F. 1999. *Langmuir* 15: 7432–37.
52. Adler, J. J., Singh, P. K., Patist, A., Rabinovich, Y. I., Shah, D. O., and Moudgil, B. M. 2000. *Langmuir* 16: 7255–62.
53. Rutland, M. W., and Senden, T. J. 1993. *Langmuir* 9: 412–18.
54. Dixit, S. G., Vanjara, A. K., Nagarkar, J., Nikoorazm, M., and Desai, T. 2002. *Colloids Surfaces* 205: 39–46.
55. Patrick, H. N., Warr, G. G., Manne, S., and Aksay, I. A. 1997. *Langmuir* 13: 4349–56.
56. Kiraly, Z., and Findenegg, G. H. 2000. *Langmuir* 16: 8842–49.
57. Findenegg, G. H., Pasucha, B., and Strunk, H. 1989. *Colloids Surfaces* 37: 223–33.

58. Giordano-Palmino, F., Denoyl, R., and Rouquerol, J. 1994. *J. Colloid Interface Sci.* 165: 82–90.
59. Hartley, G. S. 1936. *Aqueous solutions of paraffin chain salts.* Paris: Herman.
60. Mittal, K. L., ed. 1997. *Micellization, solubilization and microemulsion.* New York: Plenum Press.
61. Fairhurst, C. E., Fuller, S., Gray, J., Holmes, M. C., and Tiddy, G. J. 1996. *Lyotropic surfactant liquid crystals.* Vol. 3 of *Handbook of liquid crystals.* Ed. D. Demus, J. W. Goodby, G. W. Gray, H. W. Spiess, and V. Vill. Weinheim: Wiley-VCH.
62. Elworthy, P. H., and Florence, A. T. 1964. *Kolloid Z. Z. Polym.* 195: 23.
63. Elworthy, P. H., and MacFarlane, C. B. 1962. *J. Pharm. Pharmacol. Suppl.* 14: 100.
64. Clunie, S. J., and Ingram, B. T. 1983. In *Adsorption from solution at the solid/liquid interface.* Ed. G. D. Parfitt and C. H. Rochester. New York: Academic Press.
65. Portet, F., Desbene, P. L., and Treiner, C. 1997. *J. Colloid Interface Sci.* 194: 379.
66. Baljon, A. R. C., and Robbins, M. O. 1997. In *Micro/nanotribology and its applications.* Ed. B. Bhushan. Dordrecht: Kluwer.
67. Israelachvili, J., McGuiggan, P., Gee, M., Homola, A., Robbins, M., and Thompson, P. A. 1990. *J. Phys. Condens. Matter* 2: SA89.
68. Dan, N. 1996. *Curr. Opinion Colloid Interface Sci.* 1: 48.
69. Thompson, P. A., Grest, G. S., and Robbins, M. O. 1992. *Phys. Rev. Lett.* 68: 3448.
70. Gee, M. L., McGuiggan, P. M., and Israelachvili, J. N. 1990. *J. Chem. Phys.* 93: 1895.
71. Berman, A. D., Ducker, W. A., and Israelachvili, J. N. 1996. *Langmuir* 12: 4559.
72. Persson, B. N. J. 1995. *J. Chem. Phys.* 103: 3849.
73. Gupta, S. A., Cochran, H. D., and Cummings, P. T. 1997. *Chem. Phys.* 107: 10327.
74. Rabin, Y., and Hersht, I. 1993. *Physica A.* 200: 708.
75. Granick, S. 1991. *Science* 253: 1374.
76. Wasilewski, T., and Sulek, M. W. 2006. *Wear* 261: 230–34.
77. Biresaw, G., Adhvaryu, A., and Erhan, S. Z. 2003. *J. Am. Oil Chem. Soc.* 80: 697–704.
78. Kurth, T. L., Biresaw, G., and Adhvaryu, A. 2005. *J. Am. Oil Chem. Soc.* 82: 293–99.
79. Boschkova, K., Kronberg, B., Rutland, M., and Imae, T. 2001. *Tribol. Int.* 34: 815.
80. Penfold, J., Staples, E., Lodhi, A. K., Tucker, I., and Tiddy, G. J. T. 1997. *J. Phys. Chem.* 101: 66.
81. Kabalnov, A., and Wennerström, H. 1996. *Langmuir* 12: 276.
82. Holmes, M., Leader, M., and Smith, A. 1995. *Langmuir* 11: 356–65.
83. Ducker, W. A., and Grant, L. M. 1996. *J. Chem. Phys.* 100: 1507–11.
84. Fuerstenau, D. W., and Colic, M. 1999. *Colloids Surfaces* 146: 33–47.
85. Piekoszewski, W., Szczerek, M., and Tuszynski, W. 2001. *Wear* 240: 183–93.

18 Aqueous Solutions of Mixtures of Nonionic Surfactants as Environment-Friendly Lubricants

Tomasz Wasilewski

ABSTRACT

This chapter presents the results of physicochemical and tribological investigations of lubricating substances composed of water and a mixture of two nonionic surfactants: sorbitan monolaurate (SML) and ethoxylated sorbitan monoisostearate (ESMIS). Particular attention was paid to proper selection of surfactant mixture composition (SML/ESMIS ratio) ensuring high effectiveness in reducing friction coefficient and wear as well as seizure prevention.

Aqueous solutions of various SML/ESMIS mixtures (concentration range—0.1% to 4 wt%) were used as lubricants in this investigation. Tribological tests were carried out for steel-polyamide 6 and steel-steel friction pairs using the pin-on-disc or four-ball testers. Based on the tests the following values were determined: friction coefficient, scuffing load, seizure load and changes in roughness of friction pair surfaces. Wear was determined on the basis of measurements of wear scar diameters and wear trace profiles.

Viscosity and surface tension were measured for individual lubricant compositions. The properties of the mixtures of the surfactants used were also evaluated in terms of their ability to produce microemulsions and to form liquid crystalline structures (polarized light microscopy was used).

Based on the tests it has been concluded that solutions of the mixtures of the nonionic surfactants studied result in a threefold decrease in the coefficient of friction and wear of friction pair elements compared with water. Besides, changes in tribological properties as a function of SML/ESMIS ratio are not monotonic. The lowest values for the coefficient of friction and wear were observed in the case of the mixtures with SML/ESMIS ratios 3:7 and 5:5.

Interesting results were also obtained with regard to antiseizure properties of the compositions tested. None of the 1 wt% aqueous solutions caused seizure of friction pair elements in the load range of up to 8 kN. The lowest values of scuffing loads (when lubricant film is initially broken) were also observed for the mixtures with SML/ESMIS ratios 3:7 and 5:5.

On the basis of literature data and the tests performed, it can be stated that the effective action of the mixtures with SML/ESMIS ratios 3:7 and 5:5 results from the most favorable packing of surfactant molecules in the interfacial film at these ratios. As a result, when the stability of the adsorbed film is the highest, it leads to the highest load-carrying capacity as well as the largest decrease in motion resistance and wear.

18.1 INTRODUCTION

Interest in tribological studies has increased rapidly in recent years. The principal aim of these studies is the optimization of friction processes, which usually leads to a reduction in energy and material losses and to a longer service life of elements of machines and devices. Work is continually being done to improve the existing friction couples and develop new ones.

The application of an appropriate lubricating substance to a friction pair is an essential factor controlling its proper operation. The lubricant must be suitable for both the surface type and the material of which the mating elements are made. Moreover, friction couples often operate at various pressures, sliding velocities, and temperatures. The lubricating substance must perform its intended functions, regardless of the operating conditions involved.

The most important characteristics that a lubricating substance must exhibit are an efficient reduction in resistance to motion; prevention of wear, scuffing, and seizure of mating friction-pair elements; protection of mating elements against corrosion; and suitably high thermal conductivity and dispersibility of wear products. In recent years special attention has been paid to ecological considerations: lubricating substances should be nontoxic, able to undergo biodegradation, and should not be hazardous in contact with the human body. Lubricant compositions based on renewable resources are in particular demand. The high quality of these types of products should also be correlated with their low cost, availability, and user friendliness.

A lubricating substance is normally a multicomponent mixture in which individual components perform different functions. In the process of developing a lubricant composition, particular attention is paid to the compounds, which ensure the formation, during friction, of a lubricant film protecting friction-pair elements against direct contact and effecting a reduction in resistance to motion. The compounds are called FM (friction modifier) additives, AW (antiwear) additives, and EP (extreme pressure) additives. Additives affecting rheological properties of lubricants and anti-corrosion additives are also relatively important.

Despite the huge number of lubricants currently in use, studies are still being carried out to better understand the action mechanisms of additives for a friction pair, particularly at the molecular level [1–19]. The contributing factors are the development of modern measurement methods, including atomic force microscopy

(AFM) [17], and devices for tribological investigations, such as a device for measuring lubricant film thickness or relatively low friction forces [2, 3, 16].

During the friction process, lubricant additives interact with the surfaces of friction-pair elements: they may undergo physical adsorption or chemisorption [8]. This may result in the modification of the original surfaces of a friction pair or in the formation of the so-called operational top layer [20, 21].

Amphiphilic compounds play a special role in this process. A number of commonly used additives are composed in such a way that there are hydrophobic and hydrophilic groups within a single molecule. Depending on the structure of a compound (appropriate arrangement of hydrophilic and hydrophobic groups in the molecule, size of the groups) and on the kind of base they are in, they may exhibit surface activity. Examples of this kind of compound are zinc dialkyldithiophosphate (ZDDP), sulfonated olefins, and carboxylic acids. These compounds contain alkyl chains (hydrophobic part) in their molecules, which determine their solubility in the oil base. Apart from the alkyl chains, there appear hydrophilic groups containing heteroatoms O, S, P, and N. (These atoms have free electron pairs facilitating the formation of hydrogen bonds.) Due to such structure of the additives, they are "pushed" out from the bulk phase toward the interface (they may undergo adsorption at the interfaces). Besides, under suitable conditions (temperature, pressure) the compounds adsorbed may undergo tribochemical changes or reactions with friction-pair metals. This results in the formation of products that effectively change the character of the top layer of friction-pair elements [13].

In the literature there are a number of suggestions involving various action mechanisms of additives, particularly under boundary lubrication [1–19]. One of the assumptions is that the adsorbed additive molecules may form an ordered protective film on the surfaces that separates the mating surfaces. Such a film may exhibit low shear strength when subjected to the action of shearing forces, or it may act as a physical barrier inhibiting contacts between friction-pair elements, where its structure may resemble a very hard solid.

The ecological aspects of lubricating substances are an extremely important issue [22–33]. The problems resulting from environmental pollution by mineral or synthetic lubricants, often containing toxic additives, have been encountered for a number of years. There are a number of literature reports presenting investigation results for lubricating substances that are environmentally friendly at the stages of manufacture, application, and recycling. Examples are the studies on the application of various vegetable oils or the products of their processing [27–33]. It should be noted that raw materials of plant origin frequently differ in their composition, depending on the cultivation method and place or the variety used. This necessitates development of suitable procedures and taking those factors into account when designing lubricants based on vegetable oils.

In the present study, particular attention has been paid to aqueous solutions of surfactants as a new kind of lubricating substances. This subject has been discussed in the literature [34–49]. The surfactant–water compositions that have already been investigated include some systems that meet certain criteria that lubricants should fulfill: suitable lubricity, reduced wear, and antiseizure properties. Besides, it is very important that such compositions provide a way to eliminate problems resulting from the application of lubricants based on petroleum industry products. It is also important that aqueous solutions of surfactants can be applied to lubricate machines used in the food and pharmaceutical industries and to lubricate open friction couples from which lubricating substances find their way directly to the environment.

This chapter discusses possible applications of aqueous solutions of surfactants as potential lubricants. It also presents the aspects that need to be considered when selecting the composition of surfactants. The model lubricants proposed contain water and a mixture of two nonionic surfactants. A series of physicochemical and tribological tests have been carried out. Their aim was to demonstrate the applicability of these compositions to the lubrication of friction couples.

18.2 AQUEOUS SOLUTIONS OF SURFACTANTS AS POTENTIAL ENVIRONMENT-FRIENDLY LUBRICATING SUBSTANCES

Water is a compound that is commonly available in the natural human environment. It does not need to be treated; it is nonflammable and very cheap. It also has a number of drawbacks, such as low viscosity, tendency for metal corrosion, and inadequate lubricating properties. However, the application of water as a lubricant base was such an attractive idea that detailed studies were undertaken on its applicability. The other contributing factors were the knowledge of the mechanisms operating within friction pairs and the possibility of making use of modern measuring devices. A number of papers have been published that deal with the lubrication of friction pairs whose elements were made of ceramics [50–53], plastics, or composite materials [54–58]. The other solution is the application of appropriate coatings on friction-pair surfaces [59–62] or adding corrosion inhibitors to water [63].

The analysis of the data presented in the literature [50–66] indicates that pure water does not always satisfy the requirements expected of lubricating substances. Hence, there is a need to utilize appropriate additives that would reduce motion resistance and wear, prevent seizure, inhibit the corrosive action of water, and increase the viscosity of water. Surfactants can be used as such substances, primarily due to their adsorptivity at the interface and their ability to produce ordered structures in solutions.

It has been assumed in this chapter that aqueous solutions of surfactants will meet the criteria for modern and functional lubricating substances. However, to do so, they must perform several functions within the friction couple simultaneously. A discussion on a possible application of surfactants as effective additives to water used as a lubricant base will be presented below.

18.2.1 SOLUBILITY OF ADDITIVES

The solubility of a surfactant is the first criterion categorizing it as a potential lubricant additive. In the case of water-based lubricants, the application of FM, AW, and EP additives in oils is impossible or considerably limited. Despite the fact that ZDDP-type compounds have a hydrophilic group, its proportion in the whole molecule is too small to ensure water solubility of the whole molecule.

The so-called HLB (hydrophilic–lipophilic balance) constant has been introduced to easily assess the affinity of a surfactant for polar solvents (including water). The HLB values for the best-known surfactants can be found in the literature [67–71]. The higher the HLB value for a given compound, the more readily the surfactant dissolves in water (HLB values range from 0 to 20). The value for the surfactants that can be potential water-soluble additives must be close to 20. It should be noted here, however, that the higher the HLB constant for a given compound, the lower its ability to migrate toward the interface. Therefore, the best approach would be to use water-soluble compounds with a sufficiently large hydrophobic part.

18.2.2 Adsorption of Surfactants on the Surface of a Friction-Pair Element: Formation of a Boundary Lubricant Film to Reduce Motion Resistance and Wear

In many publications dealing with the application of aqueous solutions of surfactants as lubricants, attempts are made to demonstrate the key importance of adsorption in dictating specific tribological characteristics [34–49]. Both the structure (HLB constant) and the size of a surfactant have a considerable influence on the rate of formation and the stability of the adsorbed film. The ability of the surfactants to migrate from the bulk phase to the surface phase is essential, as it determines the formation of a lubricant film on mating surfaces. The film will protect the surfaces against direct contact and will exhibit a load-carrying capacity.

In water solutions, the surfactants characterized by a low HLB value and small molecular sizes will be able to form an adsorbed film very quickly. An increase in size or a specific structure of a surfactant molecule allowing for a high degree of packing at the interface will result in an increase in the interactions between individual molecules in the adsorbed film and, thus, the stability of the film produced. To obtain a lubricant that can effectively reduce the coefficient of friction and prevent wear, the surfactant chosen should be able to rapidly produce a film of suitable stability [34–49].

It should be mentioned here that the formation, structure, and stability of the surface phase have been the subject of a number of scientific publications [72–88]. This has resulted mainly from the fact that the formation of an adsorbed film at the liquid–solid, liquid–liquid, and liquid–gas interfaces is of crucial importance in many areas. It seems that the results of these studies can be particularly helpful in explaining phenomena occurring within a friction couple.

The knowledge of the surface-phase structure may be particularly important for the friction process. The latest data indicate that aggregates with structure different from the one observed in the bulk phase may form in the surface phase. The fact that such aggregates may form at concentrations below the critical micelle concentration

(CMC) is also important [79]. There are no tribological studies giving consideration to this matter.

18.2.3 Rheology

At room temperature, the viscosity of pure water is relatively low (about 1 mPa·s). Addition of surfactants results in a viscosity increase, though the increase is not linear as a function of additive concentration. In a range of low concentrations (depending on the compound structure, these are on the order of 0.001% to 10%), the surfactant molecules in water occur in the form of monomers. After exceeding the threshold CMC characteristic for a specific system, some monomers are present in the form of spherical aggregates called micelles. The presence of micelles in a solution considerably affects its viscosity. Increasing the concentration of a compound in solutions of nonionic surfactants usually leads to an increase in the number of micelles and, to a small extent, to a change in the shape of the micelles. In the case of ionic surfactants, increases in concentration lead to changes in the shapes of micelles and, hence, their sizes. In addition, the structure of surfactant molecules significantly affects the shape and size of micelles. The compounds with a small hydrophobic group and a relatively large hydrophilic group will produce spherical micelles. Increasing the share of the hydrophobic part in the molecule results in the formation of cylindrical micelles by surfactants, even at low concentrations.

An increase in both the number of micelles and in their size leads to a significant increase in the viscosity of the solution. The phenomenon of aggregation of surfactants into micelles may be used to obtain suitable rheological characteristics of potential lubricant compositions. The investigation results presented by Sulek and Wasilewski [47] may be quoted as an example. The viscosity of 1% water solutions of alkylpolyglucosides containing, on average, nine carbon atoms in the alkyl chain was 8 mPa·s, while the viscosity of 1% water solutions of alkylpolyglucosides with a similar structure but having 13 carbon atoms, on average, in the alkyl chain was 16 mPa·s. It is interesting that increasing the concentration up to 4% in the first case led to an increase in viscosity of only up to 10 mPa·s, while in the latter case the increase was up to 130 mPa·s. The data presented above demonstrate how even a small change in the structure of a surfactant molecule makes it possible to control the rheological properties of compositions.

Another interesting approach seems to be the application of solutions containing about 40%–80% of surfactants as a lubricant composition. Such systems may exist in a liquid crystalline form. The liquid crystalline phases can have regular, hexagonal, or lamellar structures depending, among others, on the concentration, structure, and temperature of the compound. The lamellar phase is particularly interesting from the viewpoint of its application as a lubricant composition. It has a layer structure, and the action of shearing forces may easily result in the relative displacement of individual layers of the liquid. As a result, it causes a significant decrease in friction coefficient values [26, 46, 68, 69].

18.2.4 Corrosion Prevention

A reduction in the corrosive action of water is an extremely important property of surfactants. The mechanism of this phenomenon has been presented in a work by Somasundaran and Huang [89]. The authors assume that the adsorbed film formed at the interface efficiently separates the water from the metal surface. Experiments confirming the anticorrosive action of surfactants have been presented in several works [90–94]. The following compounds were used in the experiments as additives: cationic surfactants (cetyl pyridinium chloride and trimethyl ammonium bromide) [89], ethoxylated fatty acids [90], ethoxylated fatty alcohols [91], and ethoxylated sorbitan esters [92–94].

18.2.5 Ecological Consideration

The overriding aim of the development of water-based lubricants is their environmental safety. Therefore, the surfactants used in food and pharmaceutical products seem to be of particular interest. In this respect, the best compounds will be those prepared from raw materials of natural origin: vegetable oils, animal fats, or sugars, e.g., alkylpolyglucosides [95, 96].

18.3 COMPOSITIONS CONTAINING SORBITAN ESTERS, ETHOXYLATED SORBITAN ESTERS, AND WATER AS POTENTIAL LUBRICATING SUBSTANCES

Sorbitan esters and ethoxylated sorbitan esters are an interesting group of surfactants from the viewpoint of their application to lubricating substances. These compounds do not exhibit harmful effects on living organisms. Therefore, they are commonly used as components in foodstuffs, cosmetics, and medicines [70, 71]. They are described as anticorrosion additives in literature reports and patents [92–94]. Another factor weighing in favor of the choice of sorbitan esters and ethoxylated sorbitan esters as lubricant additives is their ability to produce liquid crystalline structures [48]. There are a number of scientific publications that claim that beneficial tribological properties frequently depend on the ability of additive molecules or molecular aggregates to arrange themselves into structures exhibiting a long-term order [87–110].

Sorbitan esters are prepared by esterification of dehydrated sorbitol with fatty acids. In most cases, raw materials of plant origin are used. The esters that are most commonly obtained contain one or three alkyl chains from lauric, palmitic, stearic, isostearic, and oleic acids. The compounds obtained exhibit different polarities, depending on the acid used and on the degree of substitution. Most sorbitan esters readily dissolve in organic solvents.

Sorbitan esters can undergo ethoxylation in order to increase their hydrophilicity. An increase in the degree of ethoxylation improves the water solubility of the compounds. In practice, the most commonly synthesized compounds are the ones containing 20 moles of ethylene oxide per molecule. All of them are produced on an industrial scale.

The application of sorbitan esters and/or ethoxylated sorbitan esters as lubricant components has been discussed in the literature. It has been shown, among others, that sorbitan monoesters, monolaurate and monooleate, efficiently improve the lubricating properties of paraffin oil. When sorbitan trioleinate was used as an additive, no improvement in the lubricating properties of paraffin oil was observed [111].

Tribological test results for aqueous solutions of ethoxylated sorbitan esters have been presented in a publication [48]. It has been demonstrated that addition to water of such esters as ethoxylated sorbitan monolaurate, ethoxylated sorbitan monostearate, or ethoxylated sorbitan monooleate results in a significant reduction in the coefficient of friction and wear and prevents seizure of mating friction-couple elements. It has been found that the concentration of the additive in water, and not the compound's structure, has the decisive effect on the improvement in tribological characteristics [48].

Interesting investigation results were also obtained when mixtures of sorbitan ester with ethoxylated sorbitan ester were used in a model lubricant base (paraffin oil) [112]. It was shown that the mixtures were very efficient in reducing motion resistance and wear. The most beneficial action was observed for mixtures of sorbitan monolaurate and ethoxylated sorbitan monolaurate at ratios of 5:5, 3:7, and 1:9 [112].

The results presented in this chapter are a continuation of the studies on mixtures of this type of surfactant. This time, however, water was chosen as the lubricant base. In view of the investigation results presented by Pilpel and Rabbani [112], particular attention was paid in this study to the sorbitan-ester/ethoxylated-sorbitan-ester ratio. An attempt was made to find out how to prepare lubricating substances having the desired parameters (characteristics).

The decision to conduct the study on the mixtures also resulted from literature reports claiming that the application of a suitable sorbitan-ester/ethoxylated-sorbitan-ester composition affects the possibility of forming very stable adsorbed films at the liquid–liquid interface. The mechanisms of emulsion stabilization by such mixtures have been explained in the literature [113–121]. Models presenting the patterns of arrangement of surfactant molecules at the interfaces have also been suggested. However, the main conclusion resulting from this study is that, for a given hydrophilic-solvent–hydrophobic-solvent system, the mixture of surfactants should be selected in a way that would ensure the highest possible coefficient of packing at the interface [113, 115, 118]. The arrangement of molecules relatively close to each other will lead to an increase in interactions between them. This results in an improvement in the stability of the adsorbed film, which translates into an improvement in the stability of emulsions, foams, etc. [122–125].

In view of these investigation results, it has been assumed in this chapter that an appropriate sorbitan-ester/ethoxylated-sorbitan-ester composition will also be capable of producing stable adsorbed films at the lubricant/friction-pair material interface. It is predicted that an optimal arrangement of surfactant molecules in the surface phase can result in a very effective reduction in resistance to motion and in wear and seizure prevention.

The two compounds—sorbitan monolaurate (SML) and ethoxylated sorbitan monoisostearate (ESMIS)—were selected for the investigation out of a wide range of sorbitan esters and ethoxylated sorbitan esters produced on an industrial scale. The chemical structures of the selected compounds are shown in fig. 18.1.

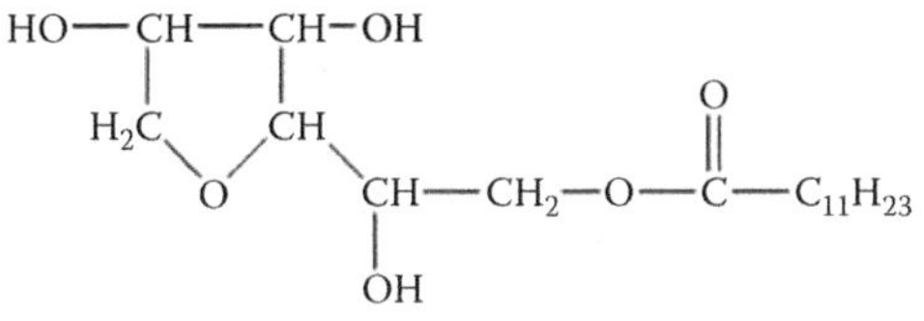

(a) Sorbitan monolaurate (SML)

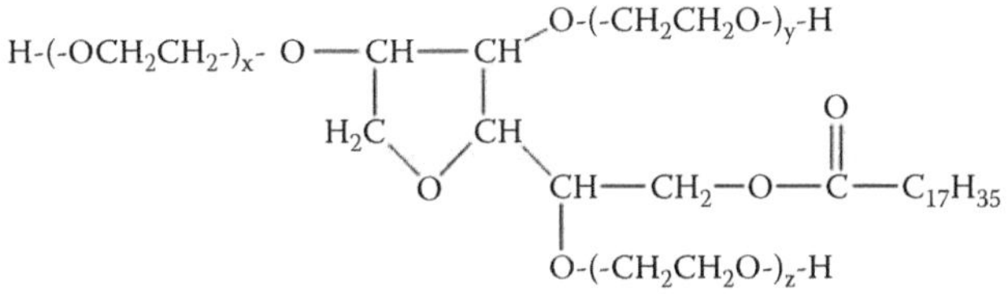

(b) Ethoxylated sorbitan monoisostearate (ESMIS)
(with x + y + z = 20)

FIGURE 18.1 Chemical structures of sorbitan monolaurate (SML) and ethoxylated sorbitan monoisostearate (ESMIS).

Sorbitan monolaurate is an effective additive for improving the lubricating properties of hydrocarbon bases [111, 112]. However, it cannot be used on its own as an additive to improve the lubricating properties of water, because it does not dissolve in it. Therefore, it is necessary to add another component to the water and SML in order to obtain a stable solution. The component selected was ESMIS, because it exhibits very high water solubility and forms solutions in combination with SML.

The choice of ethoxylated sorbitan monoisostearate (ESMIS) and sorbitan monolaurate (SML) can be justified by the fact that they do not contain unsaturated bonds susceptible to oxidation. Besides, they are liquid at room temperature. Such a molecular structure facilitates their utilization, as they can be supplied to customers in the form of easily solubilized liquids.

From the viewpoint of fundamental research, pure chemical entities should be used as additives. However, taking into account the application of the results obtained, the tendency was to use low-cost lubricating substances having certain performance characteristics. Therefore, the decision was made to use products manufactured on a large industrial scale. Such products are mixtures of compounds of similar structure. This results from the fact that the reactants in the case of such compounds are, for example, vegetable oils whose composition can vary. However, such an approach to investigations requires each time a detailed physicochemical analysis of the compositions obtained.

Physicochemical and tribological test results for lubricants composed of water and SML/ESMIS mixtures are presented in sections 18.4 and 18.5 of this chapter, respectively. Particular attention was paid to the effect of the composition of the mixture on reducing resistance to motion and prevention of wear and seizure of the mating elements. An attempt was made to show that proper selection of the composition of the surfactants in a lubricating substance makes it possible to obtain lubricants capable of forming a stable adsorbed film during friction.

18.4 PHYSICOCHEMICAL PROPERTIES OF AQUEOUS SOLUTIONS OF SML/ESMIS MIXTURES

Physicochemical studies were carried out for aqueous solutions of SML/ESMIS mixtures with the following ratios: 1:9, 3:7, 5:5, 7:3, and for an aqueous solution of ESMIS (SML/ESMIS ratio = 0:10). The ratios stand for the weight ratio of the components. The selection of SML/ESMIS ratios was determined by the water solubilities of the mixtures obtained. The compositions with a large content of SML did not dissolve in water; therefore, they were not examined.

Aqueous solutions of sorbitan-ester/ethoxylated-sorbitan-ester mixtures are an interesting research topic. Both SML and ESMIS belong to a group of nonionic surfactants. As such, they will demonstrate an ability to migrate from the bulk phase toward the surface phase and then they will undergo adsorption at the interface. This results in a decrease in interfacial tension. The behavior of SML/ESMIS mixtures in the bulk phase is also interesting. Various types of solutions can be obtained, depending on the SML/ESMIS ratio and on the mixture concentration. Monomers are observed in the solutions at the lowest concentrations. An increase in the content of a surfactant in the solution means that, apart from monomers, there will also be micelles in the solution. Whereas at concentrations on the order of 40%–80% in the bulk phase, micelles may form a certain ordering, producing various mesophases (lyotropic liquid crystals) [122–125].

As commercial products will be used in the tribological experiment, a number of physicochemical tests were conducted to confirm the presence of various types of solutions. They will form the basis for interpretation of the tribological test results.

The first test was an assessment of surface activity of SML/ESMIS mixtures. Surface tension (σ) measurements of their aqueous solutions at concentrations of 0.001%, 0.01%, 0.1%, 1%, 2%, and 4 wt% were made. A TD1 Lauda tensiometer was used. The test was carried out at 20°C. The dependences of surface tension of aqueous SML/ESMIS solutions vs. concentration are presented in fig. 18.2.

The results obtained confirm that the addition of SML/ESMIS mixtures to water causes a reduction in surface tension. For each of the mixtures tested, increasing the concentration in the range of 0.001% to 1% resulted in a linear decrease in σ. No significant changes in the value of σ were observed at concentrations of 1%, 2%,

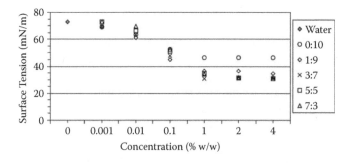

FIGURE 18.2 Surface tension vs. concentration for aqueous solutions of SML/ESMIS mixtures (0:10, 1:9, 3:7, 5:5, and 7:3).

and 4%. The ESMIS solutions (0:10) showed the poorest ability to reduce surface tension. A σ value of 46 mN/m was obtained for the highest analyzed concentration. The use of SML/ESMIS 1:9 mixture resulted in a considerable reduction in surface tension to a value of about 36 mN/m. Increasing the SML content in the mixtures led to a further reduction in surface tension. In the case of 3:7, 5:5, and 7:3 mixtures at concentrations of 1%, 2%, and 4%, the σ values ranged from 31 to 34 mN/m.

The results obtained also confirm the higher surface activity of SML. Increasing the share of SML (with very low affinity for water) in mixtures causes a more effective reduction in σ in comparison to mixtures containing ESMIS alone. Stabilization of the surface tension value at around 1% concentration indicates that micelles begin to appear in solutions at this concentration.

Determination of the effect of the SML/ESMIS ratio on the formation of microemulsions was the next test to confirm the ability of the mixtures to reduce interfacial tension. The following model systems were used in the study: water–kerosene–SML/ESMIS and water–vaseline oil–SML/ESMIS. Compositions containing a hydrophobic solvent (kerosene or vaseline oil) and an SML/ESMIS mixture were prepared at the weight ratios of 1:1 and 1:3 by adding portions of water to the resulting mixtures. Observations were carried out to find the water content at which microemulsions transformed into an emulsion. The test was carried out at 20°C. The results obtained can be seen in figs. 18.3 and 18.4.

It was found that microemulsions formed much more readily in the case of a surfactant-mixture/hydrophobic-solvent composition with the 3:1 ratio. Besides, in both systems containing kerosene and in those containing vaseline oil, the largest share of water that did not lead to a transition of microemulsions into an emulsion occurred for an SML/ESMIS mixture with the 3:7 ratio. The fact that a maximum can be observed on the dependences obtained is also important. This shows that the most beneficial arrangement of molecules of both surfactants at the oil/water interface occurs only at a specific SML/ESMIS ratio.

Viscosity is an important quantity from the viewpoint of application of aqueous solutions of mixtures to lubrication technology. The effects of the concentration,

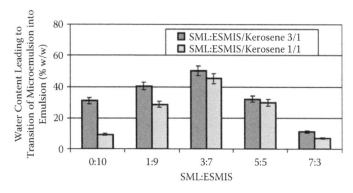

FIGURE 18.3 Effect of SML/ESMIS ratio on microemulsion-forming ability in the system: (SML/ESMIS)/kerosene/water. The following SML/ESMIS ratios were investigated: 0:10, 1:9, 3:7, 5:5, and 7:3.

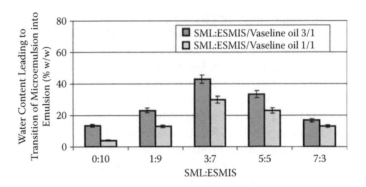

FIGURE 18.4 Effect of SML/ESMIS ratio on microemulsion-forming ability in the system: (SML/ESMIS)/vaseline oil/water. The following SML/ESMIS ratios were investigated: 0:10, 1:9, 3:7, 5:5, and 7:3.

SML/ESMIS ratio, and temperature on the kinematic viscosity of aqueous solutions of specific mixtures were analyzed. An Ubbelohde capillary viscometer was used in the tests. The measurements were carried out at temperatures from 25°C to 60°C (every 5°C). The concentrations of the solutions examined were 0.001%, 0.01%, 0.1%, 1%, 2%, and 4 wt%. The results obtained are given in table 18.1 and figs. 18.5–18.7.

Two tendencies concerning changes were observed in the case of solutions of all the mixtures. There were no visible changes in the value of the kinematic viscosity in the 0.001% to 0.1% concentration range. However, there was an increase in viscosity starting at concentration of 1% (table 18.1, fig. 18.5). The extent of these changes depends on the SML/ESMIS ratio (table 18.1, fig. 18.6). A relatively significant increase in kinematic viscosity was observed with an increase in the ratio of SML in the SML/ESMIS mixture. As an example, at 25°C, the kinematic viscosity of a 4% aqueous solution of a 7:3 mixture was about 60% higher than the one for a 4% solution of the 1:9 mixture. The results obtained indicate that it is possible to affect the kinematic viscosity of a lubricating substance by selecting a suitable SML/ESMIS ratio. In a range of relatively low concentrations (up to 4%), it is possible to obtain kinematic viscosity that is as much as two times higher than the one for pure water.

The changes in kinematic viscosity as a function of concentration indirectly prove the presence of micelles in solutions of SML/ESMIS mixtures. A noticeable increase in kinematic viscosity relative to water was not observed until the concentration reached 1%. In combination with the data from the surface-tension measurements, the CMC value can be expected around this concentration. It should also be added that there may be problems with an accurate determination of the CMC value when commercial products are used in the tests. This kind of material may contain unreacted reactants, e.g., fatty acids. As a result, changes in the quantities being measured are not very pronounced. In this case, it seems justified to quote a range or approximate concentration values at which CMC occurs.

A relatively high effect of temperature was also discovered on the basis of measurements of kinematic viscosity of aqueous solutions of SML/ESMIS mixtures (table 18.1, fig. 18.7). Kinematic viscosity decreased by about 50% in all cases from 25°C to 60°C.

TABLE 18.1

Kinematic Viscosity for Aqueous Solutions of SML/ESMIS Mixtures at Various Temperatures

Concentration (%w/w)	Kinematic viscosity (mm²/s)							
	25°C	30°C	35°C	40°C	45°C	50°C	55°C	60°C
Water	0.89	0.80	0.73	0.66	0.61	0.56	0.52	0.48
SML/ESMIS 0:10								
0.001	0.90	0.79	0.73	0.66	0.61	0.56	0.52	0.49
0.01	0.90	0.80	0.73	0.66	0.61	0.56	0.52	0.49
0.1	0.89	0.81	0.74	0.66	0.61	0.56	0.52	0.49
1	0.94	0.84	0.76	0.69	0.65	0.59	0.56	0.51
2	0.98	0.89	0.81	0.74	0.69	0.63	0.57	0.55
4	1.14	1.02	0.92	0.83	0.78	0.72	0.66	0.62
SML/ESMIS 1:9								
0.001	0.87	0.79	0.73	0.66	0.61	0.56	0.52	0.48
0.01	0.90	0.80	0.73	0.66	0.61	0.56	0.52	0.48
0.1	0.87	0.79	0.72	0.66	0.61	0.56	0.52	0.49
1	0.94	0.84	0.77	0.71	0.64	0.59	0.55	0.51
2	1.01	0.92	0.83	0.76	0.71	0.64	0.59	0.57
4	1.15	1.04	0.97	0.88	0.84	0.77	0.72	0.70
SML/ESMIS 3:7								
0.001	0.87	0.78	0.73	0.65	0.60	0.55	0.51	0.48
0.01	0.87	0.78	0.72	0.65	0.60	0.55	0.51	0.48
0.1	0.87	0.78	0.73	0.65	0.61	0.56	0.51	0.48
1	0.94	0.85	0.80	0.73	0.67	0.61	0.55	0.52
2	1.07	0.96	0.86	0.78	0.72	0.68	0.62	0.60
4	1.23	1.11	1.01	0.93	0.85	0.81	0.75	0.72
SML/ESMIS 5:5								
0.001	0.87	0.78	0.72	0.65	0.62	0.57	0.51	0.47
0.01	0.87	0.79	0.72	0.65	0.63	0.58	0.52	0.49
0.1	0.87	0.78	0.72	0.66	0.61	0.56	0.52	0.49
1	0.95	0.86	0.77	0.72	0.65	0.61	0.56	0.52
2	1.05	0.94	0.86	0.78	0.72	0.66	0.63	0.59
4	1.27	1.11	1.02	0.95	0.88	0.84	0.77	0.71
SML/ESMIS 7:3								
0.001	0.86	0.77	0.71	0.64	0.59	0.55	0.50	0.47
0.01	0.86	0.77	0.69	0.65	0.59	0.55	0.50	0.47
0.1	0.90	0.79	0.72	0.65	0.60	0.55	0.51	0.48
1	0.99	0.90	0.81	0.68	0.67	0.63	0.57	0.53
2	1.17	1.06	0.95	0.87	0.79	0.69	0.62	0.57
4	1.87	1.67	1.49	1.30	1.14	1.01	0.86	0.76

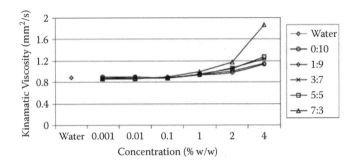

FIGURE 18.5 Dependence of kinematic viscosity on concentration of aqueous solutions of SML/ESMIS mixtures (25°C). The following SML/ESMIS ratios were investigated: 0:10, 1:9, 3:7, 5:5 and 7:3.

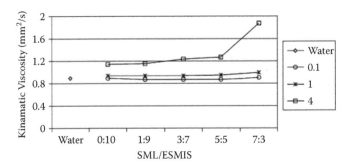

FIGURE 18.6 Dependence of kinematic viscosity on SML/ESMIS ratio for 0.1%, 1%, and 4% solutions (25°C).

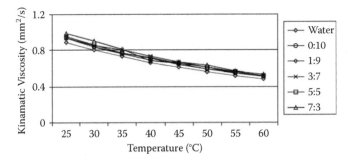

FIGURE 18.7 Dependence of kinematic viscosity on temperature for 1% aqueous solutions of SML/ESMIS mixtures. The following SML/ESMIS ratios were investigated: 0:10, 1:9, 3:7, 5:5 and 7:3.

It has been assumed in this chapter that lubricant compositions will contain relatively low concentrations of additives, only up to a few percent. However, properties of aqueous solutions of SML/ESMIS mixtures at concentrations of about 40%–80% seem to be interesting. Liquid crystalline structures may form in such solutions.

To confirm the presence of mesophases, dynamic viscosity measurements were made and microscopic photographs in polarized light were taken for aqueous solutions of SML/ESMIS mixtures at the concentrations of 10%, 20%, 30%, 40%, 50%, 60%, 70%, 80%, and 90 wt%.

A Brookfield RV DVI+ viscometer was used for the viscosity measurements. The measurements were carried out at 25°C and at a rotational speed of the spindle of 100 rpm. Some of the dependences obtained are presented in fig. 18.8.

The observed course of viscosity change as a function of concentration is characteristic for aqueous solutions of compounds capable of producing mesophases. It shows that the existence of mesophases can be expected in the case of solutions of the SML/ESMIS 1:9 mixture in the 40%–80% concentration range, and in the case of the 3:7 mixture in the range of 50%–80%.

Microscopic photographs in polarized light were taken in the concentration ranges given above. A Polar polarizing microscope manufactured by PZO Warszawa (Poland) and equipped with a digital camera was used. The measurements were conducted at 25°C and the magnification used was 150×.

It has been found that lyotropic liquid crystals form in the solutions, depending on their concentration and on the SML/ESMIS ratio. The liquid crystals have lamellar and hexagonal structures. Some of the photographs taken are shown in fig. 18.9.

Another extremely interesting aspect resulting from the investigation of highly concentrated solutions is the possibility of easily controlling lubricant viscosity only

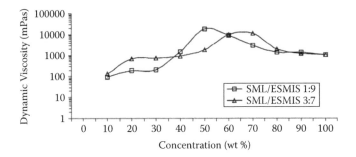

FIGURE 18.8 Dependence of dynamic viscosity on concentration of aqueous solutions of SML/ESMIS mixtures with 1:9 and 3:7 ratios (Brookfield RV DVI+, temperature 25°C).

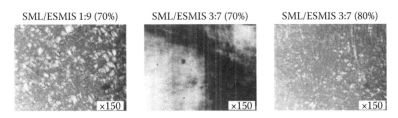

FIGURE 18.9 Microscope photographs in polarized light of aqueous solutions of SML/ESMIS mixtures with 1:9 (70%) and 3:7 (70% and 80%) ratios; magnification 150×; measurement temperature 25°C.

by increasing the concentration of additives without the need to add other viscosity modifiers. The solutions with a viscosity maximum are of particular interest. The reason why such systems are characterized by high viscosities is the presence of cylindrical micelles whose length frequently reaches as much as 0.2 μm. They can form a certain order, but can also become intertwined. The structure then produced resembles a polymer, with a polymer chain being replaced by a micelle composed of several hundred to several thousand surfactant molecules [126–129].

18.5 TRIBOLOGICAL INVESTIGATIONS OF AQUEOUS SOLUTIONS OF SML/ESMIS MIXTURES

Aqueous solutions of 1:9, 3:7, 5:5, and 7:3 SML/ESMIS (sorbitan monolaurate/ethoxylated sorbitan monoisostearate) mixtures and aqueous ESMIS solutions (SML/ESMIS = 0:10) were used as model lubricants. Two devices were used to carry out the tribological studies:

- T-11 tribometer (pin-on-disc contact)
- T-02 four-ball tester

18.5.1 T-11 TRIBOMETER (PIN-ON-DISC)

The tests were conducted on aqueous solutions of SML/ESMIS mixtures at the concentrations of 0.01%, 0.1%, 1%, and 4 wt%. A T-11 tribometer (pin-on-disc) produced by ITeE Poland (Institute for Sustainable Technologies, National Research Institute, Radom, Poland) was used in the tests. The pins used were 3.0 mm in diameter and were made of LH15 steel or polyamide-6, while the discs were 25 mm in diameter and were made of polyamide-6. Before the tests, all components of the friction couples were thoroughly chemically cleaned. An ultrasonic cleaner was employed. The steel elements were cleaned in n-hexane, acetone, ethyl alcohol, and distilled water. The polyamide-6 elements were cleaned in ethyl alcohol and distilled water. After the cleaning process all friction-couple components were dried (50°C, 30 min). The time of the test run was 900 s, the friction-couple load was 10 N, and the sliding velocity was 0.1 m/s. The coefficient of friction was calculated from friction force measurements using

$$\mu = \frac{F_T}{P} \tag{18.1}$$

where
F_T = friction force (N)
P = load (N)

In the diagrams illustrating friction coefficient changes with time (figs. 18.10 and 18.15), the points on the graphs are friction coefficients averaged in 30-s intervals. Figures 18.11–18.14 and figs. 18.16–18.19 show the mean friction coefficient values from the complete 900-s test and three independent tests. The errors were evaluated using Student's t-test at a confidence level of 0.90.

18.5.1.1 The Couple: Steel-Pin–Polyamide-Disc

The results obtained from tests using a steel-pin–polyamide-disc couple are given in figs. 18.10–18.14.

Figure 18.10 presents examples of dependences of friction coefficient changes as a function of time for water and 0.01% aqueous solutions of SML/ESMIS mixtures. The highest friction coefficient values were observed for the tests in which pure water was the lubricant. Over the test duration, the coefficient of friction increased from a value of 0.2 to about 0.3. Over the course of measurement, in the case of all the solutions containing the SML/ESMIS mixtures, only a slight reduction in resistance to motion occurred, and no changes in friction coefficient values were observed. However, the compositions analyzed differed in their ability to reduce μ. A set of mean friction coefficient values from the whole 900-s test can be seen in fig. 18.11.

The friction coefficient value for water was about 0.3. For a solution containing ESMIS (SML/ESMIS 0:10), the measured value dropped to 0.2. Increasing the SML content in the mixture lead to a decrease in the parameter tested: for the 1:9 and 3:7 mixtures, the μ values were 0.14 and 0.09, respectively. In the case of the 5:5 mixture, there was a slight increase in the coefficient of friction up to 0.1. A change in the mixture ratio to 7:3 resulted in an increase in μ of up to 0.18.

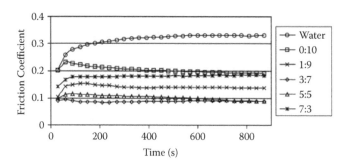

FIGURE 18.10 Examples of dependences of friction coefficient changes on time for a steel-pin–polyamide-disc pair for water and 0.01% aqueous solutions of SML/ESMIS mixtures. The following SML/ESMIS ratios were investigated: 0:10, 1:9, 3:7, 5:5, and 7:3.

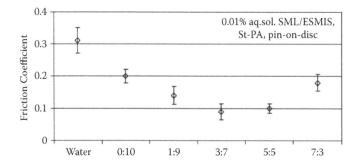

FIGURE 18.11 Mean friction coefficient values for a steel-pin–polyamide-disc pair for water and 0.01% aqueous solutions of SML/ESMIS mixtures. The following SML/ESMIS ratios were investigated: 0:10, 1:9, 3:7, 5:5, and 7:3.

Tests were then carried out for 0.1% solutions. The results are presented in fig. 18.12. The results obtained for 0:10 and 1:9 SML/ESMIS mixtures show similar values of the coefficient of friction ranging from 0.13 to 0.16. The highest resistance to motion was observed for the composition containing the highest share of SML (7:3); the mean μ value equaled 0.24.

A relatively high scatter in mean values of μ was observed when 1% aqueous solutions of SML/ESMIS mixtures were examined (fig. 18.13). A friction coefficient value of 0.22 was obtained for the composition containing only ESMIS (0:10). Then the μ value decreased to 0.2 for a 1:9 mixture. The lowest resistance to motion (μ = 0.1) was observed for the 3:7 ratio. The solutions of the other two mixtures with 5:5 and 7:3 SML/ESMIS ratios exhibited higher resistances to motion; the μ values obtained equaled 0.16 and 0.22, respectively.

Figure 18.14 shows the friction coefficient values obtained in tests carried out in the presence of 4% aqueous solutions of SML/ESMIS mixtures. No significant differences in friction coefficient values were observed; for solutions of all the mixtures, the values ranged from 0.16 to 0.18.

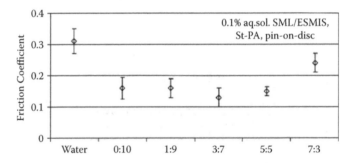

FIGURE 18.12 Mean friction coefficient values for a steel-pin–polyamide-disc pair for water and 0.1% aqueous solutions of SML/ESMIS mixtures. The following SML/ESMIS ratios were investigated: 0:10, 1:9, 3:7, 5:5, and 7:3.

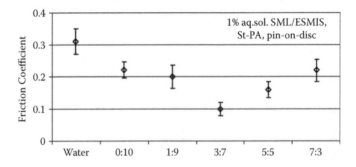

FIGURE 18.13 Mean friction coefficient values for a steel-pin–polyamide-disc pair for water and 1% aqueous solutions of SML/ESMIS mixtures. The following SML/ESMIS ratios were investigated: 0:10, 1:9, 3:7, 5:5, and 7:3.

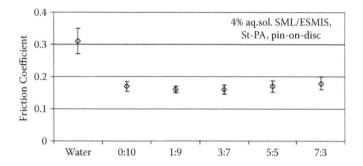

FIGURE 18.14   Mean friction coefficient values for a steel-pin–polyamide-disc pair for water and 4% aqueous solutions of SML/ESMIS mixtures. The following SML/ESMIS ratios were investigated: 0:10, 1:9, 3:7, 5:5, and 7:3.

18.5.1.2 The Couple: Polyamide-Pin–Polyamide-Disc

The test results for the polyamide-pin–polyamide-disc couple are presented in figs. 18.15–18.19.

Figure 18.15 shows examples of dependences of μ on time for water and 0.01% aqueous solutions of SML/ESMIS mixtures. In the case of water, during the course of the test the coefficient of friction increased slightly from the value of 0.12 to 0.16. The dependences obtained for solutions of all mixtures were similar: the μ values practically did not change during the course of the test.

For the polyamide–polyamide pair, the mean values of friction coefficients obtained for water and all the SML/ESMIS solutions studied are quite similar (figs. 18.16–18.19). The relatively low μ value (0.14) observed during tests in which pure water was the lubricant is worth noting. Addition of an SML/ESMIS mixture to water reduces the resistance to motion to some degree. Regardless of the SML/ESMIS ratio or mixture concentration, mean μ values were close to the value obtained for water and ranged from 0.09 to 0.16.

18.5.2 T-02 Tribometer (the Four-Ball Tester)

The tests were conducted for 1% aqueous solutions of SML/ESMIS mixtures. The T-02 tester was manufactured at the Institute for Sustainable Technologies–National Research Institute in Poland. The balls of ½-in. diameter were made of LH 15 bearing steel. Surface roughness was $R_a = 0.032$ μm. The device was employed to measure motion resistance, wear, and antiseizure abilities in the presence of a lubricant. The methodology of the tests is described in the literature [130]. The values of the quantities measured are arithmetic means of three independent measurements. The errors were evaluated with Student's t-test at a confidence level of 0.90.

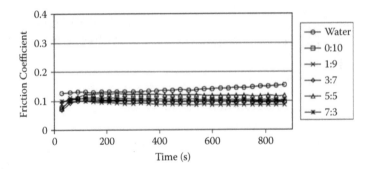

FIGURE 18.15 Examples of dependences of friction coefficient changes on time for a poly-amide-pin–polyamide-disc pair for water and 0.01% aqueous solutions of SML/ESMIS mixtures. The following SML/ESMIS ratios were investigated: 0:10, 1:9, 3:7, 5:5, and 7:3.

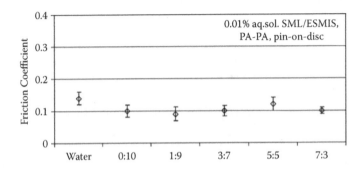

FIGURE 18.16 Mean friction coefficient values for a polyamide-pin–polyamide-disc pair for water and 0.01% aqueous solutions of SML/ESMIS mixtures. The following SML/ESMIS ratios were investigated: 0:10, 1:9, 3:7, 5:5, and 7:3.

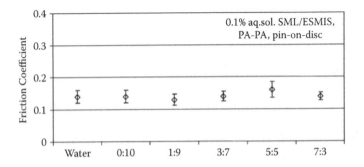

FIGURE 18.17 Mean friction coefficient values for a polyamide-pin–polyamide-disc pair for water and 0.1% aqueous solutions of SML/ESMIS mixtures. The following SML/ESMIS ratios were investigated: 0:10, 1:9, 3:7, 5:5, and 7:3.

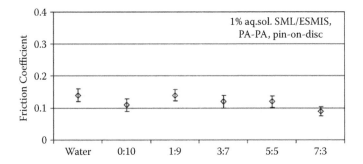

FIGURE 18.18 Mean friction coefficient values for a polyamide-pin–polyamide-disc pair for water and 1% aqueous solutions of SML/ESMIS mixtures. The following SML/ESMIS ratios were investigated: 0:10, 1:9, 3:7, 5:5, and 7:3.

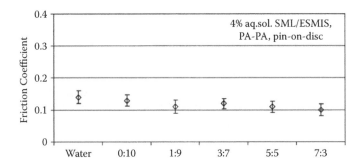

FIGURE 18.19 Mean friction coefficient values for a polyamide-pin–polyamide-disc pair for water and 4% aqueous solutions of SML/ESMIS mixtures. The following SML/ESMIS ratios were investigated: 0:10, 1:9, 3:7, 5:5, and 3:7.

18.5.2.1 Tests under Constant Load

The influence of aqueous solutions of SML/ESMIS mixtures on the coefficients of friction, wear, and temperature in the friction pair was evaluated under the following conditions: rotational speed of the spindle, 200 ± 20 rpm; constant applied load, 3.0 kN (in the case of water, the tests were conducted at the load of 2.0 kN); test duration, 900 s. Data were automatically acquired every second.

The friction coefficient (μ) was evaluated from the friction torque measurement using

$$\mu = 222.47 \frac{M_T}{P} \tag{18.2}$$

where
M_T = friction torque (N·m)
P = load (N)

The coefficient 222.47 is imposed by forces acting in the four-ball tester.

Changes in the coefficient of friction as a function of time observed for 1% aqueous solutions of SML/ESMIS mixtures are shown in fig. 18.20. Out of the lubricant compositions tested, the highest resistance to motion was observed for water. Its lubricating properties were so poor that, just a few seconds after starting the tester under the load of 3.0 kN, welding of the balls took place. Therefore, tests with pure water were carried out at a load of 2.0 kN. Under these conditions, the coefficient of friction was relatively high, fluctuating around 0.3.

The addition of SML/ESMIS mixtures to water significantly improved its lubricity. At a load of 3.0 kN, seizure of balls was not observed. In the case of solutions of all the mixtures analyzed, a similar behavior of friction coefficient changes as a function of time was observed. In the initial phase of the test, the μ values were the highest (about 0.2–0.3). Then, after about 100 s, the μ value decreased to about 0.1. It is interesting that in the case of SML/ESMIS mixtures with the 3:7, 5:5, and 7:3 ratios, the friction coefficient value was practically constant after the 100th second of the test. There were fluctuations in the μ values for the other compositions (0:10 and 9:1).

When analyzing the mean values of μ (fig. 18.21) it can be seen that resistance to motion decreases with an increased share of SML in the mixture. However, the differences in the mean μ values are small. The μ value for an ESMIS solution (0:10)

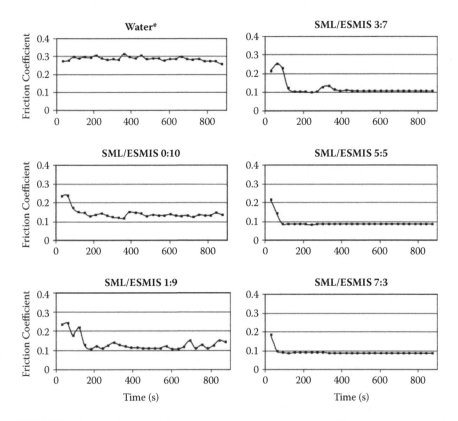

FIGURE 18.20 Friction coefficient changes vs. time for water and 1% aqueous solutions of SML/ESMIS mixtures at a load of 3000 N (four-ball tester). (*Note*: Test for water at 2000 N.)

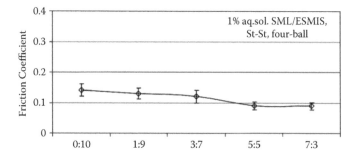

FIGURE 18.21 Effect of SML/ESMIS mixture ratio on mean friction coefficient value (1% solutions) using a four-ball tester at a friction-pair load of 3000 N. The following SML/ESMIS ratios were investigated: 0:10, 1:9, 3:7, 5:5, and 7:3.

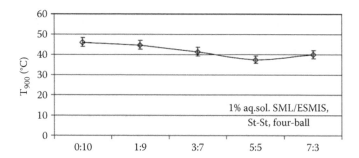

FIGURE 18.22 Effect of SML/ESMIS mixture ratio on lubricant temperature after a 900-s test (1% solutions) using a four-ball tester at a friction-pair load of 3000 N. The following SML/ESMIS ratios were investigated: 0:10, 1:9, 3:7, 5:5, and 7:3.

was 0.14, whereas it was about 0.09 for a solution of the SML/ESMIS mixture with the 7:3 ratio.

Temperature measurements were carried out during each test. A thermocouple used was in direct contact with the lubricating substance. Figure 18.22 shows the temperatures of the lubricants immediately after the tests.

One can observe a minimum in the dependence obtained (fig. 18.22). The 1% aqueous solutions of the 5:5 mixture had the lowest temperature (37.4°C). Slightly higher temperature (40°C) was measured for a solution of the 7:3 mixture. A reduction in the share of SML in the mixtures increased the temperature at the end of the friction tests. The values obtained for the solutions of 3:7, 1:9, and 0:10 mixtures were 41.4°C, 44.8°C, and 46.2°C, respectively.

To assess the antiwear properties of the solutions tested, the wear-scar diameter of the balls was measured after each experiment. The measurements were carried out in two directions: parallel and perpendicular to the motion. The average of the two values was taken as the estimate of the parameter. The results obtained are shown in fig. 18.23. Moreover, a wear-scar profile was analyzed after each test. A TOPO 01 profilometer produced by IZTW Kraków (Institute of Advanced Manufacturing Technology, Krakow, Poland) was used in the tests, in which a needle scanning at a speed of 0.5 m/s covered a measuring distance of 2.5 mm. The results obtained are presented in fig. 18.24.

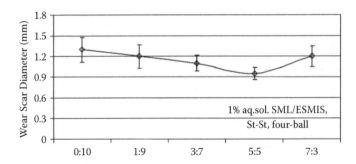

FIGURE 18.23 Wear-scar diameters after four-ball tests at a load of 3000 N in the presence of 1% aqueous solutions of SML/ESMIS mixtures. The following SML/ESMIS ratios were investigated: 0:10, 1:9, 3:7, 5:5, and 7:3.

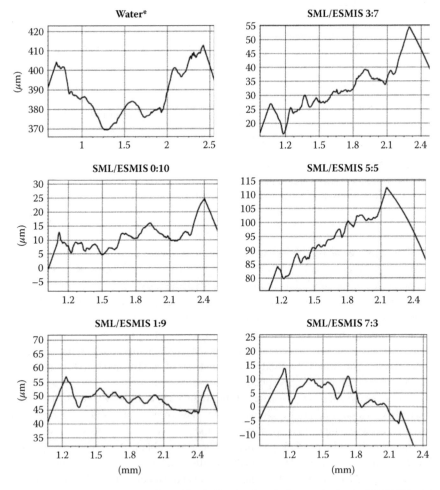

FIGURE 18.24 Wear-scar profiles of the balls after four-ball tests at a load of 3000 N in the presence of 1% aqueous solutions of SML/ESMIS mixtures. (*Note*: Tests for water at 2000 N.)

The character of changes in wear of the balls as a function of the SML/ESMIS ratio (fig. 18.23) is similar to the one observed for temperature (fig. 18.22). The lowest wear-scar value (0.95 mm) was obtained for the tests in the presence of a 1% aqueous solution of the SML/ESMIS mixture with the 5:5 ratio. Relatively low wear scar was also found for the 7:3 ratio. In the case of the other mixtures, wear scars of the balls ranged from 1.2 mm to 1.3 mm. In the tests for water (friction-pair load of 2.0 kN), the wear-scar diameter was 1.8 mm.

Figure 18.24 shows wear-scar profiles of the balls obtained after tests. The results obtained indicate a similar kind of wear behavior during tests carried out in the presence of solutions of individual mixtures.

18.5.2.2 Tests under Variable Loads

In order to assess antiseizure properties, tests were performed at a linearly increasing load [130]. During the measurements, the load increased from zero to 8000 N at a rate of 409 N/s, and the rotational velocity was 500 rpm. After each test, wear was evaluated as an average of a wear-scar diameter measured in two directions: parallel and perpendicular to the friction direction, d. Measurements of wear-scar profiles were also made in each case.

The schematic of the dependence of friction torque on the applied load is presented in fig. 18.25. The following denotations were adopted:

- P_t: scuffing load, in N, at which the boundary layer is broken; it can be characterized by a rapid increase in friction torque
- P_{oz}: seizure load, in N, which corresponds to a friction torque equal to 10 N·m

Dependences of friction torque on time are presented in fig. 18.26, and the mean values of scuffing load, wear-scar diameters of the balls, and wear-scar profiles obtained after the tests in the presence of 1% aqueous solutions of SML/ESMIS mixtures are shown in figs. 18.27–18.29.

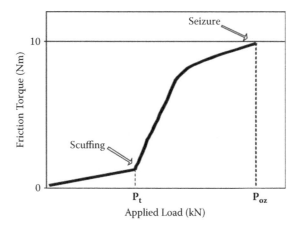

FIGURE 18.25 Simplified dependence of friction torque on load.

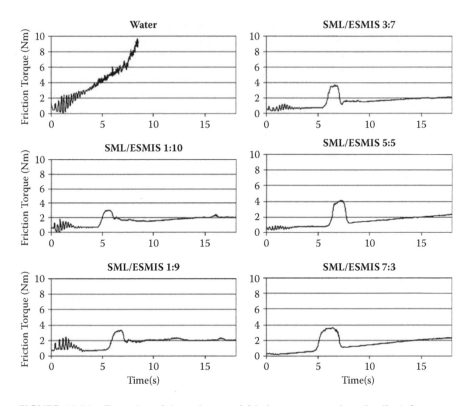

FIGURE 18.26 Examples of dependences of friction torque on time (loading) for water and 1% aqueous solutions of SML/ESMIS mixtures (four-ball tester, rotational speed of the spindle 500 rpm, load increase rate 409 N/s).

It was observed in the case of water that, almost immediately after starting the test, there occurred a sudden increase in friction torque (fig. 18.26). This resulted in a rapid seizure of friction-pair elements. The value of scuffing load P_t was estimated to be at a level of 200 N, while seizure load (P_{oz}) was about 3200 N. The wear-scar diameter of the balls after the tests was 2.6 mm.

Aqueous solutions (1%) of SML/ESMIS mixtures exhibited entirely different antiseizure properties than water. For SML/ESMIS mixtures, the obtained dependences of friction torque on time (a change in time is equivalent to increasing load) had a similar course. No sudden increase in friction torque was observed within the first seconds of the test. This shows that a relatively stable adsorbed film that effectively reduced the motion resistance was produced on the mating elements. Scuffing load was estimated from the dependences obtained (fig. 18.26). The results obtained are presented in fig. 18.27. The highest P_t values were found for 1% aqueous solutions of mixtures with the 5:5, 3:7, and 1:9 ratios. These equaled, respectively, 2500 N, 2350 N, and 2300 N. Lower P_t values were obtained for the other compositions. For a solution containing the 7:3 SML/ESMIS mixture, the P_t value was 2000 N, while for the ESMIS solution (0:10) the value was 1900 N.

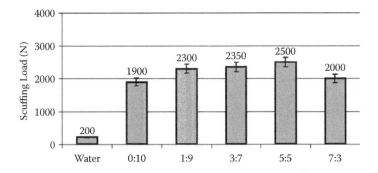

FIGURE 18.27 Dependence of scuffing load (P_t) on SML/ESMIS ratio in 1% aqueous solutions. The following SML/ESMIS ratios were investigated: 0:10, 1:9, 3:7, 5:5, and 7:3.

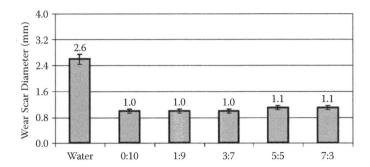

FIGURE 18.28 Wear-scar diameters after four-ball tests at a linearly increasing load (409 N/s) in the presence of water and 1% aqueous solutions of SML/ESMIS mixtures. The following SML/ESMIS ratios were investigated: 0:10, 1:9, 3:7, 5:5, and 7:3.

It was also interesting to observe that no seizure occurred in subsequent stages of the tests. Increasing the load of a friction pair resulted, admittedly, in an expected increase in friction torque, but after 2 or 3 s, resistance to motion decreased again. Seizure of the balls was not observed before the test ended (up to 8000 N). In such cases, according to the accepted methodology, the seizure load assumes the conventional value of 8000 N.

A characteristic "bulge" can be seen in the dependences obtained for the solutions of individual mixtures. One might expect that the larger the area under the "bulge," the higher would be the wear of friction-pair elements. However, the obtained values of wear-scar diameters were similar (fig. 18.28).

Profilograms of wear scars of the balls provided important information on wear intensity (fig. 18.29). Individual scars differed considerably in their depths. The highest depth was observed after tests in the presence of aqueous solutions of mixtures with the ratios of 5:5 and 7:3. Furthermore, the surface of the scars was very smooth. A reduction in the SML content in the SML/ESMIS mixture caused the scars to be less deep but more rough. The smallest depth of the scars was observed after tests in the presence of ESMIS (0:10) alone.

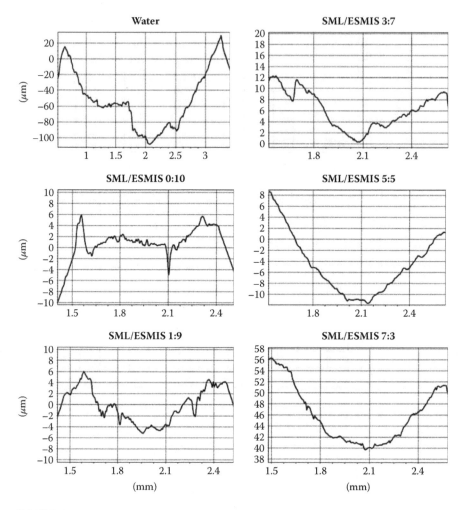

FIGURE 18.29 Wear-scar profiles of the balls after four-ball tests at a linearly increasing load (409 N/s) in the presence of water and 1% aqueous solutions of SML/ESMIS mixtures.

18.6 DISCUSSION

In this chapter, aqueous solutions of SML/ESMIS mixtures were proposed as model lubricants. They are environmentally friendly. Particular attention was paid to the proper selection of the SML/ESMIS ratio, which is essential for high efficiency in reducing resistance to motion as well as preventing wear and seizure of friction-couple elements.

Physicochemical studies have confirmed that the mixtures used in the tests exhibit surface activity (fig. 18.2). The compositions with a significant proportion of SML are more capable of reducing interfacial tension. However, the results of the tests in which the ability to form microemulsions was assessed indicate that the 3:7 and 5:5 compositions of the mixtures are the optimal ones (figs. 18.3 and 18.4).

The importance of the composition of the mixtures defined by the weight ratio of the esters is clear from the viewpoint of application of the mixtures in tribology. An attempt at a molecular interpretation requires an estimation of molar proportions. Taking into account molecular masses of the esters: $M_{SML} = 346$ g/mole, $M_{ESMIS} = 1310$ g/mole, and including the solutions containing pure ESMIS as an additive (SML/ESMIS ratio 0:10), the weight ratios correspond to the following molar ratios: 1:9 → 0.42:1, 3:7 → 1.6:1, 5:5 → 3.8:1, and 7:3 → 8.8:1.

Thus, despite the fact that SML/ESMIS weight ratios indicate a comparable or predominant proportion of the ethoxylated form (ESMIS) in the mixture, there are, in fact, more SML molecules in the solutions in both cases. It should also be assumed that the composition of the surface phase will be comparable to the composition of the bulk phase or that there will be more SML molecules in it. This results from the fact that, compared with ESMIS, an SML molecule has a considerably smaller hydrophilic part (fig. 18.1) and, therefore, it will much more readily leave an aqueous solution.

The fact that increasing the SML proportion in the mixture does not result in easier production of microemulsions is an extremely interesting observation. This may show that, in the case of the mixtures tested, there exists an optimal SML/ESMIS ratio at which interactions of surfactant molecules in the adsorbed film are the strongest. This directly results in the stability of such a film. It seems that one of the models suggested in the literature [113, 115, 118] can be applied to describe the ordering of molecules of the SML/ESMIS mixture at the interface.

The physicochemical test results presented in this chapter also indicate that micelles form in aqueous solutions of SML/ESMIS mixtures when the concentration is around 1% (figs. 18.2 and 18.5). The presence of aggregates in the bulk phase increases the viscosity of the compositions (figs. 18.5–18.7, table 18.1). The possibility of producing liquid crystalline structures in aqueous solutions of SML/ESMIS mixtures is a consequence of micelle formation (figs. 18.8 and 18.9).

One of the essential properties of lubricating substances should be their ability to reduce resistance to motion. A pin-on-disc device was used to carry out tribological tests employing steel–polyamide and polyamide–polyamide friction pairs. In the former case it was found that all the SML/ESMIS mixtures tested caused a reduction in motion resistance relative to pure water (figs. 18.10–18.14). An effect of the SML/ESMIS ratio on the coefficient of friction was observed at the lowest concentration examined (0.01%). The lowest resistances to motion were observed, as expected, for the mixtures with the 3:7 and 5:5 ratios. A similar effect was not observed at higher concentrations. With a few exceptions, all the mixtures had a similar friction coefficient.

The explanation for this phenomenon can be sought in the structure of the surface phase. The 0.01% concentration is much lower than the CMC value. The adsorbed film composed of surfactant molecules can be a monolayer or it may be quite thin. Furthermore, surface aggregates may not form in the surface phase at such a low concentration. The arrangement of individual molecules relative to each other may be very important in this type of layer. At appropriately high concentrations of additives in a lubricant composition, the surface phase is so thick that it efficiently protects the mating elements against direct contact, regardless of the type of SML/ESMIS mixture.

In the case of a polyamide–polyamide pair, no difference was found between the lubricating properties of water and those of aqueous solutions of SML/ESMIS mixtures (figs. 18.15–18.19). The reason for such behavior of the lubricants may be the relatively high wettability of polyamide by pure water. Similar results were also obtained in earlier investigations in which aqueous solutions of anionic surfactants were the lubricants [49].

The four-ball tests were carried out using a steel–steel pair. In this case, the addition of an SML/ESMIS mixture to water strongly reduced motion resistance and wear (figs. 18.20, 18.21, 18.23, and 18.24). An increased SML content in the mixture results in a decrease in μ. As far as wear is concerned, the lowest wear-scar diameter value was observed for the 5:5 mixture.

The SML/ESMIS mixture with the ratio of 5:5 turned out to be the most effective one in preventing seizure (figs. 18.26–18.29). Based on a test in which friction-couple load was increased linearly, it was found that the lubricant film broke only when the load reached 2500 N. These data confirm earlier assumptions that the selection of an appropriate ratio of a sorbitan ester/ethoxylated sorbitan ester composition is extremely important for the effective protection of friction couples against detrimental friction effects.

The experiments provided one more extremely valuable piece of information. None of the compositions analyzed allowed seizure of mating elements. This fact needs to be emphasized, as the conditions within a friction couple during such tests are extreme (rotational speed of the spindle 500 rpm, load increase 409 N/s, maximum loading of friction couple 8000 N). Many of the commonly used lubricants based on conventional bases and additives cannot withstand such severe experimental conditions [130]. A schematic of changes in friction torque for commercial products is shown in fig. 18.25.

In the case of aqueous solutions of SML/ESMIS mixtures, the dependence of friction torque on load is different. After exceeding the load at which the lubricant film breaks (P_t), relatively intense wear of the balls occurs. When the elements are not protected by a lubricant, adhesion bonding and seizure may easily occur. As the results obtained indicate (fig. 18.26), this is prevented by the action of aqueous solutions of the mixtures. After a short time, there is another increase in friction torque and, as one might expect, the lubricant film is reconstructed.

The profilograms of wear scars of the balls after the tests are extremely interesting (fig. 18.29). Although the wear-scar diameters are quite similar, the scars are very deep in the case of the 5:5 and 7:3 SML/ESMIS mixtures. The tendency is that the larger the SML proportion in the mixture, the deeper is the scar. The surface roughness of the scars is also important. The mixtures with the largest proportion of SML leave the surface relatively smooth, while in the case of the compositions with a high ESMIS content, the scars are less deep but their surface has a considerably higher roughness.

These results show that there may be differences in the ability to prevent wear under extreme loading conditions, depending on the SML/ESMIS ratio employed and, hence, on the stability of the adsorbed film. This behavior of the SML/ESMIS mixtures can be explained on the basis of the so-called Rehbinder effect [130–133].

According to this theory, surfactant molecules penetrate microirregularities and microcracks. Subsequently, as a result of adsorption inside such depressions, they exert pressure on them. This leads to the formation of localized stresses in the top layer of the friction-pair material. Under the influence of such stresses and also other factors, such as temperature, localized softening of the friction-pair surface occurs, and relatively easy wear of the material results from its loss of hardness. Referring to the results obtained, it seems that SML molecules will have a much higher capability of surface softening. This may be due to their relatively small size and much higher surface activity.

Another reason why SML/ESMIS mixtures with the ratios of 5:5 and 7:3 display a very good ability to soften surfaces is the possibility that there exist strong interactions between individual molecules in the films produced by the mixtures. As mentioned before, this may result from a high coefficient of packing. The effect of this may also be a higher probability of more complete filling of microcracks by surfactants.

It would be extremely interesting to compare the softening action of surfactant molecules having the same HLB value but of different sizes. However, this type of investigation was beyond the scope of this study.

The conclusion is that SML/ESMIS mixtures can be successfully used as antiseizure additives. However, there is a danger that such compounds may contribute to excessive wear of friction-couple elements under the conditions of high loads and high sliding velocities.

18.7 SUMMARY

The research results and discussion presented in this chapter clearly indicate that the issue of proper selection of surfactants for water-based lubricant compositions is extremely important. It has been demonstrated that a proper selection of the structure of surfactant compositions is extremely important for controlling the properties of the lubricants being developed.

The results obtained form the basis for a claim that, in the near future, water-based lubricating substances may eliminate most of the products currently in use. However, further research is necessary to gain insight into the action mechanisms of surfactants in friction pairs and then to apply the results of this research to practical applications.

ACKNOWLEDGMENT

This research was financed by the State Committee for Scientific Research, grant no. T08A-022-30. The paper was prepared when the author was a Foundation for Polish Science scholar.

REFERENCES

1. Spikes, H. A. 1997. *Lubrication Sci.* 9: 221–53.
2. Guangteng, G., Cann, P. M., Olver, A. V., and Spikes, H. A. 2000. *Tribol. Int.* 33: 183–89.
3. Ratoi, M., Anghel, V., Bovington, C., and Spikes, H. A. 2000. *Tribol. Int.* 33: 241–47.
4. Hsu, S. M., and Gates, R. S. 2005. *Tribol. Int.* 38: 305–12.
5. Hsu, S. M. 2004. *Tribol. Int.* 37: 553–59.
6. Hsu, S. M. 2004. *Tribol. Int.* 37: 537–45.

7. Zhang, Ch. 2005. *Tribol. Int.* 38: 443–48.
8. Zhang, Y. 2006. *Mol. Liquids* 128: 56–59.
9. Zhang, S., and Lan, H. 2002. *Tribol. Int.* 35: 321–27.
10. Matveevskii, R. M. 1995. *Tribol. Int.* 28: 51–54.
11. Yoshizawa, H., Chen, Y. L., and Israelachvili, J. 1993. *Wear* 168: 161–66.
12. Cho, Y. K., Cai, L., and Granick, S. 1997. *Tribol. Int.* 30: 889–94.
13. Kajdas, Cz. 2005. *Tribol. Int.* 38: 337–53.
14. Morina, A., Neville, A., Priest, M., and Green, J. H. 2006. *Tribol. Int.* 39: 1545–57.
15. Vicente, J., Stokes, J. R., and Spikes, H. A. 2006. *Food Hydrocolloids* 20: 483–91.
16. Choo, J. H., Spikes, H. A., Ratoi, M., Glovnea, R., and Forrest, A. 2007. *Tribol. Int.* 40: 54–159.
17. Gao, Ch., and Bhushan, B. 1995. *Wear* 190: 60–75.
18. Wu, Y. L., and Dacre, B. 1997. *Tribol. Int.* 30: 445–53.
19. Myshkin, N. K., Petrokovets, M. I., and Kovalev, A. V. 2005. *Tribol. Int.* 38: 910–21.
20. Marczak, R. 2002. *Tribologia* 183: 939–56 (in Polish).
21. Burakowski, T., and Wierzchon, T. 1999. *Surface engineering of metals principles, equipment, technologies.* Boca Raton, FL: CRC Press LLC.
22. Bartz, W. J. 1998. *Tribol. Int.* 31: 35–47.
23. Havet, L., Blouet, J., Robbe Valloire, F., Brasseur, E., and Slomka, D. 2001. *Wear* 248: 140–46.
24. Durak, E. 2004. *Industrial Lubrication Tribol.* 56: 23–37.
25. Wilson, B. 1998. *Industrial Lubrication Tribol.* 50: 6–15.
26. Biresaw, G., ed. 1990. *Tribology and the liquid-crystalline state.* Oxford: Oxford University Press.
27. Biresaw, G., Adhvaryu, A., and Erhan, S. Z. 2003. *Am. Oil Chemists Soc.* 80: 697–704.
28. Kurth, T. L., Biresaw, G., and Adhvaryu, A. 2005. *Am. Oil Chemists Soc.* 82: 293–99.
29. Kodali, D. 2002. *Ind. Lubr. Tribol.* 54: 165–70.
30. Feldman, D. G., and Kessler, M. 2002. *Ind. Lubr. Tribol.* 54: 117–29.
31. Qing-Ye, G., Lai-Gui, Y., and Cheng-Feng, Y. 2002. *Wear* 253: 558–62.
32. Jiusheng, L., Yanyan, Z., Tianhui, R., Weimin, L., and Xingguo, F. 2002. *Wear* 253: 720–24.
33. Krzan, B., and Vizintin, J. 2003. *Tribol. Int.* 36: 827–33.
34. Ratoi, M., and Spikes, H. A. 1999. *Tribol. Trans.* 42: 479–86.
35. Wei, J., and Xue, Q. 1993. *Wear* 162–164: 229–33.
36. Wei, J., and Xue, Q. 1996. *Wear* 199: 157–59.
37. Ono, H., Okabe, H., and Masuko, M. 1995. *Tribol. Trans.* 38: 693–99.
38. Boschkova, K., Kronberg, B., Rutland, M., and Imae, T. 2001. *Tribol. Int.* 34: 815–22.
39. Boschkova, K., Kronberg, B., Stalgren, J. J. R., Persson, K., and Ratoi-Salagean, M. 2002. *Langmuir* 18: 1680–87.
40. Boschkova, K., Feiler, A., Kronberg, B., and Stalgren, J. J. R. 2002. *Langmuir* 18: 7930–35.
41. Duffy, D. C., Friedmann, A., Boggis, A., and Klenerman, D. 1998. *Langmuir* 14: 6518–27.
42. Plaza, S., Margielewski, L., Celichowski, G., Wesolowski, R. W., and Stanecka, R. 2001. *Wear* 249: 1077–89.
43. Gao, Y., Jing, Y., Zhang, Z., Chen, G., and Xue, Q. 2002. *Wear* 253: 576–78.
44. Misra, S. K., and Skold, R. O. 2000. *Colloids Surfaces A* 170: 91–106.
45. Ratoi, M., Bovington, C., and Spikes, H. 2003. *Tribol. Lett.* 14: 33–40.
46. Sulek, M. W., and Wasilewski, T. 2005. *Tribol. Lett.* 18: 197–205.
47. Sulek, M. W., and Wasilewski, T. 2006. *Wear* 260: 193–204.
48. Sulek, M. W., and Wasilewski, T. 2002. *Appl. Mech. Eng.* 7: 189–95.

49. Sulek, M. W., and Wasilewski, T. 2006. *Tribologia* 208: 123–37 (in Polish).
50. Basu, B., Vleugels, J., and van der Biest, O. 2001. *Wear* 250: 631–41.
51. Chen, M., Kato, K., and Adachi, K. 2002. *Tribol. Int.* 35: 129–35.
52. Zhang, H., Hu, X., Yan, J., and Tang, S. 2005. *Materials Lett.* 59: 583–87.
53. Saito, T., Imada, Y., and Honda, F. 1999. *Wear* 236: 153–58.
54. Lee, S., and Spencer, N. D. 2005. *Tribol. Int.* 38: 922–30.
55. Zhu, C., Jacobs, O., Jaskulka, R., Koller, W., and Wu, W. 2004. *Polymer Testing* 23: 665–73.
56. Li, X. Y., Dong, H., and Shi, W. 2001. *Wear* 250: 553–60.
57. Jia, J., Chen, J., Zhou, H., Hu, L., and Chen, L. 2005. *Composites Sci. Technol.* 65: 1139–47.
58. Jia, J., Lu, J., Zhou, H., and Chen, J. 2004. *Mater. Sci. Eng.* 381: 80–85.
59. Masuko, M., Suzuki, A., Sagae, Y., Tokoro, M., and Yamamoto, K. 2006. *Tribol. Int.* 39: 1601–8.
60. Yamamoto, K., and Matsukado, K. 2006. *Tribol. Int.* 39: 1609–14.
61. La, P., Xue, Q., and Liu, W. 2000. *Tribol. Int.* 33: 469–75.
62. Wu, P. Q., Mohrbacher, H., and Celis, J.-P. 1996. *Wear* 201: 171–77.
63. Sato, T., Besshi, T., Sato, D., and Tsutsui, I. 2001. *Wear* 249: 50–55.
64. Gao, Y., Jing, Y., Zhang, Z., Chen, G., and Xue, Q. 2002. *Wear* 253: 576–78.
65. Huang, W., Dong, J., Li, F., and Chen, B. 2002. *Wear* 252: 306–10.
66. Lai, Ch. L., Harwell, J. H., O'Rear, E., Komatsuzaki, S., Arai, J., Nakakawaji, T., and Ito, Y. 1995. *Colloids Surfaces A* 104: 231–41.
67. Meyers, D. 1999. *Surface, interface and colloids.* Weinheim, Germany: Wiley-VCH.
68. Tsujii, K. 1998. *Surface activity: Principles, phenomena & applications.* New York: Academic Press.
69. Laughlin, R. G. 1994. *Aqueous phase behavior of surfactants.* New York: Academic Press.
70. Kuo-Yann, L., ed. 1997. *Liquid detergents.* New York: Marcel Dekker.
71. Broze, G., ed. 1999. *Handbook of detergents.* New York: Marcel Dekker.
72. Cases, J. M., Villieras, F., Michot, L. J., and Bersilon, J. L. 2002. *Colloids Surfaces A* 205: 85–99.
73. Patrick, H. N., and Warr, G. G. 2000. *Colloids Surfaces A* 162: 149–57.
74. Wanless, E. J., Davey, T. W., and Ducker, W. A. 1997. *Langmuir* 13: 4223–28.
75. Wall, J. F., and Zukoski, Ch. F. 1999. *Langmuir* 15: 7432–37.
76. Dixit, S. G., Vanjara, A. K., Nagarkar, J., Nikoorazm, M., and Desai, T. 2002. *Colloids Surfaces A* 205: 39–46.
77. Patrick, H. N., Warr, G. G., Manne, S., and Aksay, I. A. 1997. *Langmuir* 13: 4349–56.
78. Patrick, H. N., Warr, G. G., Manne, S., and Aksay, I. A. 1999. *Langmuir* 15: 1685–92.
79. Levitz, P. E. 2002. *Colloids Surfaces A* 205: 31–38.
80. Fan, A., Somasundaran, P., and Turro, N. J. 1997. *Langmuir* 13: 506–11.
81. Denoyel, R. 2002. *Colloids Surfaces A* 205: 61–71.
82. Adler, J. J., Singh, P. K., Patist, A., Rabinovich, Y. I., Shah, D. O., and Moudgil, B. M. 2000. *Langmuir* 16: 7255–62.
83. Warr, G. G. 2000. *Curr. Opinion Colloid Interface Sci.* 5: 88–94.
84. Atkin, R., Craig, V. S. J., Wanless, E. J., and Biggs, S. 2003. *Adv. Colloid Interface Sci.* 103: 219–304.
85. Tiberg, F., Brinck, J., and Grant, L. 2000. *Curr. Opinion Colloid Interface Sci.* 4: 411–19.
86. Paria, S., and Khilar, K. C. 2004. *Adv. Colloid Interface Sci.* 110: 75–95.
87. Klitzing, R., and Muller, H. J. 2002. *Curr. Opinion Colloid Interface Sci.* 7: 42–49.
88. Somasundaran, P., and Huang, L. 2000. *Adv. Colloid Interface Sci.* 88: 179–208.
89. Free, M. L. 2002. *Corrosion Sci.* 44: 2865–70.

90. Osman, M. M., and Shalaby, M. N. 2003. *Mater. Chem. Phys.* 77: 261–69.
91. Abdallah, M. 2003. *Corrosion Sci.* 45: 2705–16.
92. Mu, G., Li, X., and Liu, G. 2005. *Corrosion Sci.* 47: 1932–52.
93. Osman, M. M., and Shalaby, M. N. 1997. *Anti-Corrosion Methods and Materials* 44: 318–22.
94. Batton, C. B., Chen, T.-Y., and Towery, C. Ch. 1998. U.S. Patent 5,849,220, issued 1998.
95. Balzer, D. 2000. In *Nonionic surfactants: Alkyl polyglucosides*. Ed. D. Balzer and H. Luders. New York: Marcel Dekker.
96. Nickel, D., Forster, T., and von Rybinski, W. 1996. In *Alkyl polyglucosides*. Ed. K. Hill, W. von Rybinski, and G. Stoll. New York: VCH.
97. Bermudez, M. D., Martinez-Nicolas, G., and Carrion-Vilches, F. J. 1997. *Wear* 212: 188–194.
98. Yao, J., Wang, Q., Xu, Z., Yin, J., and Wen, S. 2000. *Lubrication Eng.* 56: 21–25.
99. Kupchinov, B. I., Rodnenkov, V. G., Ermakov, S. F., and Parkalov, V. P. 1991. *Tribol. Int.* 24: 25–28.
100. Vekteris, V., and Murchachver, A. 1995. *Lubrication Eng.* 51: 851–53.
101. Mori, S., and Iwata, H. 1996. *Tribol. Int.* 29: 35–39.
102. Fischer, T. E., Bhattacharya, S., Salher, R., Lauer, J. L., and Ahn, Y. J. 1988. *Tribol. Trans.* 31: 442–48.
103. Tichy, J. A. 1990. *Tribol. Trans.* 33: 363–70.
104. Kimura, Y., Nakano, K., Kato, T., and Morishita, S. 1994. *Wear* 175: 143–49.
105. Fuller, S., Li, Y., Tiddy, T. J., Wyn-Jones, E., and Arnell, R. D. 1995. *Langmuir* 11: 1980–83.
106. Lee, H. S., Winoto, S. H., Winer, W. O., Chiu, M., and Friberg, S. E. 1990. In *Tribology and the liquid crystalline state*. Ed. G. Biresaw. Washington, DC: Oxford University Press.
107. Lockwood, F. E., Benchaita, M. T., and Friberg, S. E. 1987. *ASLE Trans.* 30: 539–48.
108. Kumar, K., and Shah, D. O. 1990. in *Tribology and the liquid crystalline state*. Ed. G. Biresaw. Washington, DC: Oxford University Press.
109. Boschkova, K., Elvesjo, J., and Kronberg, B. 2000. *Colloids Surfaces A* 166: 67–77.
110. Friberg, S., Ward, A., Gunsel, S., and Lockwood, F. 1990. In *Tribology and the liquid crystalline state*. Ed. G. Biresaw. Washington, DC: Oxford University Press.
111. Wasilewski, T., and Sulek, M. W. 2006. *Wear* 261: 230–34.
112. Pilpel, N., and Rabbani, M. E. 1988. *Colloid Interface Sci.* 122: 266–73.
113. Kirikou, M., and Sherman, P. 1979. *Colloid Interface Sci.* 71: 51–54.
114. Lu, D., Burgess, D. J., and Rhodes, D. G. 2000. *Langmuir* 16: 10329–33.
115. Singhal, S., Moser, C. C., and Wheatley, M. A. 1993. *Langmuir* 9: 2426–29.
116. Davies, R., Graham, D. E., and Vincent, B. 1987. *Colloid Interface Sci.* 116: 88–99.
117. Lu, D., and Rhodes, D. G. 2000. *Langmuir* 16: 8107–12.
118. Pilpel, N., and Rabbani, M. E. 1987. *Colloid Interface Sci.* 119: 550–58.
119. Gullapalli, R. P., and Sheth, B. B. 1999. *European Pharmaceutics Biopharmaceutics* 48: 233–38.
120. Gurkov, T. D., Horozov, T. S., Ivanov, I. B., and Borwankar, R. P. 1994. *Colloids Surfaces A* 87: 81–92.
121. Ogino, K., and Abe, M. eds. 1993. *Mixed surfactant system*. New York: Marcel Dekker.
122. Chattopadhyay, A. K., and Mittal, K. L. eds. 1996. *Surfactants in solutions*. New York: Marcel Dekker.
123. Friberg, S. E., and Lindman, B. eds. 1992. *Organized solutions*. New York, Marcel Dekker.

124. Holmberg, K., ed., 2001. *Handbook of applied surface and colloid chemistry.* New York: John Wiley & Sons.
125. Walker, L. M. 2001. *Curr. Opinion Colloid Interface Sci.* 6: 451–56.
126. Yang, J. 2002. *Curr. Opinion Colloid Interface Sci.* 7: 276–81.
127. Platz, G., Polike, J., Thuning, Ch., Hofmann, R., Nickiel, D., and von Rybinski, W. 1995. *Langmuir* 11: 4250–55.
128. Boden, N. 1994. In *Micelles, membranes, microemulsions and monolayers.* Ed. W. M. Gelbart, A. Ben-Shaul, and D. Roux. New York: Springer-Verlag.
129. Piekoszewski, W., Szczerek, M., and Tuszynski, W. 2001. *Wear* 240: 183–193.
130. Savenko, V. I., and Schukin, E. D. 1996. *Wear* 194: 86–94.
131. Gorbatkina, Y. A., and Ivanova-Mumjieva, V. G. 1999. *Colloids Surfaces A* 160: 155–62.
132. Butyagin, P. 1999. *Colloids Surfaces A* 160: 107–115.
133. Summ, B. D., and Samsonov, V. M. 1999. *Colloids Surfaces A* 160: 63–77.

19 Surfactant Adsorption and Aggregation Kinetics
Relevance to Tribological Phenomena

Rico F. Tabor, Julian Eastoe, and Peter J. Dowding

19.1 INTRODUCTION

The action of surfactant adsorption at interfaces and aggregation in bulk solutions is central to many processes, among them biological, pharmaceutical, and industrial. A frequently cited biological example is the case of natural surfactants in the lungs that keep the alveoli from collapsing; furthermore, cells themselves are encased in a surfactant bilayer. In industrial processes, surfactants are key additives for oil recovery and pharmaceuticals and for use as dispersants for mineral recovery and engineering oils, as vesicles for drug delivery, as foams, and as surface coatings, among countless other examples. Due to the wide structural variety of surfactants and different interfaces of interest, it is not surprising that the physicochemical characterization of surfactants remains at the forefront of research.

Surfactant adsorption/self-assembly at solid–liquid interfaces not only affects solution properties, but also the tribological aspects of the solid surface, in particular lubrication of interacting surfaces through mediation of friction and wear [1].

Friction can be considered the resistance to motion of one body in contact with another, and it is quantified by the determination of friction coefficients using techniques such as pin-on-disc or Cameron–Plint [2]. Lubrication of surfaces occurs in various regimes (involving different mechanisms), dependent upon the film thickness of the lubricating species. If two solid surfaces are in relatively close contact involving a thin monolayer of lubricant film (i.e., where fluid film formation is not possible), then boundary lubrication occurs where the chemi- and/or physisorption of surfactants influences the friction coefficient. As the film thickness increases, a hydrodynamic region is entered (via a mixed hydrodynamic–boundary regime), where the number of solid–solid contacts (resulting from asperities in the surface) diminishes and the friction coefficient also decreases. Thicker films beyond this point result in higher friction coefficients due to solvent viscosity and internal friction [3].

Surfactants have the greatest influence on frictional characteristics in the boundary lubrication regime, with both head and tail structures influencing the resultant friction coefficient. Such molecules adsorb onto metal surfaces via covalent or hydrogen bonding, with the magnitude of interaction mediated through bond strength. For example, a surfactant with an acidic head group (such as stearic acid) is believed to form a (strong) covalent bond with iron (via deprotonation). However, if the acid species is replaced by ester or hydroxyl functionality (e.g., methyl stearate or stearyl alcohol), interaction occurs via a weaker hydrogen-bonding mechanism [3–6]. Further molecules can be adsorbed, resulting in the formation of multimolecular clusters. Such layers are hard to compress but readily undergo shear (i.e., low shear strength), which results in a low friction coefficient.

The tail group not only imparts oil solubility, but when the head group is adsorbed onto the steel surface, tail groups align perpendicular to the surface and interact cohesively (through van der Waals interactions) to effectively produce a hydrocarbon layer. The production of multilayers can also occur through van der Waals interactions of tail groups and polar interactions of the head groups. The chemical nature of the tail group also influences the resultant friction [3]. Beltzer and Jahanmir [4] investigated the variations in the tail structures of C_{18} carboxylic acids and concluded that changes that increase tail–tail interactions, such as an increase in chain length (to promote van der Waals interactions), or the inclusion of polar groups, such as hydroxyls (to promote hydrogen bonding), reduce the resultant friction coefficient. Similarly, changes that reduce tail–tail interaction, such as unsaturation or branching, increase the friction coefficient [4, 7]. The friction modification is a temperature-dependent behavior that is controlled by both the chain length of the surfactant and the lubricating solvent [7]. Amphiphilic materials are used in lubricants to mediate the effects of friction and improve vehicle fuel economy (termed "friction modifiers"). These generally comprise structures based on carboxylic acids, phosphoric/phosphonic acids, amines, or amides [1, 2].

Wear of solid surfaces, which involves physical loss of material, can occur via adhesion (bonding of surface asperities, which then break under shear), abrasion (due to the presence of heterogeneous particles in the lubricating film), fatigue (which initiates below the solid surface, propagating to a point of weakness in the surface, which then results in cracking), and corrosive wear. The formation of low-shear-strength surfactant films on metal surfaces (as described above) can help reduce adhesive wear in the boundary lubrication regime by preventing bonding of the tips of asperities of the solid surfaces in contact [1].

The study of surfactant kinetics is vital in explaining the fundamentals of adsorption and aggregation. Determining the rates of these processes can increase mechanistic understanding and guide the design of new molecules.

The purpose of this chapter is (a) to present currently held kinetic models for various important interfaces and aggregation structures and (b) to discuss recent developments in the study of surfactant aggregation and adsorption kinetics.

19.2 AIR–LIQUID INTERFACES

19.2.1 ADSORPTION AT THE AIR–WATER INTERFACE

One of the most fundamentally accepted and understood properties of surfactants is their propensity to adsorb at interfaces. The kinetics of adsorption at the air–water interface has been far more comprehensively researched and modeled than at any other interfacial boundary.

The generally accepted air–water dynamic model posits that when a fresh interface is created between a surfactant solution and air, a nonequilibrium state is generated, with surfactant molecules diffusing from the subsurface layer and adsorbing and orienting themselves at the interface. This is thought to be an activation-controlled process, which depletes the concentration in the subsurface, setting up a concentration gradient with the bulk solution. Molecules then diffuse toward the subsurface, and this is a diffusion-controlled process. Either of these two steps can be rate-determining for adsorption.

In 1907, Milner [8] first suggested that the variation of surface tension of a surfactant solution could be mediated by molecules diffusing to the interface. Some considerable time later, Langmuir and Schaeffer [9] made a significant advance when they looked at the diffusion of ions into monolayers and proposed a mathematical model of the diffusion process. However, it was not until the seminal 1946 paper of Ward and Tordai [10] that the first complete model for diffusion-based kinetics emerged. The Ward–Tordai model accounts for three variables: the bulk concentration, the subsurface concentration, and the surface tension. This led to the celebrated Ward–Tordai equation:

$$\Gamma_t = 2C_0 \left(\frac{Dt}{\pi} \right)^{\frac{1}{2}} - 2 \left(\frac{D}{\pi} \right)^{\frac{1}{2}} \int_0^{t^{\frac{1}{2}}} C_s (t - \tau) d\tau^{\frac{1}{2}} \tag{19.1}$$

where

    Γ = surface excess
    C_0 = bulk concentration
    D = diffusion coefficient
    t = time
    C_s = subsurface concentration
    τ = dummy integration variable

They suggested that if the transport from the bulk solution to the interface was purely a diffusion-controlled process, then calculated, model-derived diffusion coefficients should match experimentally determined values. However, they found that the experimental diffusion coefficients were much lower than the predictions, implying that diffusion was not the rate-determining process, and leading to the conclusion that there is an (additional) activation barrier.

As the Ward–Tordai equation contains two independent variable functions (surface excess and subsurface concentration), its application requires a further equation relating the two functions. The first attempt at this was by Sutherland [11], who incorporated a linear adsorption isotherm. This, however, proved to be quite limiting, and so various other isotherms were employed [12, 13]. Even so, these extended theorems accurately matched experimental results only in the case of some nonionic surfactants.

Diamant and Andelman [14] cited the three primary drawbacks of these approaches:

1. The second relation between surface excess and subsurface concentration is effectively an external boundary condition, and not intrinsic to the model.
2. The calculated dynamic surface tension uses an equilibrium equation of state, and assumes it works in nonequilibrium conditions.
3. The theories cannot be applied to ionic surfactant solutions. For these reasons, they developed a free-energy approach to solving the adsorption kinetics [15].

For nonionic surfactants, their theory yields an equation similar in form to that of Ward and Tordai:

$$\phi_0(t) = \left(\frac{D}{\pi a^2}\right)^{1/2}\left[2\phi_b t^{1/2} - \int_0^t \frac{\phi_1(\tau)}{(t-\tau)^{1/2}}\,d\tau\right] + 2\phi_b - \phi_1 \tag{19.2}$$

where

ϕ_0 = volume fraction of surfactant at interface (surface coverage)
ϕ_b = volume fraction of surfactant in bulk
ϕ_1 = volume fraction of surfactant in subsurface
D = diffusion coefficient
a = surfactant molecular size
τ = dummy integration variable

In their work, Diamant and Andelman [14] concluded that most nonionic surfactants follow diffusion-limited adsorption. In the case of ionic surfactants (for which the resulting expression is somewhat involved), a pause is noted in the surface adsorption (representing a free-energy minimum), during which domains of the denser surface phase are nucleated, and after which further surface adsorption can occur. In systems where an added salt charge screens the ionic surfactant, the behavior is similar to the case of nonionic surfactants, and diffusion control takes over.

The first attempt to provide a kinetic treatment incorporating both diffusion and activation processes was made by Liggieri and coworkers [16]. A further development along similar lines was again proposed by Liggieri et al. [17], resulting in the extended "master" equation:

$$\int_0^\Gamma \frac{1}{g(\Gamma)}\,d\mu = \left(\frac{D_a}{\pi}\right)^{1/2}\left[2C^\infty t^{1/2} - \int_0^t \frac{C_0(\tau)}{(t-\tau)^{1/2}}\,d\tau\right] \tag{19.3}$$

where the apparent diffusion coefficient D_a is defined as

$$D_a = D\exp\left(\frac{-2\varepsilon_A}{k_B T}\right) \tag{19.4}$$

and where

 Γ = surface excess

 μ = surfactant chemical potential

 C^∞ = concentration of bulk solution

 C_0 = concentration at $x = 0$ (where x represents distance from the interface)

 τ = dummy integration variable

 ε_A = activation energy for adsorption

 k_B = Boltzmann constant

 T = temperature

It should be noted that for the limiting case of $\varepsilon_A = 0$ the expression tends to a Ward–Tordai-type equation. Liggieri et al. [17] also make the distinction that, even when adsorption appears to be diffusion controlled, there may still be an adsorption barrier present, and although it is not the rate-determining process, its existence should not be discounted entirely.

Various experimental studies have been conducted to substantiate activation-diffusion kinetics. Eastoe et al. [18] were able to determine the size of the activation barrier as 5–12 kJ·mol^{-1} for a range of alkylpolyglycol ether and di-chain glucamide surfactants. They attained a similar value of $\approx$10 kJ·mol^{-1} for a cationic surfactant, n-hexylammonium n-dodecyl sulfate [19]. It should be noted here for clarity that there is little consistency in the literature between the units of rates and rate constants due to their dependence on the various models and analyses used. In this chapter, values are reported and discussed in the units given in those original publications.

Lee and coworkers [20] determined the model-derived adsorption-rate constants for a range of polyglycol ethers to be around 5.6×10^{-6} cm^3·mol^{-1}·s^{-1}, regardless of molecular size. However, they found that the desorption rate constant differed greatly with molecular size. Conversely, Bleys and Joos [21] determined rate constants for adsorption and desorption for a homologous series of α,ω-diol bolaform surfactants and found that desorption ($\approx$10^{-6} mol·m^{-2}·s^{-1}) was independent of chain length, whereas adsorption ($\approx$10 to 10^{-2} m·s^{-1}) was not. They were the first to suggest a mechanism for the activation process, via formation of activated complex. They surmised that in order for a surfactant molecule to adsorb, it must first shed its sheath of structured water. (This is an entropic process that would depend on chain length, leaving a higher energy "unsolvated" activated complex that is then able to adsorb.) Desorption is via the same activated complex, but as the molecule moves to the activated complex state, and then gains its structured water, the rate should not depend on chain length.

There is still some discussion as to the exact nature of the adsorption barrier, and whether or not molecular reorientation at the interface is an important process. In 1983, van den Tempel and Lucassen-Reynders [22] published a review of advances up to that point. They noted that for small molecules, any orientation at the interface was fast, and so diffusion would still be the rate-determining step. A more significant aspect was the distinction made between a kinetic process resulting in surfactant molecules being (a) adsorbed from the subsurface layer to the interface (usually faster than diffusion from the bulk to the subsurface) and (b) reoriented when at the interface, usually a slower process involving cooperative motions. Serrien and Joos [23] explained the slow

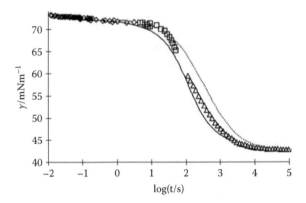

FIGURE 19.1 Dynamic surface-tension data for n-alkyldimethylphosphine oxides, as measured by maximum bubble-pressure technique (◇), drop-volume tensiometry (□), and de Noüy ring tensiometry (△), and model fit ignoring reorientation (dotted line) and incorporating reorientation (solid line). (From Fainerman, V. B., et al. 2000. *Adv. Colloid Interface Sci.* 86 (1–2): 83–101. With permission.)

surface-tension relaxation of sodium di-octyl sulfosuccinate (a straight-chain analogue of the much-researched AOT [Aerosol-OT Sodium bis (2-ethylhexyl) sulfosuccinate]) with such a cooperative surface reorientation of molecules.

Fainerman et al. [24] proposed a model for reorientation at the interface (based on an earlier suggestion by Langmuir) in which molecules could adsorb parallel to the interface at low surface pressure, and then adsorb normally at higher surface pressures. Later work substantiates the model, showing that it accurately describes experimental evidence from n-alkyldimethyl phosphine oxides (fig. 19.1) and polyglycol ethers [25]. Miller et al. [26] suggested that reorientation can either accelerate or decelerate the decrease of surface tension, depending on the adsorption characteristics of the surfactant in its two adsorbed states.

19.2.2 ADSORPTION AT THE AIR–WATER INTERFACE FROM MICELLAR SOLUTIONS

The classic picture of adsorption from a micellar solution is that when fresh surface is created, molecules adsorb at the interface from the subsurface layer. This depletion at the subsurface sets up a concentration gradient, which drives free surfactant from the bulk to diffuse toward the subsurface. The decrease in concentration results in micellar breakdown; hence the micelles act as a "reservoir" of surfactant monomers.

Using this model, it is possible to determine rates for micellar breakdown from dynamic surface-tension measurements. The first attempt to do so was by Lucassen in 1975 [27], with mixed conclusions about the results and the validity of the model. Later, Rillaerts and Joos [28] were able to determine rate constants for micelle breakdown with cationic ammonium bromide surfactants, found to be on the order of ≈200 s^{-1} for cetyltrimethylammonium bromide. In a significant review of the area, Noskov [29] concludes that due to the large number of experimental parameters associated with the model, numerical solution is difficult, and hence the rate information gained realistically only represents a rough approximation. A better approach was proposed to simplify the mathematical model using relaxation methods. It was also noted that due to the intrinsic assumptions made, the model is accurate only for small deviations from equilibrium.

While the classical model is probably correct for ionic micelles, Colegate and Bain [30] showed that for a nonionic glycol ether, entire micelles can adsorb at the interface at a diffusion-controlled rate without first breaking down. This adds an additional pathway to the kinetic model for nonionics. Song et al. [31] provided a substantial theoretical study of similar nonionic surfactants and suggested an addition to the model whereby, at low aggregate concentrations, depletion in the subsurface as monomer adsorbs leads to a "micelle-free zone" that extends into the bulk [31]. Conversely, if the aggregate concentration is high enough to overcome this, a depletion layer cannot be established even in the subsurface layer, leading to the possibility that entire micelles may directly adsorb.

19.2.3 Adsorption at the Air–Oil Interface

While many theoretical studies and models consider the general case of adsorption at a liquid–air interface, the vast majority of experimental work has been carried out with aqueous solutions. Data for other air–solvent interfaces are sparse, with most results being related to crude-oil systems. Some equilibrium adsorption and aggregation studies have involved nonaqueous solvents, but kinetic data are lacking.

Bauget et al. [32] looked at the kinetics of asphaltenes (multiaromatic surface-active compounds found in crude oil) at organic solvent–air interfaces and found that adsorption was activation barrier-limited. They postulated that this could be due to adsorption at the interface from the subsurface or from molecular rearrangement when adsorbed at the interface. As asphaltenes comprise planar sheets that can stack via the interaction of their conjugated π-orbitals, there is evidence that these stacks can self-assemble to form "micellar" structures. It is the agglomeration of these species that results in phase separation or aggregation. Later work by Jeribi and coworkers [33] found similar results, in that asphaltene adsorption followed diffusion-controlled kinetics at the air interface with toluene (a good solvent) and apparent mixed kinetics in toluene-heptane (a poorer solvent). They cited the reason being that the asphaltene molecules have a greater propensity to aggregate in poorer solvents, and so mobility and diffusivity are decreased.

19.2.4 Mixed Surfactants at the Air–Water Interface

Ariel, Diamant, and Andelman [34] extended their free-energy formalism approach to look at mixed surfactant systems. Consider a mixture of two molecules that differ in molecular weight (diffusion coefficient) and surface activity (hydrophobicity). Their model predicts four stages of kinetics, each with an associated rate constant. Firstly, rapid adsorption occurs from the subsurface layer to the surface, depleting the subsurface. The second stage is diffusion from bulk to subsurface, during which the more mobile species will diffuse more rapidly, and hence preferentially populate the surface. If the less mobile species is also the more surface active, a third stage could occur, as it displaces the less active species that has already adsorbed. The final stage is then diffusion-limited relaxation to a state in which the surface, subsurface,

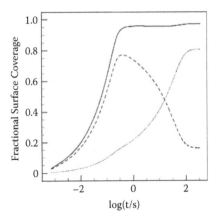

FIGURE 19.2 Competitive surfactant adsorption at the air–water interface. The more mobile species (dashed line) rapidly covers the surface; this is then replaced more slowly by the more surface-active species (dotted line). The solid line represents overall coverage. (From Ariel, G., Diamant, H., and Andelman, D. 1999. *Langmuir* 15 (10): 3574–81. With permission.)

and bulk are in mutual equilibrium. These stages can be seen in fig. 19.2, where the dashed line represents the more mobile species, which rapidly covers the surface. This is then replaced more slowly by the more surface-active species, represented by the dotted line. The solid line represents overall coverage.

Fainerman and Miller [35] found that displacement of an initially adsorbed surfactant by a second, more surface-active species allowed measurement of the desorption rate of the former. For example, competitive adsorption of sodium decyl sulfate and the nonionic Triton X-165 gave a desorption rate constant for the former of 40 s^{-1}. Mulqueen and coworkers [36] recently developed a diffusion-based model to describe the kinetics of surface adsorption in multicomponent systems, based upon the Ward–Tordai equation. Experimental work with a binary mixture of two nonionic alkyl ethoxylate surfactants [37] showed good agreement with the model, demonstrating a similar temporal adsorption profile to that found by Diamant and Andelman [34].

Frese et al. [38] studied the adsorption kinetics of mixed anionic/cationic surfactant systems, finding that net charge-neutral mixed micelles were formed. They found that the dynamic surface tension was higher than expected for small-surface ages, possibly due to an activation barrier associated with the mixed-micelle complexes.

19.3 LIQUID–LIQUID SYSTEMS

19.3.1 SINGLE SURFACTANT SYSTEMS

Moving from air–water to liquid–liquid interfaces introduces added complications, since the surfactant always has some degree of solubility in both phases, and hence partitioning effects become important. This means that not only will there be processes associated with adsorption and rearrangement at the interface (potentially from both phases), but also with molecules transferring from one bulk phase to an equilibrium partition state (fig. 19.3).

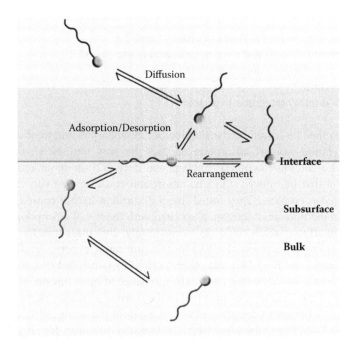

FIGURE 19.3 A summary of dynamic processes that may occur at the liquid–liquid interface.

As Ravera et al. [39] point out in an extensive review of surfactant kinetics at liquid–liquid interfaces, the partitioning between phases should be included as a partition coefficient, k_p, derived from the equilibrium concentrations of the surfactant in the two liquid phases, c_1^{eq} and c_2^{eq}:

$$k_p = \frac{c_1^{eq}}{c_2^{eq}} \tag{19.5}$$

If k_p is zero (or small), the systems may be considered analogous to the air–water interface. However, due to the amphiphilic nature of surfactants, the more realistic case is where k_p is nonzero, and there is an appreciable surfactant solubility in both phases. Accounting for this is difficult, and this complication is often overlooked. Note that potentially there may be fluxes to the interface from both phases.

It is then possible to consider the problem as essentially mass-transport dominated. This was first done for the water–hexane interface by van Hunsel and Joos [40] based on earlier work by Hansen [41] considering a water–air interface, where the solute itself was volatile. This approach resulted in an extended Ward–Tordai diffusion equation, of the form

$$\Gamma_t = \frac{2c_{01}}{\pi^{1/2}}\left[D_1^{1/2} + \frac{c_{02}D_2^{1/2}}{c_{01}}\right]t^{1/2} - \frac{1}{\pi^{1/2}}\int_0^t\left[D_1^{1/2} + k_pD_2^{1/2}\right]\frac{c_{1s}(\tau)}{(t-\tau)^{1/2}}\,d\tau \tag{19.6}$$

where:

D_1, D_2 = diffusion coefficients in phases 1 and 2

c_{01}, c_{02} = initial surfactant concentrations in phases 1 and 2

c_{1s} = subsurface concentration in phase 1

t = time

τ = dummy integration variable

This can then be solved by the addition of a boundary isotherm, just as is done for the air–liquid case. If adsorption barriers are now introduced, the problem becomes yet again more complex. England and Berg [42] set up an activation–diffusion model that introduced a kinetic adsorption equation as a barrier condition for the diffusion problem. They found that a desorption barrier caused a transient minimum in the interfacial tension, associated with the initial adsorption of surfactant at the interface, which could also significantly slow mass transfer. Rubin and Radke [43] attempted a physical model for this minimum, experimentally observed by both England and Berg [42] and others [44]. They concluded that an interfacial tension minimum would be observed if the resistance to mass transfer (diffusion) in the adjacent phase were greater than that in the starting saturated phase, or if there were an energy barrier to desorption into the adjacent phase (fig. 19.4). Ferrari and coworkers [45, 46] found the same interfacial tension minimum for alkylphosphine surfactants at the water/hexane boundary, citing the same reason and determining the equilibrium concentration to be 34 times higher in the organic phase.

Ravera and coworkers extended their previously mentioned diffusion-barrier model [16] to a situation where the solute is soluble in both phases (i.e., for real liquid–liquid systems) [47]. They suggested that the model of England and Berg [42] is

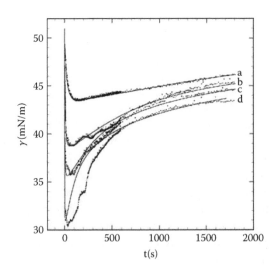

FIGURE 19.4 Dynamic interfacial tension data showing a minimum for C_{13}-dimethylphosphine oxide at the water–hexane interface. (Experimental data, increasing concentration a→d [symbols] and theoretical data [lines].) (From Liggieri, L., et al. 1997. *J. Colloid Interface Sci.* 186 (1): 46–52. With permission.)

in fact invalid, as it assumes an instant equilibrium between the adsorbed molecules and those that have diffused to the surface; this is unphysical and contradicts the presence of an adsorption barrier. The model of Ravera et al. [47] allows for diffusion and adsorption at the interface from either phase, as well as accounting for an adsorption barrier, and hence is one level more sophisticated.

The work described so far has, at least theoretically, dealt with "semi-infinite" solutions, i.e., those whose volume is considered large enough to outstrip the concentration gradient during the timescale of the observation. It is interesting to note that when this is not the case, kinetics can vary dramatically, and when considering droplet size in emulsions and microemulsions, this is extremely relevant. This dependence on phase volume was mentioned by Ravera and coworkers [45] in the previously cited work with alkylphosphine surfactants. Logically, if surfactant is diffusing across a liquid–liquid interface from a semi-infinite reservoir of surfactant solution to a small surfactant-devoid phase, in the absence of any activation (energy) barrier to desorption, then the new phase will become rapidly saturated. The equilibrium partition is rapidly set up, and the interfacial tension should decrease rapidly to a fixed value and remain steady, as molecules desorbing from the interface to the new phase are rapidly replaced by those in the bulk.

Alternatively, if a small reservoir of surfactant solution is used against a large surfactant-devoid phase, the molecules will rapidly diffuse to the interface and then disperse into the larger phase. Because there is a very limited supply of molecules to replace them, only over time will the equilibrium distribution be established. In this case, the rapid diffusion to the interface results in a transient minimum in interfacial tension. The interfacial tension then increases again as the molecules desorb from the interface to the larger phase.

Conversely, for small volume changes of a hexane/water system with binary mixtures of nonionic surfactants, Mulqueen and Blankschtein [48] found no change in interfacial tension. It should be noted in this case, though, that the difference in phase volumes is probably not enough to show the effect described above, which is only marked when there is a difference of at least several orders of magnitude between phase volumes.

19.3.2 Surfactant Mixtures

The whole area of mixed surfactants at liquid–liquid interfaces has direct relevance for industrially important areas such as emulsions and detergents; surprisingly, there are few publications including such studies. Campanelli and Wang [49] provided some interesting results and modeling by considering systems with only one surfactant dissolved in the water phase, and one in the oil. They found that when one surfactant was significantly more surface active than the other, it was the former that dominated the rate of adsorption, and the effect of the less active species could be considered as negligible. Conversely, if both species were highly surface active, they both had significant adsorption rates, and the dynamic interfacial tension could be explained by a desorption-corrected version of the model suggested by Miller et al. [50]. Mulqueen and Blankschtein [48] provide a useful model based on single-component parameters for predicting the equilibrium interfacial tensions of mixed systems, although they do not extend this to dynamic effects.

19.4 SOLID–LIQUID INTERFACES

19.4.1 SINGLE-SURFACTANT SYSTEMS

The process of surfactant adsorption from a solution onto a solid surface is perhaps the most relevant in the field of tribology. It is the process by which all surfactant-mediated lubrication occurs. The physical model for this can be described by a series of steps. As with other systems, there will be diffusion from the bulk to a subsurface layer, adsorption of (initially) a monolayer of surfactant molecules, and the possibilities of subsequent rearrangement of molecules on the surface, deposition of a bilayer and subsequent further layers, and also desorption. The interactions that govern the adsorption and desorption processes can be hydrophilic, hydrophobic, and/or electrostatic, depending on the nature of the surfactant and surface concerned. Now the situation is very complex, and because of this, it is sensible to tackle the kinetics of nonionic and ionic surfactants separately.

The majority of kinetic studies of nonionic surfactant adsorption at solid interfaces have been carried out using model surfaces, particularly silica, cellulose, and polystyrene. Much of the work on the adsorption kinetics of nonionic surfactants onto solids has been carried out by Tiberg and coworkers [51–53]. The nature of the surface attraction between nonionic surfactants and solid surfaces is still under some debate. Behl and Moudgil [54] concluded that, in the case of silica, a hydrogen-bonding interaction between the surfactant head group and surface hydroxyls results in attractions. Tiberg's model sets up three zones within a micellar solution: a subsurface layer in direct proximity with the solid surface (usefully defined as the region in which surface attractions are important); a stagnant layer outside this, in which surfactant diffusion is significant; and the bulk solution, in which convection is important. The two steps governing adsorption can then be considered as diffusion from the stagnant layer to the subsurface, and then the subsequent adsorption onto the surface itself. Adsorption is then seen to be diffusion controlled, and it is assumed that micelles only act as monomer reservoirs, and cannot adsorb directly onto the solid surface. This model, using resolved mass-transfer equations, was found to accurately describe experimental data for poly(ethylene glycol) monoalkyl esters at the silica/water interface [55].

Surfactant adsorption at the water–cellulose interface is of particular interest for applications in papermaking, and hence has attracted considerable attention. Torn et al. [56] have made significant advances, finding adsorption and desorption rate constants for poly(ethylene oxide) alkylethers on the orders of $2–5 \times 10^{-6}$ m·s^{-1} and $0.02–0.2 \times 10^{-6}$ mol·m^2·s^{-1}, respectively. They suggested that a rearrangement of surface molecules takes places during adsorption, to form more favorable surface micelles or aggregates. While the adsorption rate is initially attachment controlled, the desorption rate is controlled by the dissociation of the surface aggregates.

Paria and coworkers [58] suggest that the cellulose has two different types of available adsorption sites, of hydrophilic and hydrophobic nature. In later work, their model predicted that nonionics only adsorb at the hydrophobic sites, which fits experimental data well, suggesting an initial adsorption rate constant of 3×10^2 m^3·mol^{-1}·min^{-1} [58].

Working with alkyl ethylene oxides at the water–polystyrene interface, Geffroy et al. [59] have recently gained some useful insights into the kinetic behavior. In this case, it is the surfactant tail group that is adsorbed to the surface, leaving the polar head group presented to the solution. They employed the local equilibrium model suggested by Dijt [60], whereby it is assumed that molecules adsorbed at the interface are in local equilibrium with those in the subsurface layer. Interestingly, they found that surfactant adsorption was partly cooperative, and so adsorption and desorption processes within a certain concentration range were dependent on the strength of surfactant–surfactant interactions. This was deduced from kinetic measurements of desorption, where the initial dissociation of surfactants was fast, but a smaller amount of molecules were much slower to dissociate.

In all of these cases, the adsorption profiles are classical (expected). There is an initial fast, linear rate of adsorption, which gradually decreases to a plateau after a characteristic time (which depends on the specific surfactant, concentration, and surface) when the equilibrium adsorbed amount is achieved.

When ionic surfactants are considered, additional complications arise. As well as electrostatic forces between surfactant molecules and also with counterions (and also potentially the surface substrate), interactions with the electrical double layer must also be considered. Many solid surfaces will spontaneously develop a (usually negative) surface charge when immersed in water, attracting an atmosphere of counterions, the two together being the so-called electrical double layer. As Koopal and Avena [61] point out when modeling the case of aggregation at solid interfaces, a charged surfactant molecule will experience an effect from the double layer in both diffusion and adsorption processes.

The adsorption of cationic surfactants onto silica surfaces has been researched for over 35 years, and so this represents a well-understood model system. There are several available explanations for the changes in rate of adsorption seen. Rupprecht and Gu [62] provided perhaps the first explanation for the experimental surface excess profile. They suggested that at a fresh silica interface (sporting negative surface charges) there would be an initial rapid adsorption of surfactant ions to these surface-charged sites owing to electrostatics. Once surface neutrality is attained, there is a brief plateau in surface concentration of surfactant ions. Additional adsorption comes from hydrophobic interactions between surfactant tail groups nucleating around those already adsorbed at the surface, forming so-called hemimicelles with hydrophilic head groups presented to the solution.

A similar, alternative model, originally proposed by Somasundaran and Fuerstenau [63], suggests that surfactant ions first adsorb onto charged sites on the interface, as for Rupprecht and Gu's model [62], but that strong hydrophobic interactions between the tails of adsorbed molecules then cause them to aggregate, forming so-called hemimicelles (summarizing earlier work). Yeskie and Harwell [64] modified this model to include the possibility of both hemimicelle and admicelle formation. (The former is a local region of surface monolayer that presents a hydrophobic film to the bulk aqueous solution, and the latter a region of bilayer on top of this, hence

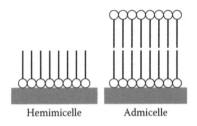

Hemimicelle   Admicelle

FIGURE 19.5 The structure of hemimicelles and admicelles.

presenting a hydrophilic surface to the solution [see fig. 19.5].) They found that this tied in well with the decrease in mineral particle flotation efficiency on increasing surfactant concentration.

Intuitively, even when the equilibrium adsorbed amount is reached, there will still be surfactant exchange between the interface and solution. This equilibrium exchange is, as at other interfaces, the least studied of kinetic processes. For the case of the solid–liquid interface, pulsed field-gradient nuclear magnetic resonance (PFG-NMR) is often used. This technique, first suggested in the 1950s, has been well documented and reviewed with respect to its usefulness in measuring diffusion coefficients [65–68]. Schonhoff and Soderman [69] were the first to directly derive kinetic parameters from PFG-NMR data, in studies of poly(ethylene oxide) surfactants on a polystyrene latex surface. They found residence times for surfactant molecules at the interface to be around 10 ms, decreasing slightly with increasing bulk concentration.

More recently, an interesting technique has been used by Clark and Ducker [70] to measure kinetics, in the form of attenuated total reflectance Fourier transform infrared (ATR-FTIR) spectroscopy. They found that total surface exchange of a cationic ammonium bromide surfactant on a silica surface occurred in slightly less than 10 s (fig. 19.6). This technique had been used previously by Couzis and Gulari [71, 72] to look at the adsorption kinetics of anionic surfactants at the alumina–water interface with apparent timescales in the region of tens of hours.

Another novel approach is the streaming potential optical reflectometer, developed by Theodoly et al. [73] (effectively a continuous optical reflectometer combined with a flow cell), with which they suggest kinetic measurements can be made.

Electron spin resonance (ESR) spectroscopy has also been used to look at adsorption kinetics at the solid–liquid interface. Malbrel and coworkers [74] found, when looking at the adsorption of Aerosol-OT (AOT, sodium bis(ethyl-hexyl)sulfosuccinate), that 40% of total adsorption had occurred within the first 5 s [74].

19.4.2 SURFACTANT MIXTURES

For practical reasons, and to facilitate data analysis, single-surfactant systems are normally studied. Importantly, the thermodynamic and kinetic models are easier to handle when compared with multicomponent systems, and there are only a few reliable methods for the unequivocal measurement of mixed systems. However, in "real-world" applications, mixed and blended surfactants are normally employed for commercial reasons and to improve performance.

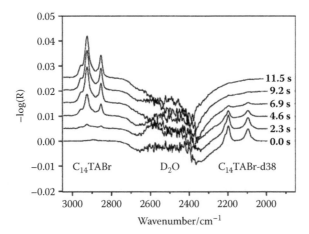

FIGURE 19.6 Exchange over time of previously adsorbed deuterated tetradecyltrimeth-ylammonium bromide for its protonated counterpart at the solid/aqueous interface, as measured by ATR-FTIR spectroscopy. (From Clark, S. C., and Ducker, W. A. 2003. *J. Phys. Chem. B* 107 (34): 9011–21. With permission.)

There are a few experimental studies involving the kinetics of mixed surfactant systems at solid–liquid interfaces. Brinck et al. [52] looked at binary systems of nonionic poly(ethylene glycol) surfactants at the silica–water interface with ellipsometry. As they point out, the large number of parameters involved makes deducing quantitative factors very difficult, but some global features can be understood. They found that mixtures with significantly differently sized head groups behaved in a similar fashion to one-component systems, whereas when tail groups were of different sizes, unusual desorption kinetics were seen. A theoretical model was published two years later, reemphasizing the particular importance of desorption kinetics of surfactant mixtures [53].

There have been a number of studies concerning mixtures of anionic surfactants with either nonionics or cationics, but only a very few have addressed the kinetics of these complex systems [75, 76]. When looking at enhancement of anionic surfactant adsorption at the water–cellulose interface, Paria et al. [75] found that the greatest rate increase could be achieved by pretreating the surface with cationic surfactant, rather than using a mixed solution.

19.5 AGGREGATE DYNAMICS

19.5.1 MICELLAR SOLUTIONS

The first meaningful studies on the kinetics of dynamic processes within micellar solutions, which looked at dissociation rates via temperature-jump (T-jump) and related techniques, were carried out in the 1960s. The first notable attempt at a complete theory of micellization kinetics was the work of Kresheck et al. [77], who proposed a stepwise surfactant aggregation model based on a monodisperse system, where all micelles have the same aggregation number, n. However, this model found

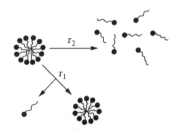

FIGURE 19.7   Relaxation processes in micellar solutions: τ_1 represents loss of a single surfactant molecule from a micelle; τ_2 represents complete micellar breakdown.

poor agreement with data from real systems. The significance of early results was not fully recognized, and it was not until work by Muller [78] in 1972 that a critical appreciation of the processes involved began to emerge. It had been found that there were two relaxation processes associated with micellar kinetics, with characteristically different relaxation times (designated τ_1 and τ_2) that differed by up to three orders of magnitude ($\tau_2 \gg \tau_1$).

The work of Zana and coworkers [79] (1974) gave a useful evaluation of available models and experiments to explain this discrepancy along with their own data. Their conclusion was that the fast process, τ_1, could be qualitatively assigned to represent the exchange of surfactant molecules between micelles and the bulk solution (fig. 19.7), i.e.,

$$A_n - A \leftrightarrow A_{n-1} \tag{19.7}$$

The slow process was determined to be complete dissolution of the micelles (fig. 19.7), such as might be particularly noted after perturbation, e.g., T-jump experiments:

$$A_n \leftrightarrow nA \tag{19.8}$$

The real breakthrough in terms of kinetic theory was published in 1973 by Aniansson and Wall [80, 81], who provided much more applicable kinetic equations for stepwise micelle formation using a polydisperse model. In a substantial paper two years later they were able to predict the first-order rate constants for the dissociation/association of surfactant ions to and from micelles (and hence residence times/lifetimes of surfactant monomers within micelles) [82]. They found values for the association and dissociation of surfactants into/from micelles (k^- and k^+, respectively) for sodium dodecyl sulfate (SDS) as 1×10^7 s^{-1} and 1.2×10^9 mol^{-1}·s^{-1}. Their kinetic model still remains essentially unchanged as a basis for the kinetics of micellar formation and breakdown. Modifications made to existing theory also allowed them to offer a significant thermodynamic explanation for the low enthalpy change upon micellization.

As it has been so comprehensively documented and reviewed [82–84], Aniansson and Wall's full derivation is not duplicated here. However, the most significant results are the expressions for the two relaxation times, τ_1 and τ_2:

$$\frac{1}{\tau_1} = \left(\frac{k^-}{\sigma^2}\right)\left[1 + \left(\frac{\sigma^2}{N}\right)a\right]$$ (19.9)

$$\frac{1}{\tau_2} \cong \frac{N^2}{cmc\left[1 + \left(\frac{\sigma^2}{N}\right)a\right]R}$$ (19.10)

where the equivalent concentration a is defined as

$$a = \frac{(C - cmc)}{cmc}$$ (19.11)

The function, R, is effectively a measure of the resistance to the movement of monomers between the bulk solution and micelles, and is calculated as

$$R = \sum_{s_1+1}^{s_2}\left(k_s^-\left[A_s\right]^{eq}\right)^{-1}$$ (19.12)

where
k^- = rate constant for dissociation of one surfactant molecule from a micelle
k_s^- = rate constant for dissociation of one surfactant molecule from an aggregate (and hence equal to k^- in the range of monodisperse micelles)
σ = width of the Gaussian micellar aggregation number distribution
N = average micelle aggregation number
C = total surfactant concentration
cmc = critical micelle concentration
$[A_s]^{eq}$ = equilibrium concentration of aggregates (micelles)

Lessner and coworkers [85, 86] suggested that micelles could form via coagulation of smaller micellar aggregates (which had previously been thought energetically impossible), and vice versa that micelles could fragment to smaller aggregates, particularly under the influence of added salt. A similar modification to make the theory more applicable to ionic systems was derived independently by Hall [87]. The work of Rharbi and Winnik [88] has since shown that in salt-screened solutions, the slow process that Lessner et al. had attributed to the fusion of two smaller micelles was in fact due to the fragmentation of the micelles, which then grow by the addition of surfactant monomer from the bulk solution.

More recently, Dushkin [89] (1998) derived a model interposed between the idealized monodisperse and the more realistic polydisperse. This pseudo-polydisperse system allowed for two types of micelles, one at each side of the bell-shaped size distribution.

An alternative new approach was that of Kuni and coworkers [90] using nucleation theory to derive a new kinetic model accounting fully for the size distribution of micelles. Their work follows the theory that nucleation is an activated process, although this has yet to become widely accepted.

Other recent, interesting studies include that of Nyrkova and Semenov [84, 91] on block copolymers. They showed that the activation barrier to micelle formation may be so large that it is kinetically unfeasible at the "conventional" *cmc*, as the timescale would be too great. Instead, micelles form at an increased "apparent" *cmc*, emphasizing the fact that micellization is indeed an activation-controlled process, and that thermodynamics as well as kinetics must be considered for a complete understanding.

So far, all of the studies detailed have dealt primarily with nonequilibrium states, i.e., micelle formation or breakdown stimulated by an external trigger, e.g., a jump in pressure, temperature, or concentration. It has of course been shown, however, that even at equilibrium, micelles are by their nature transient species with millisecond-scale lifetimes, and that dynamic processes exchange surfactant molecules between aggregates, interfaces, and bulk solutions. Studies of the kinetics of these processes in equilibrium solutions are much more scarce, mainly owing to experimental difficulties.

The main technique used to look at exchange processes in equilibrium systems employs labeled surfactants, particularly with ESR spectroscopy. Fox's ESR study [92] of a paramagnetic surfactant in micellar solution was the first of its kind, and yielded a solution–micelle monomer exchange rate of $\approx 10^5$ s^{-1} at room temperature for 2,2,6,6-tetramethylpiperidine-oxidedodecyldimethylammonium bromide. These techniques, along with time-resolved luminescence quenching, have shown that the entry of surfactant molecules into micelles is near-diffusion controlled, whereas loss from micelles is rate limiting, and hence kinetically controlled [93]. A decade later (1981), Bolt and Turro [94] were able to find the separate exit and reentry rate constants for 10-(4-bromo-1-naphthoyl)decyltrimethylammonium bromide as 3.2 × 10^3 s^{-1} and 5.7 × 10^7 mol^{-1}·s^{-1}, respectively.

19.5.2 Vesicle Systems

As has been shown above, the kinetics of micelle formation, breakdown, and associated dynamic processes has been documented. However, much less is known about the kinetic processes involved with transformations between other aggregate structures.

Surfactant vesicle systems have attracted much research and attention in recent years due to their potential applications in drug delivery, although their kinetics are just starting to be understood. Egelhaaf and Schurtenberger [95] suggested that vesicles can form from a micellar solution upon dilution via an intermediate structure, the whole process exhibiting first-order kinetics and a time constant of around 90 min. This transition state was later elucidated as a nonequilibrium system of disc-shaped micelles [96]. Leng, Egelhaaf, and Cates [97] proposed and tested a mechanism for vesicle formation by which disclike micelles rapidly form, grow to a critical radius, and then coalesce (close) to form a vesicle. They derived expressions for the rate of initial growth (coalescence of two disclike micelles) and the subsequent disc→vesicle transition (closure) that fit experimental results from their aqueous lecithin/bile salt system well. Whether vesicles ever truly represent an equilibrium (energy minimum) state is still a matter of some debate [98, 99].

19.5.3 MICROEMULSIONS

Microemulsions consist of either three or four components: two solvents, a surfactant, and sometimes an alcohol/cosurfactant. This complexity of composition means that there are potentially many relaxation processes. Despite this, microemulsion kinetics has been relatively well researched due to sustained interest in their structure and optimization. There have been several important reviews of the area, including summaries of work on the dynamic processes in such systems [100, 101].

The dynamic processes that can occur can be summarized as: (a) motion of interfacial molecules, (b) exchange, and (c) coalescence/fission of drops. Motion of interfacial molecules refers to the mobile hydrophobic chains of surfactant molecules adsorbed at the interface. Ahlnäs et al. [102] determined that this occurs on the picosecond timescale. Exchange processes can be of surfactant, cosurfactant, or dispersed solvent. For a dispersed phase of water, the exchange between immobilized or "bound" water (in a layer associated with the polar surfactant head groups) and free water was determined to occur on the millisecond timescale by NMR [103].

Exchange rates of cosurfactant molecules between interface and bulk solution seem to vary with molecular size (mobility) and environment, but a typical rate of $\approx 10^8$ s^{-1} was found by Lang et al. [104] for butan-1-ol in water/toluene/SDS microemulsions. They also found that surfactant exchange could be on a similar timescale, but may be greatly affected by many other parameters [105]. It also seems that the exchange times found in different studies are dependent on the method of measurement, so the picture is not yet complete.

Exchange of material in droplets seems to occur via a process of coalescence and then rapid fission, with nice data supporting this theory coming from Abe and coworkers in 1987 [106]. They found that, when mixing two three-component microemulsions with different sized droplets, a unimodal distribution was spontaneously reached without stirring, demonstrating that there must be an exchange process occurring more rapidly than Ostwald ripening, which can only be interpreted as fusion-fission (fig. 19.8).

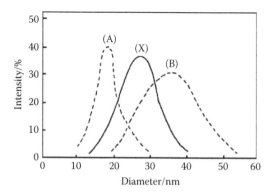

FIGURE 19.8 Spontaneous readjustment of microemulsion droplet size. Microemulsions with droplet sizes A and B are contacted. A single phase with droplet size X is spontaneously reached. (From Abe, M., et al. 1987. *Chem. Lett.* (8): 1613–16. With permission.)

The seminal paper in this field came from Fletcher, Howe, and Robinson [107], who emphasized that droplet exchange (via coalescence/fission) is rapid and continuous. They were even able to determine the activation barrier to water exchange for AOT-stabilized water-in-oil microemulsions (on the order of 70-100 kJ·mol^{-1}), with an associated rate constant for water exchange of 10^6-10^8 mol^{-1}·dm^3·s^{-1}.

19.6 CONCLUSION

Although there has been much effort spent in developing and improving our understanding of surfactant adsorption and aggregation kinetics, there are still important issues left to resolve. Much research has been conducted at the air–water, liquid–liquid, and at certain solid–liquid interfaces, and this has provided a valuable insight into dynamic surface and interfacial tensions and into rates of surface coverage. It has afforded understanding into the way molecules diffuse in solutions and also how they aggregate and rearrange at interfaces.

However, the area of exchange kinetics in equilibrium systems is still heavily underresearched, mostly due to experimental difficulties associated with observing and measuring such processes. Equally, systems involving organic solvents, and the complications these pose in terms of different solvophilic and -phobic interactions, and partitioning between solvents, are not fully understood. These have special relevance for the field of tribology, in terms of grease and lubricant formulation and action. How quickly molecules can arrange and rearrange is vital to understanding the science of lubrication at interfaces.

Using both classical techniques to look at new systems, and exciting novel experimental methods, there is still great scope for advancement within this expansive and continually evolving field.

ACKNOWLEDGMENT

R.T. thanks Infineum UK Ltd. and the University of Bristol Doctoral Training Grant for Ph.D. funding.

REFERENCES

1. Bovington, C. H. 1992. In *Chemistry and technology of lubricants*. Ed. R. M. Mortier and S. T. Orszulik. 2nd ed. London: Blackie.
2. Crawford, J., Psaila, A., and Orszulik, S. T. 1992. In *Chemistry and technology of lubricants*. Ed. R. M. Mortier and S. T. Orszulik. 2nd ed. London: Blackie.
3. Davidson, J. E., Hinchley, S. L., Harris, S. G., Parkin, A., Parsons, S., and Tasker, P. A. 2006. *J. Molecular Graphics Modelling* 25: 495–506.
4. Beltzer, M., and Jahanmir, S. 1985. *ASLE Trans.* 29: 423–30.
5. Frey, M., Harris, S. G., Holmes, J. M., Nation, D. A., Parsons, S., Tasker, P. A., and Winpenny, R. E. P. 2000. *Chem. Eur. J.* 6: 1407–15.
6. Frey, M., Harris, S. G., Holmes, J. M., Nation, D. A., Parsons, S., Tasker, P. A., and Winpenny, R. E. P. 1998. *Angewandte Chemie International Edition* 37: 3245–48.
7. Bowden, F. P., and Tabor, D. 1986. *The friction and lubrication of solids*. 2nd ed. New York: Oxford University Press.
8. Milner, R. S. 1907. *Phil. Mag.* 13: 96.

9. Langmuir, I., and Schaefer, V. J. 1937. *J. Am. Chem. Soc.* 59: 2400–14.

10. Ward, A. F. H., and Tordai, L. 1946. *J. Chem. Phys.* 14 (7): 453–61.

11. Sutherland, K. L. 1952. *Aust. J. Sci. Res. A – Phys. Sci.* 5 (4): 683–96.

12. Miller, R. 1981. *Colloid Polym. Sci.* 259 (3): 375–81.

13. McCoy, B. J. 1983. *Colloid Polym. Sci.* 261 (6): 535–39.

14. Diamant, H., and Andelman, D. 1996. *Europhys. Lett.* 34 (8): 575–80.

15. Diamant, H., and Andelman, D. 1996. *J. Phys. Chem.* 100 (32): 13732–42.

16. Ravera, F., Liggieri, L., and Steinchen, A. 1993. *J. Colloid Interface Sci.* 156 (1): 109–16.

17. Liggieri, L., Ravera, F., and Passerone, A. 1996. *Colloids Surf. A* 114: 351–59.

18. Eastoe, J., Dalton, J. S., Rogueda, P. G. A., Crooks, E. R., Pitt, A. R., and Simister, E. A. 1997. *J. Colloid Interface Sci.* 188 (2): 423–30.

19. Eastoe, J., Dalton, J., Rogueda, P., Sharpe, D., Dong, J. F., and Webster, J. R. P. 1996. *Langmuir* 12 (11): 2706–11.

20. Lee, Y. C., Stebe, K. J., Liu, H. S., and Lin, S. Y. 2003. *Colloids Surf. A* 220 (1–3): 139–50.

21. Bleys, G., and Joos, P. 1985. *J. Phys. Chem.* 89 (6): 1027–32.

22. van den Tempel, M., and Lucassen-Reynders, E. H. 1983. *Adv. Colloid Interface Sci.* 18 (3–4): 281–301.

23. Serrien, G., and Joos, P. 1990. *J. Colloid Interface Sci.* 139 (1): 149–59.

24. Fainerman, V. B., Miller, R., Wustneck, R., and Makievski, A. V. 1996. *J. Phys. Chem.* 100 (18): 7669–75.

25. Fainerman, V. B., Miller, R., Aksenenko, E. V., Makievski, A. V., Kragel, J., Loglio, G., and Liggieri, L. 2000. *Adv. Colloid Interface Sci.* 86 (1–2): 83–101.

26. Miller, R., Aksenenko, E. V., Liggieri, L., Ravera, F., Ferrari, M., and Fainerman, V. B. 1999. *Langmuir* 15 (4): 1328–36.

27. Lucassen, J. 1975. *Faraday Disc.* 59: 76–87.

28. Rillaerts, E., and Joos, P. 1982. *J. Phys. Chem.* 86 (17): 3471–78.

29. Noskov, B. A. 2002. *Adv. Colloid Interface Sci.* 95 (2–3): 237–93.

30. Colegate, D. M., and Bain, C. D. 2005. *Phys. Rev. Lett.* 95 (19): Article 198302, 1–4.

31. Song, Q., Couzis, A., Somasundaran, P., and Maldarelli, C. 2006. *Colloids Surf. A* 282: 162–82.

32. Bauget, F., Langevin, D., and Lenormand, R. 2001. *J. Colloid Interface Sci.* 239 (2): 501–8.

33. Jeribi, M., Almir-Assad, B., Langevin, D., Henaut, I., and Argillier, J. F. 2002. *J. Colloid Interface Sci.* 256 (2): 268–72.

34. Ariel, G., Diamant, H., and Andelman, D. 1999. *Langmuir* 15 (10): 3574–81.

35. Fainerman, V. B., and Miller, R. 1995. *Colloids Surf. A* 97 (1): 65–82.

36. Mulqueen, M., Stebe, K. J., and Blankschtein, D. 2001. *Langmuir* 17 (17): 5196–207.

37. Mulqueen, M., Datwani, S. S., Stebe, K. J., and Blankschtein, D. 2001. *Langmuir* 17 (24): 7494–500.

38. Frese, C., Ruppert, S., Sugar, M., Schmidt-Lewerkuhne, H., Wittern, K. P., Fainerman, V. B., Eggers, R., and Miller, R. 2003. *J. Colloid Interface Sci.* 267 (2): 475–82.

39. Ravera, F., Ferrari, M., and Liggieri, L. 2000. *Adv. Colloid Interface Sci.* 88 (1–2): 129–77.

40. van Hunsel, J., and Joos, P. 1987. *Langmuir* 3 (6): 1069–74.

41. Hansen, R. S. 1960. *J. Phys. Chem.* 64 (5): 637–41.

42. England, D. C., and Berg, J. C. 1971. *AIChE J.* 17 (2): 313.

43. Rubin, E., and Radke, C. J. 1980. *Chem. Eng. Sci.* 35 (5): 1129–38.

44. Mansfield, W. W. 1952. *Aust. J. Sci. Res. A – Phys. Sci.* 5 (2): 331–38.

45. Ferrari, M., Liggieri, L., Ravera, F., Amodio, C., and Miller, R. 1997. *J. Colloid Interface Sci.* 186 (1): 40–45.

46. Liggieri, L., Ravera, F., Ferrari, M., Passerone, A., and Miller, R. 1997. *J. Colloid Interface Sci.* 186 (1): 46–52.
47. Ravera, F., Liggieri, L., Passerone, A., and Steinchen, A. 1994. *J. Colloid Interface Sci.* 163 (2): 309–14.
48. Mulqueen, M., and Blankschtein, D. 2002. *Langmuir* 18 (2): 365–76.
49. Campanelli, J. R., and Wang, X. H. 1999. *J. Colloid Interface Sci.* 213 (2): 340–51.
50. Miller, R., Lunkenheimer, K., and Kretzschmar, G. 1979. *Colloid Polym. Sci.* 257 (10): 1118–20.
51. Brinck, J., Jonsson, B., and Tiberg, F. 1998. *Langmuir* 14 (5): 1058–71.
52. Brinck, J., Jonsson, B., and Tiberg, F. 1998. *Langmuir* 14 (20): 5863–76.
53. Brinck, J., and Tiberg, F. 1996. *Langmuir* 12 (21): 5042–47.
54. Behl, S., and Moudgil, B. M. 1993. *J. Colloid Interface Sci.* 161 (2): 443–49.
55. Tiberg, F., Jonsson, B., and Lindman, B. 1994. *Langmuir* 10 (10): 3714–22.
56. Torn, L. H., Koopal, L. K., de Keizer, A., and Lyklema, J. 2005. *Langmuir* 21 (17): 7768–75.
57. Paria, S., Manohar, C., and Khilar, K. C. 2005. *Colloids Surf. A* 252 (2–3): 221–29.
58. Paria, S., Manohar, C., and Khilar, K. C. 2005. *Industrial Eng. Chem. Res.* 44 (9): 3091–98.
59. Geffroy, C., Stuart, M. A. C., Wong, K., Cabane, B., and Bergeron, V. 2000. *Langmuir* 16 (16): 6422–30.
60. Dijt, J. C., Stuart, M. A. C., and Fleer, G. J. 1992. *Macromolecules* 25 (20): 5416–23.
61. Koopal, L. K., and Avena, M. J. 2001. *Colloids Surf. A* 192 (1–3): 93–107.
62. Rupprecht, H., and Gu, T. 1991. *Colloid Polym. Sci.* 269 (5): 506–22.
63. Somasundaran, P., and Fuerstenau, D. W. 1966. *J. Phys. Chem.* 70 (1): 90.
64. Yeskie, M. A., and Harwell, J. H. 1988. *J. Phys. Chem.* 92 (8): 2346–52.
65. Johnson, C. S. 1999. *Prog. Nucl. Magn. Reson. Spectrosc.* 34 (3–4): 203–56.
66. Stilbs, P. 1987. *Prog. Nucl. Magn. Reson. Spectrosc.* 19 1–45.
67. Stejskal, E. O., and Tanner, J. E. 1965. *J. Chem. Phys.* 42 (1): 288.
68. Soderman, O., and Stilbs, P. 1994. *Prog. Nucl. Magn. Reson. Spectrosc.* 26: 445–82.
69. Schonhoff, M., and Soderman, O. 1997. *J. Phys. Chem. B* 101 (41): 8237–42.
70. Clark, S. C., and Ducker, W. A. 2003. *J. Phys. Chem. B* 107 (34): 9011–21.
71. Couzis, A., and Gulari, E. 1992. *ACS Symp. Ser.* 501: 354–65.
72. Couzis, A., and Gulari, E. 1993. *Langmuir* 9 (12): 3414–21.
73. Theodoly, O., Cascao-Pereira, L., Bergeron, V., and Radke, C. J. 2005. *Langmuir* 21 (22): 10127–39.
74. Malbrel, C. A., Somasundaran, P., and Turro, N. J. 1990. *J. Colloid Interface Sci.* 137 (2): 600–3.
75. Paria, S., Manohar, C., and Khilar, K. C. 2004. *Colloids Surf. A* 232 (2–3): 139–42.
76. Penfold, J., Staples, E., Tucker, I., and Thomas, R. K. 2002. *Langmuir* 18 (15): 5755–60.
77. Kresheck, G. C., Hamori, E., Davenport, G., and Scheraga, H. A. 1996. *J. Am. Chem. Soc.* 88 (2): 246.
78. Muller, N. 1972. *J. Phys. Chem.* 76 (21): 3017.
79. Lang, J., Zana, R., Bauer, R., Hoffmann, H., and Ulbricht, W. 1975. *J. Phys. Chem.* 79 (3): 276–83.
80. Aniansson, E. A. G., and Wall, S. N. 1974. *J. Phys. Chem.* 78 (10): 1024–30.
81. Aniansson, E. A. G., and Wall, S. N. 1975. *J. Phys. Chem.* 79 (8): 857–58.
82. Aniansson, E. A. G., Wall, S. N., Almgren, M., Hoffmann, H., Kielmann, I., Ulbricht, W., Zana, R., Lang, J., and Tondre, C. 1976. *J. Phys. Chem.* 80 (9): 905–22.
83. Zana, R. 2005. In *Dynamics of surfactant assemblies.* Vol. 125. Ed. R. Zana. Boca Raton, FL: Taylor & Francis.

84. Nyrkova, I. A., and Semenov, A. N. 2005. *Macromolecular Theory Simulations* 14 (9): 569–85.
85. Lessner, E., Teubner, M., and Kahlweit, M. 1981. *J. Phys. Chem.* 85 (11): 1529–36.
86. Lessner, E., Teubner, M., and Kahlweit, M. 1981. *J. Phys. Chem.* 85 (21): 3167–75.
87. Hall, D. G. 1981. *J. Chem. Soc., Faraday Trans. II* 77: 1973–2006.
88. Rharbi, Y., and Winnik, M. A. 2002. *J. Am. Chem. Soc.* 124 (10): 2082–83.
89. Dushkin, C. D. 1998. *Colloids Surf. A* 143 (2–3): 283–99.
90. Rusanov, A. I., Kuni, F. M., and Shchekin, A. K. 2000. *Colloid J.* 62 (2): 167–71.
91. Nyrkova, I. A., and Semenov, A. N. 2005. *Faraday Disc.* 128: 113–27.
92. Fox, K. K. 1971. *Trans. Faraday Soc.* 67 (585): 2802.
93. Brigati, G., Franchi, P., Lucarini, M., Pedulli, G. F., and Valgimigli, L. 2002. *Res. Chem. Intermediates* 28 (2–3): 131–41.
94. Bolt, J. D., and Turro, N. J. 1981. *J. Phys. Chem.* 85 (26): 4029–33.
95. Egelhaaf, S. U., and Schurtenberger, P. 1997. *Physica B* 234: 276–78.
96. Egelhaaf, S. U., and Schurtenberger, P. 1999. *Phys. Rev. Lett.* 82 (13): 2804–7.
97. Leng, J., Egelhaaf, S. U., and Cates, M. E. 2003. *Biophys. J.* 85 (3): 1624–46.
98. Luisi, P. L. 2001. *J. Chem. Educ.* 78 (3): 380–84.
99. Yatcilla, M. T., Herrington, K. L., Brasher, L. L., Kaler, E. W., Chiruvolu, S., and Zasadzinski, J. A. 1996. *J. Phys. Chem.* 100 (14): 5874–79.
100. Lopez-Quintela, M. A., Tojo, C., Blanco, M. C., Rio, L. G., and Leis, J. R. 2004. *Curr. Opinion Colloid Interface Sci.* 9 (3–4): 264–78.
101. Moulik, S. P., and Paul, B. K. 1998. *Adv. Colloid Interface Sci.* 78 (2): 99–195.
102. Ahlnäs, T., Soderman, O., Hjelm, C., and Lindman, B. 1983. *J. Phys. Chem.* 87 (5): 822–28.
103. Hansen, J. R. 1974. *J. Phys. Chem.* 78 (3): 256–61.
104. Lang, J., Djavanbakht, A., and Zana, R. 1980. *J. Phys. Chem.* 84 (12): 1541–47.
105. Tondre, C. 2005. In *Dynamics of surfactant assemblies*. Vol. 125. Ed. R. Zana. Boca Raton, FL: Taylor & Francis.
106. Abe, M., Nakamae, M., Ogino, K., and Wade, W. H. 1987. *Chem. Lett.* 16 (8): 1613–16.
107. Fletcher, P. D. I., Howe, A. M., and Robinson, B. H. 1987. *J. Chem. Soc., Faraday Trans. I* 83: 985–1006.

20 Effect of Surfactant and Polymer Nanostructures on Frictional Properties

P. Somasundaran and Parag Purohit

20.1 INTRODUCTION

Surfactants and polymers used in boundary lubrication systems adsorb on solid surfaces and form a protective film. The effectiveness of boundary lubricants has often been attributed to the adsorption affinity and the integrity of the adsorbed film. Such adsorption is influenced by additives incorporated into the system to reduce thermal degradation, corrosion, sludge formation, foaming, etc. There are many interactions that can take place between the additives, the surfactants, and the base oil, leading, in addition to adsorption effects, to a number of interfacial and colloidal phenomena such as micellization, precipitation, and solubilization as well as flocculation of particulate matter in fluid [1–3].

20.2 ADSORPTION PHENOMENA

Clearly, the adsorption and integrity of the adsorbed film will depend, among others, on the molecular geometry of the interacting species. For example, among the various saturated mono-, bi-, and tricyclic aromatics, alkanes, and polyethers tested as lubricating fluids by Hentschel [4], the irregularly shaped molecules yielded high traction coefficients, whereas molecules with a threadlike shape and minimal structural subunits or functional groups yielded low coefficients. Evidently the ability of molecules to undergo intramolecular as well as intermolecular motions is important for maintaining the integrity of the lubricating films.

The geometry of the molecules in the adsorbed layer and the resultant structure of the adsorbed layer on a molecular scale are fundamental to the lubrication phenomena, and knowledge of the mechanism on this scale is essential for any major technological leaps in this area. However, while there has been a significant amount of work on a macroscopic scale to study the thickness of the film, its rheology, thermal conductivity, oxidation ability [5], and corrosion, there has been only a very limited effort to determine the adsorption properties [6], and no studies have been carried out, to the authors' knowledge, to understand the molecular structure of the adsorbed film in relation to lubrication.

20.2.1 Techniques for Investigation

Conventionally, adsorption is investigated by macroscopic measurements of adsorption isotherms, zeta potential, wettability, and enthalpy of adsorption [7–9]. These studies have been helpful for developing mechanisms that govern the formation of the adsorbed layers on solids. Of particular interest here is that a major driving force for the adsorption of long-chain surfactants has been considered to result from the self-aggregation tendency of the surfactant molecules at the solid–liquid interface. Such surface aggregation leads to the formation of two-dimensional aggregates called hemimicelles or solloids in general.

Hemimicellization occurs as a result of the high lateral interactions between adsorbed species on the solid surface. When polymers and other additives are present, additional microstructures can form as a result of polymer–surfactant interactions, with marked effect on such properties as adhesion.

Long-chain fatty acids that chemisorb onto metal oxides are better lubricants than simple hydrocarbons. This is so because, possibly, the former orient themselves normal to the surface to form a hemimicelle-type structure, whereas the latter adsorb weakly and do not lead to organized structures. It must also be noted that, with an increase in temperature and pressure associated with friction, such structures can break up or reorganize due to both the effect of such variables and irregular motions as well as to possible chemical conversion of adsorption sites on the solid surfaces.

20.2.2 Surfactant Microstructure and Orientation

Clearly the adsorbed layer structure in terms of molecular packing and orientation has to be considered as the primary property that controls the lubrication behavior of adsorbed films and should receive attention. Also, a full knowledge of microviscosity or fluidity on a nanoscale and its variation with temperature and pressure is necessary along with interactive forces between the film molecules under such conditions.

20.2.2.1 Fluorescence and ESR Investigation

The difficulty in studying the above properties thus far had been the unavailability of suitable techniques for reliable in situ investigation of solid–liquid interfaces [10]. Techniques such as infrared spectroscopy and ellipsometry involve ex situ procedures such as freezing, drying in vacuo, etc., and/or have necessitated the use of ideal surfaces. Clearly, these methods have limited applicability for studying surfactant adsorption onto particulate solids in aqueous media.

Advances in fluorescence and electron spin resonance (ESR) spectroscopies have enabled the use of organic fluorescent probes for in situ characterizations of molecular environments. The utility of fluorescence methods in such studies arises from the fact that the fluorescence and ESR responses of numerous probes are highly dependent on the environment, so that specific information can be obtained by appropriate choice of probes. Fluorescence responses that have been shown to depend on a micellar environment include excitation and emission, fluorescence polarization, and quenching (or sensitization). These responses have been, in turn, related to molecular properties such as polarity, viscosity, diffusion, solute partitioning, and aggregation numbers.

Pyrene and dinaphthylpropane (DNP) fluorescence and nitroxide ESR probes have been successfully used to investigate the structure of the adsorbed layer of sodium dodecyl sulfate at the alumina–water interface [11, 12]. The fluorescence fine structure of pyrene yielded information on the polarity of the microenvironment in the adsorbed layer. Intramolecular excimer formation of DNP was used to measure the microfluidity of this environment. The results indicate the presence of highly organized surfactant aggregates at the solid–liquid interface, formed by the association of hydrocarbon chains.

Fluorescence decay methods enabled the determination of the size of these aggregates and their evolution as a function of surface coverage (see fig. 20.1). Evidently, these surfactant aggregates, leading essentially to an organic film, should contribute toward lubricity, at least on the submicroscopic scale. The integrity of the film, on the other hand, will depend on the rigidity of such films. A technique that is capable of yielding information on film integrity on a submicroscopic level is ESR spectroscopy. Toward this purpose, three nitroxide ESR probes (5, 12, 16-DOXYL stearic acid) were used to examine the nanostructure of sodium dodecyl sulfate film between water and alumina. It was found that the microviscosity within the surface aggregate varied from about 360 cP close to the solid to 120 cP near the exterior.

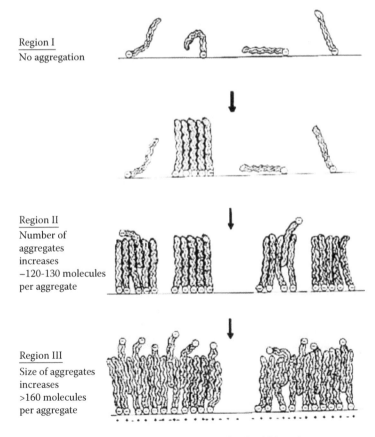

Region I
No aggregation

Region II
Number of
aggregates
increases
−120-130 molecules
per aggregate

Region III
Size of aggregates
increases
>160 molecules
per aggregate

FIGURE 20.1   Evolution of surface aggregates at solid–liquid interface.

This first report of flexibility (microviscosity) of the film suggests that the adsorbed film is sufficiently rigid and may be removed only with extreme pressure caused with application of high friction.

20.3 EFFECT OF SOLID SURFACE

Adsorption of the surfactants/polymers and the structure of the adsorbed layer will depend, to a large extent, on the nature of the solid surfaces involved and the interactions between them. Atomic force microscopy can be conveniently used to understand the effect of polymer adsorption on materials such as fibers [13]. It can be seen from fig. 20.2 that such treatment of fibers produces a smooth, uniform, and structurally relaxed surface as compared with the untreated fiber. This indicates that the silicone treatment may be responsible for changes in the microproperties of the fibers, and that it can be used to modify structural properties of fibers to induce smoothness, bounciness, and other such desirable properties.

Recently, it has been shown by Briscoe et al. [14] that the frictional stress between sliding surfaces coated with amphiphilic surfactant layers (boundary lubricants) immersed in water may be reduced by one to two orders of magnitude or more, relative to its value in dry air. Such reduction was suggested due to the shift of the slip plane during frictional sliding from the surfactant–surfactant midplane interface (in air) to the much better lubricated surfactant–substrate interfaces (underwater). The surfactant polar head groups will be hydrated upon being immersed in water [15], and the resultant surrounding fluid sheaths at the surfactant–substrate interface provide much better lubrication. Thus the nanostructure of the adsorbed surfactant layer on the substrate is very important. The higher rigidity or packing of the layer could be of benefit for lubrication in air or oil, but it might not be necessary in water, since the water molecules cannot penetrate through the layer easily to hydrate the surfactant head groups [16]. Therefore, a delicate design of surfactants is needed for specific applications.

In the case of metals subjected to friction, it would be important to simultaneously characterize the surface, particularly with respect to its oxidation state,

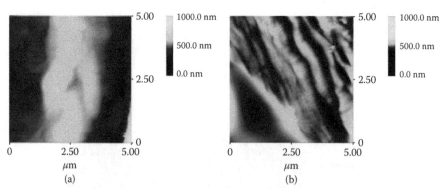

FIGURE 20.2 AFM images of (a) untreated fabric and (b) treated fabric (using modified silicone polymer). The difference in surface morphology indicates the effect of polymer interaction with solid surface, resulting in smoother fibers.

using such techniques as x-ray photoelectron spectroscopy. With the surface and the adsorbed layer well characterized, it would prove very useful to investigate the forces between surfaces and alterations in the forces when the surfaces are separated by only a few nanometers. This can now be done easily using devices that have been developed by Tabor and Israelachvili [17, 18]. Furthermore, changes in the forces within films that have been subjected to high temperatures and pressures should also be taken into account.

20.4 CONCLUDING REMARKS

Properties of adsorbed layers on a molecular scale must be recognized as critical in determining the tribological behavior of solids. The nanoscopic behavior of adsorbed layers using spectroscopic and various interfacial techniques must be taken into account while determining the role of various additives and environmental (e.g., temperature and pressure) factors in controlling frictional behaviors. Importantly, such information as a function of the chemical structure of the coating agent can serve as a guideline for selecting appropriate lubricants.

ACKNOWLEDGMENTS

The authors would like to acknowledge the financial support from the National Science Foundation (NSF) and Industry/University Center for Surfactants at Columbia University. The authors also would like to acknowledge Dr. Bingquan Li, postdoctoral research scientist, Columbia University, for his help in revising the manuscript.

REFERENCES

1. Somasundaran, P. 1980. In *Fine particles processing*. Vol. 2. Ed. P. Somasundaran. New York: American Institute of Metallurgical Engineers.
2. Somasundaran, P., Celik, M. S., and Goyal, A. 1981. In *Surface phenomena in enhanced oil recovery*. Ed. D. O Shah. New York: Plenum.
3. Celik, M. S., Manev, E. D., and Somasundaran, P. 1982. American Institute of Chemical Engineers Symposium Series, No. 78.
4. Hentschel, K. H. 1985. *Synthetic Lubrication* 2: 143–65.
5. Bartz, W. J. 1978. *Wear* 49: 1–18.
6. Jahanmir, S., and Beltzer, M. 1986. *Tribology* 108: 109–16.
7. Somasundaran, P., Healy, T. W., and Fuerstenau, D. W. 1964. *J. Phys. Chem.* 68: 3562.
8. Hough, D. B., and Rendall, H. M. 1983. In: *Adsorption from solution at the solid-liquid interface*. Ed. G. D. Parfitt and C. H. Rochester. New York: Academic Press.
9. Somasundaran, P., Chandar, P., and Chari, K. 1983. *Solloids Surfaces* 8: 121–36.
10. Somasundaran, P., and Moudgil, B. M. 1979. In *Surface contamination: Genesis, detection and control*. Vol. 1. Ed. K. L. Mittal. New York: Plenum Press.
11. Somasundaran, P., Turro, N. J., and Chandar, P. 1986. *Solloids Surfaces* 20: 145–50.
12. Chandar, P., Somasundaran, P., Waterman, K. C., and Turro, N. J. 1987. *J. Phys. Chem.* 91 (1): 148–150.
13. Purohit, P., Somasundaran, P., and Kulkarni, R. 2006. *J. Colloid Interface Sci.* 298: 987–90.
14. Briscoe, W. H., Titmuss, S., Tiberg, F., Thomas, R. K., McGillivray, D. J., and Klein, J. 2006. *Nature* 444: 191–94.

15. Williams-Daryn, S., Thomas, R. K., Castro, M. A., and Becerro, A. 2002. *J. Colloid Interface Sci.* 256: 314–24.
16. Yoshizawa, H., Chen, Y. L., and Israelachvili, J. 1993. *J. Phys. Chem.* 97: 4128–40.
17. Tabor, D., and Winterton, R. H. S. 1969. *Proc. Royal Soc. London* A312: 415–30.
18. Israelachvili, J. N., and Adams, G. E. 1976. *Nature* 262: 774–76.

21 An Extended-Drain Engine Oil Study

Joseph M. Perez

ABSTRACT

One trend in the use of engine oil in heavy duty diesel engines is to extend the time interval between oil drains. This technology requires the use of sophisticated additive packages including dispersants and detergents combined with the appropriate base stocks. In addition, adequate oil supply must be maintained over the extended use period. In the study described in this chapter, three engine oils, formulated with similar additive packages but different base oils, were evaluated in an extensive on-highway truck fleet test. The engine lubricants were formulated with a conventional API Type I petroleum base oil, an API Type III low sulfur, high paraffinic base oil and an API Type IV synthetic base oil. The change periods for the engine oils was extended to 75,000 miles for the Type I base oil formulation and 100,000 miles for the other two engine oils. All three engine oils performed satisfactory in the study.

21.1 INTRODUCTION

Engine oils are formulated to meet the specifications of the current engine and vehicle technology. As a result, the nature of base stocks, antioxidants, dispersants, detergents, viscosity improvers, and antiwear additives used has changed drastically over the past couple of decades. Most of these changes have been driven by environmental requirements to meet emission regulations, such as the reduction of sulfur, phosphorus, and aromatic compounds from both the lubricant and fuel. Reduction of phosphorus and sulfur additives is the direct result of the detrimental role they play in reducing the effectiveness of aftertreatment systems. Long-term oil drains and the use of environmentally friendly lubricants require trade-offs in the types of additives found in lubricant formulations. Also, the need for increased gas mileage has led to the use of friction modifiers and lower viscosity fluids that still perform adequately without a drastic increase in wear. This chapter reports the results of a study [1] that compares the effects of different base stocks on the performance of some extended-drain engine oils.

Extended-drain intervals, when successful, are one way to more efficiently use our petroleum resources. This also results in decreased quantities of waste oil and less downtime for long-haul diesel trucks. This is not only an environmental benefit, but also an economic benefit for the user. This study evaluates the use of three engine oils formulated for extended-drain use in heavy-duty diesel engines.

21.2 EXPERIMENTAL

21.2.1 Test Oils

Three oils are compared in this study using base oils from three different American Petroleum Institute (API) base oil categories. The oils were formulated using an API Type I petroleum base oil (engine oil A), an API Type III highly processed petroleum base oil (engine oil B), and an API Type IV synthetic base oil (engine oil C). Two of the engine oils, A and C, are commercially available top-tier oils that meet or exceed CH-4, CG-4, CF-4, and CF/SJ API requirements. The third engine oil (B) is a custom-blended prototype. The oils all contain essentially the same additive treatment. Oil A is a 15W-40, API CH-4 standard reference oil. Oil B is an SAE 5W-40 prototype oil. The additive package is similar to that of oil A but is enhanced to potentially meet the extended-drain requirements of this study and the specification for the next generation of engine oils, designated as PC-9. Oil C is an SAE 5W-40, API CH-4, custom-blended fully synthetic oil containing a poly-alpha-olefin (PAO) base oil.

21.2.2 Test Procedures

All oils were used in an over-the-road long-haul truck fleet. The fleet test consisted of 15 trucks. All trucks used in the study were new Class 8 trucks containing the same model engine from a single OEM. The trucks were initially operated on their standard factory fill engine oil for 25,000 miles prior to draining, flushing, and changing over to the test oils. Each test oil was used in five trucks in the fleet. The test ran for approximately 14 months, with the trucks accumulating some 700–800 miles per day. Approximately 300,000 miles were accumulated on each truck in the study. Used oil samples were obtained every 10,000–15,000 miles. Oil levelers were used to maintain a constant oil level in the crankcase. The standard manufacturer-recommended drain period for the vehicles was 15,000 miles. FleetGuard® series 9000 oil filters were used on all test engines and changed at the test scheduled oil periods. The drain period for oil A, a conventionally refined oil, was 75,000 miles. The drain periods for oil B, a highly refined base oil, and oil C, a synthetic base oil, were extended to 100,000 miles.

21.2.3 Oil Analysis Test Methods

The objective in this study was to compare the long-term performance of these oils. To do this, changes in chemical and physical properties of the used oils were monitored using American Society of Testing Materials (ASTM) standard test methods [2]. Typically, used engine oil analyses include measurements of viscosity, acid and base numbers, water, glycol, soot, and metals content. In addition to the standard tests, fuel economy, deposit-forming tendencies, and friction and wear characteristics were determined on new and used oil samples in this study.

21.2.3.1 Viscosity

The viscosities of new and used oil samples were determined using ASTM Standard Test Method D997.

21.2.3.2 Metals Content

Metals were determined using inductively coupled plasma–atomic emissions spectroscopy (ICP-AES).

21.2.3.3 Total Base Number (TBN)

The base number was determined using ASTM Standard Test Method D4739. The initial TBN for test oils A and B was 10 and for oil C it was 12.

21.2.3.4 Total Acid Number (TAN)

The acidity of the oil was determined using a modified ASTM Standard Test Method D664. In this test, the acid number is determined using titration to an endpoint.

21.2.3.5 RULER™ Method

A second method, the RULER (remaining useful life evaluation routine) method [3], was used as a cross-check of TAN and TBN of samples as received. The method agreed to ±1 mg KOH/g of oil with values obtained by an outside laboratory.

21.2.3.6 Soot

Particulate matter (soot) in the crankcase oil was measured using both Fourier-transform infrared (FTIR) spectroscopy and thermogravimetric analyses (TGA). A comparison of the two methods is shown in fig. 21.1. The correlation between the methods is good, but the values for the TGA method are slightly higher (10–15%) than the FTIR values.

21.2.3.7 Four-Ball Tests

Friction and wear were measured using the Penn State sequential four-ball test. The sequential test was developed to evaluate hydraulic fluids [4–6]. This test operates in the boundary lubrication regime. Test variables included time, load, speed, and temperature. Both torque (friction) and wear were measured. The sequential test con-

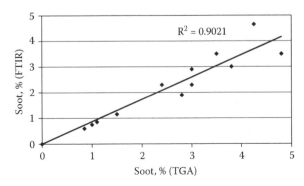

FIGURE 21.1   Comparison of FTIR and TGA soot methods.

TABLE 21.1

Penn State Microoxidation Test: Use, Test Variables, and Test Conditions

Use:

Evaluate volatility and oxidation stability using thin oil film

Measure parameters (wt%): volatility loss and deposit formation

Test variables:

Time

Temperature

Sample size (20–40 υg)

Gas flow composition (air, oxygen, nitrogen, argon)

Gas flow rate

Test coupon material (steel, aluminum, gold)

Test conditions:

	Typical	This study
Time (min)	30, 60, 90, 120, 240	30, 60, 90, 120, 240
Temperature (°C)	175–250	225
Sample size (υg)	20–40	40
Flow rate (cc/min)	20	20 (air)
Test coupon	steel, aluminum	low-carbon steel

ditions in this study included three 30-min test segments run at 40 kg (392 N) load, 75°C, and 600 rpm. The test specimens used were ANSI 52-100, 0.5-in.-diameter ball bearings. In the sequential test, the first 30-min segment was the break-in segment, the second was the steady-state segment, and the last segment was used to evaluate film formation. Approximately 10 mL of test fluid was used in the first two segments, and an 80-cSt white oil was used in the third. The ball pot was washed after each segment without removing the ball bearings. The wear scars were then measured, the fluid recharged, and the test unit reassembled for the next 30-min run.

21.2.3.8 Deposit-Forming Tendencies

The rate of formation of deposits was measured using the Penn State thin-film microoxidation (PSMO) test. The volatility and deposit-forming tendency of the three oils were determined. The conditions at which the tests were conducted in this study are shown in table 21.1. The deposit-formation induction time in the PSMO test was used as an indicator of the deposit-forming tendencies of the oil. The test method was developed to evaluate automotive engine oils [7] and later expanded to study other lubricants [8]. The test procedures are described in detail in the literature [9].

21.3 RESULTS AND DISCUSSION

21.3.1 ACID AND BASE NUMBERS

In engine oils, additives are used to control the acidity of the oil. The acidity is the result of (a) blowby of combustion oxidation products that end up in the crankcase oil and (b) oxidation of the engine oil in the crankcase over time. Total acid number

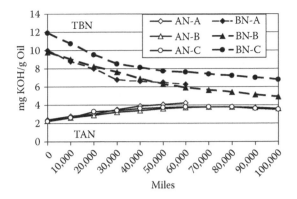

FIGURE 21.2 TAN and TBN changes of oils with use (RULER™).

(TAN) and total base number (TBN) measurements were used to track the change in acidity of the fluids using the RULER method and ASTM Standard Test Methods D664 and D4739. Normally, the rate of depletion of the additives is an indication of end of oil life. This is indicated by a crossover of the TAN and TBN values. The average values as shown in fig. 21.2 indicated that the oils in this study performed satisfactorily, and the oils contained adequate additive at the time the oil was changed (75,000 miles for oil A, 100,000 miles for oils B and C). It appears that oil A would cross over if the drain period was extended beyond 75,000 miles. The TAN and TBN curves of oil B (fig. 21.2) are approaching crossover at its oil-change period of 100,000 miles. For oil C, it appears feasible to use less TBN additive in the initial formulation, or the drain period could be extended to about 125,000 miles.

21.3.2 VISCOSITY

A significant increase or decrease in viscosity (ASTM Method D997) is a second parameter used to evaluate the performance of engine oils. The engine oils used in this study were 15W-40 and 5W-40 weight oils. Some test data and the viscosity range at 100°C for an SAE 40W oil are shown in fig. 21.3. Oils A and B remained within range throughout the test, with less than a 2% change in viscosity. Oil C had an occasional sample out of range. The viscosity change is attributed to soot accumulation and some oxidation of the crankcase oil. The stability of the oils in these tests is attributed to the constant addition of new oil to replace oil losses due to evaporation and oxidation.

21.3.3 PARTICULATE (SOOT)

Soot in the crankcase oil is the result of blowby of soot from the combustion process. When combustion occurs in the engine, some combustion gases and soot are carried past the rings and end up in the crankcase. In this study, the soot was determined by two independent methods, FTIR and a TGA method. The two methods correlated well. The TGA method produced values approximately 15% higher, as seen in fig. 21.1. The increase in particulate with time is essentially linear, as seen in fig. 21.4. This would be expected unless other factors were involved, such as microfiltration or

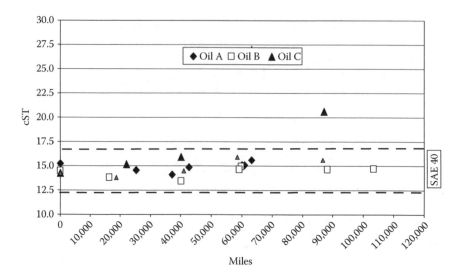

FIGURE 21.3   Oil viscosity changes of samples at 100°C.

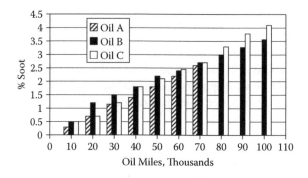

FIGURE 21.4   Crankcase soot concentration (FTIR) vs. oil miles.

loss of oil control. The increase in particulate level is similar in all three oils. The particle size, as determined by light scattering, also increased, probably as the result of particle agglomeration. Since all three additive packages were similar, the soot results are similar. The soot level is approximately 2% at 75,000 miles and 4% at 100,000 miles. This level is anticipated to increase when exhaust gas recirculation (EGR) technology is applied to the engines to meet the next level of emissions regulations.

21.3.4 VOLATILITY

The PSMO test was utilized to study the used-oil volatility and deposit-forming tendencies. The volatility loss for the new oils at 225°C is shown in fig. 21.5. The volatile losses of the oils plateau at about 80%. This is partly due to the formation of deposits, which results in a reduction in the availability of liquid product.

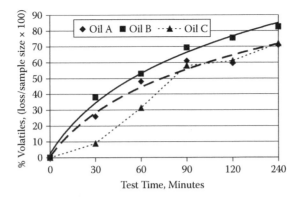

FIGURE 21.5   Volatility of new oils at 225°C (PSMO tests).

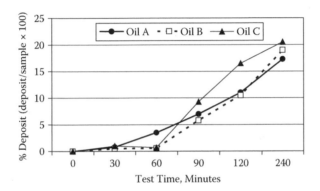

FIGURE 21.6   Deposit-forming tendencies: new oils at 225°C (PSMO).

21.3.5 DEPOSITS

The deposit-forming tendencies of the oils were determined at 225°C (fig. 21.6). The deposit induction time for oil A was about 30 min, and it was 60 min for oils B and C. All oils showed low deposit-forming tendencies throughout the test, with only slight increases in deposits due to the accumulation of soot.

21.3.6 FRICTION AND WEAR

Earlier studies suggested an increase in wear due to the presence of soot. The increase in wear was attributed to either selected adsorption of additives on the surface of the soot particles or due to an abrasiveness from the soot [10, 11]. The soot did affect the wear characteristics of the used oils. An increase in wear with time (mileage) did occur for all oils. Friction tended to vary some, but essentially stayed the same or exhibited a slight decrease with time, possibly the result of fatty acid-type oxidation products.

The three oils exhibit excellent wear levels when new. With use, wear increases with oils A and B. They have somewhat lower values than oil C. The total wear for the first two 30-min segments of sequential test, the wear-in plus the steady-state segments, are found in fig. 21.7. The wear level for oil C is high after 80,000 miles. The friction data for the first two segments are found in figs. 21.8 and 21.9, respectively. The data show a reduction in friction with time. Oil C exhibits the highest wear and lowest friction values after 80,000 miles. Overall, oil C exhibited the lowest friction values of the three oils.

To confirm these trends, the used oil samples were combined based on mileage. The results found for oil C (fig. 21.10) are typical of the findings of the three oils. The total wear shown is the combined wear for the run-in and steady-state segments. The wear increase with increased mileage is in agreement with the data found in fig. 21.7. Although the wear increases with used-oil mileage, the oils still show an acceptable wear level at the 75,000–100,000-mile level. The delta wear values shown are the increase in wear over the previous wear segment. In the case of the first segment, the break-in segment, the delta wear is the increase in wear over the Hertz indentation scar. The white-oil segment, the third 30-min segment of the sequential test, is run

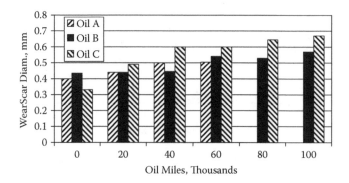

FIGURE 21.7 Four-ball sequential test: total wear at 60 min.

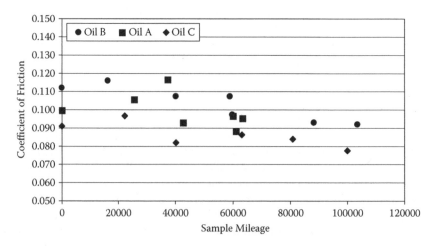

FIGURE 21.8 Four-ball sequential test: friction data, segment 1 (wear-in).

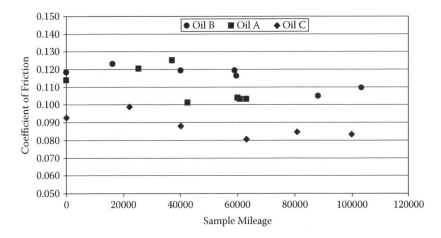

FIGURE 21.9   Four-ball sequential test: friction data, segment 2 (steady state).

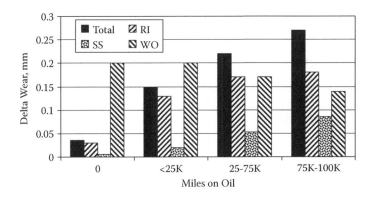

FIGURE 21.10   Four-ball incremental wear tests of used oil C samples. (RI = run-in, SS = steady state, and white oil segments of sequential test. Total wear is the sum of wear of RI and SS test segments.)

with a nonadditive white oil. The data suggest that, with use, a more lubricious film remains in the contact area at the end of the first 60 min of oil testing.

21.3.7 METALS

Aluminum, copper, chromium, and iron metal contents of the used oils were measured. The total metal contents for the oils are found in fig. 21.11. The relative quantities of the metals were copper < chromium < aluminum < iron. The end-of-test levels were less than 6 ppm for Cu, less than 16 ppm for chromium, less than 50 ppm for Al, and less than 200 ppm for iron, all acceptable levels. None of the oils exhibited an exponential increase in the amount of any of the wear metals, indicating that the additive packages were performing satisfactorily at the end of the test.

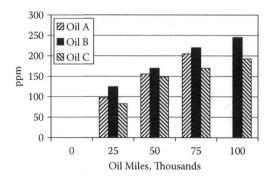

FIGURE 21.11 Total metals contents of used oils.

21.3.8 Fuel Consumption

Although fuel consumption was measured in these tests, there was no significant difference in fuel consumption. This was probably due to the variations in fuel quality, as the fleet vehicles traveled from coast to coast, refueling as needed across the country.

21.4 SUMMARY

This study demonstrates the feasibility of sustaining extended drain intervals in heavy-duty diesels for up to 100,000 miles when using lubricants formulated for extended-drain use. Three oils formulated with different base stocks but similar additive packages performed well throughout the test. The Type I, conventionally refined base oil, performed well up to 75,000 miles. The Type III, highly refined base oil, performed well up to 100,000 miles. The Type IV synthetic fluid with a slightly higher additive package performed well at 100,000 miles. With improvement in dispersancy to improve the wear characteristics, the Type IV oil could have run for at least another 25,000 miles before oil change.

Use of extended-drain lubricants is environmentally desirable. However, the drain limits found in this study may need to be reduced as soot levels increase with the use of exhaust gas recirculation (EGR), unless the effectiveness of the dispersant additives used can handle the increase in particulate soot. EGR is a factor in the emissions reduction protocol for the next generation of EPA regulations for heavy-duty diesel engines.

ACKNOWLEDGMENTS

This research is based on results found in the masters thesis of Edward Tersine [1]. The study was supported by a Tribology Consortium at Penn State University that included the Valvoline Company (Lexington, KY), the Oil Products Group (USDA Regional Laboratory, Peoria, IL), and the Tribology Group (Argonne National Laboratory, Argonne, IL). The results and conclusions as discussed in this chapter are those of the author and researchers at Penn State and are not necessarily those of the participants of the consortium.

REFERENCES

1. Tersine, E. 2002. *An Extended Drain Lubricant Study*. M.S. thesis, The Pennsylvania State University.
2. American Society of Testing Materials, 1916 Race Street, Philadelphia, PA 19103-1187.
3. RULER™. http://www.koehlerinstrument.com/tribology-index.html.
4. Perez, J. M., and Klaus, E. E. 1983. *SAE Technical Paper No. 831680.*
5. Perez, J. M., Klaus, E. E., and Hanson, R. C. 1990. *Lubrication Eng. 46*: 249–55.
6. Perez, J. M., Klaus, E. E., and Hanson, R. C. 1996. *Lubrication Eng. 52*: 416–22.
7. Klaus, E. E., Cho, L., and Dang, H. 1980. *SAE Technical Paper No. 801362.*
8. Hsu, S. M., and Perez, J. M. 1991. *SAE Technical Paper No. 910454.*
9. Erhan, S. Z., and Perez, J. M. eds. 2002. *Biobased Industrial Fluids and Lubricants*. Champaign, IL: AOCS Press.
10. Rounds, F. G. 1977. *SAE Technical Paper No. 770829.*
11. Green, D. A., and Lewis, R. 2007. *SAE Technical Paper No. 20077071.*

Index

T - #0192 - 071024 - C0 - 234/156/0 - PB - 9780367387242 - Gloss Lamination